Lecture Notes in Mathematics

Edited by A. Dold and B. Eckmann
Series: Mathematisches Institut der Universität Bonn
Adviser: F. Hirzebruch

627

Modular Functions of One Variable VI

Proceedings International Conference,
University of Bonn, Sonderforschungs-
bereich Theoretische Mathematik
July 2–14, 1976

Edited by
J.-P. Serre and D. B. Zagier

Springer-Verlag
Berlin Heidelberg New York 1977

Editors

Jean-Pierre Serre
Collège de France
75231 Paris Cedex 05/France

Don Bernard Zagier
Mathematisches Institut
der Universität Bonn
Wegelerstr. 10
53 Bonn/BRD

AMS Subject Classifications (1970): 10C15, 10D05, 10D25, 12A99, 14H45, 14K22, 14K25

ISBN 3-540-08530-0 Springer-Verlag Berlin Heidelberg New York
ISBN 0-387-08530-0 Springer-Verlag New York Heidelberg Berlin

Printed in Germany

Printing and binding: Beltz Offsetdruck, Hemsbach/Bergstr.
2140/3140-543210

PREFACE

This is the second and final volume of proceedings for the Conference on Modular Forms held in Bonn in July 1976. The first volume appeared as Lecture Notes n° 601, under the title "Modular Functions of One Variable V" (cf. Lecture Notes n° 320, 349, 350 and 476).

Jean- Pierre Serre

Don Zagier

CONTENTS

Values of Dirichlet Series at Integers in the Critical Strip

by

Michael J. Razar

§1. Introduction

This note is primarily a summary of some recent work on the values at integer points in the critical strip of Dirichlet series associated to newforms on $\Gamma_o(N)$. The first such results seem to be due to Shimura [9], who derived them for the Dirichlet series $\sum_{n=1}^{\infty} \tau(n)n^{-s}$ associated to the cusp form $\Delta(z)$ of weight 12 for the full modular group. Somewhat later, Manin [4] extended these results to cusp forms of arbitrary integral weight for the full modular group. In the interim, Birch had introduced the "modular symbol" for cusp forms of weight two on $\Gamma_o(N)$ and these were studied and used by Manin [2] and [3], Mazur and Swinnerton-Dyer [5] and others. Recently, V. Miller in his thesis [6] extended the definition of the modular symbol to $\Gamma_o(N)$. Just this year (1976), Shimura [11], using totally different methods, has extended almost everything to $\Gamma(N)$ and has obtained rationality results similar to those described below.

The main result of the present note is Theorem 4. The proof consists of two main parts. The first is based on Shimura's isomorphism between cusp forms and Eichler cohomology with real coefficients. In this respect it is similar to the techniques of Manin [4]. The second part is the interpretation of the coefficients of an Eichler cocycle as residues at poles of the Mellin transform of a multiple integral of the corresponding cusp form. This is based on the Hecke correspondence (Proposition 1) and on an additive character analog of Weil's theorem (Proposition 2). Weil has also developed such a procedure recently in a paper delivered at the Takagi conference (1976).

International Summer School on Modular Functions

Bonn 1976

One advantage of proceeding in this way is that the same methods are applicable in other settings. For example, they work for Eisenstein series. This is discussed briefly in §4.

Detailed proofs of everything discussed here will appear in [7] and [8]. It is convenient in the present note to stick to cusp forms of Hauptypus for $\Gamma_o(N)$ since the Shimura isomorphism only makes sense for cusp forms with respect to a real character. However, in [8] a modified version of the Shimura isomorphism is used to prove analogous results for arbitrary cusp forms of Nebentypus.

§2. The Shimura isomorphism

We begin with a brief discussion of the Eichler cohomology and its relationship to the space of cusp forms. Fix a Fuchsian group of the first kind $G \subset SL_2(\mathbb{R})$, an integer k, and a real character v of G. Assume G contains translations. If f is a function on the upper half plane, define

$$(f|_{k,v}\sigma)(z) = v(\sigma)(cz+d)^{-k} f\left(\frac{az+b}{cz+d}\right), \quad \sigma = \begin{pmatrix} a & b \\ c & d \end{pmatrix}. \qquad (1)$$

Let K be a subfield of $\mathbb{R}$ such that $G \subset SL_2(K)$ and denote by $X(K) = X_{k-2,v}(K)$ the space of polynomials of degree at most $k-2$ with coefficients in K and G-module structure given by (1), but with k replaced by $2-k$.

$H^1(G,X(K))$ and $Z^1(G,X(K))$ are respectively the first cohomology and first cocycle groups of G with values in $X_{k-2,v}(K)$. Let $\bar{H}^1(G,X(K))$ be the subspace of $H^1(G,X(K))$ consisting of those cohomology classes whose restriction to every (cyclic) parabolic subgroup of G is trivial. Let $\bar{Z}^1(G,X(K))$ consist of those cocycles $P: \sigma \mapsto P_\sigma(z)$ such that if $\sigma \in G$ is parabolic, then for some $Q \in X(K)$, $P_\sigma = Q|_{2-k,v}\sigma - Q$ and such that furthermore, if σ is a translation, then $P_\sigma = 0$. The only coboundaries in $\bar{Z}^1(G,X(K))$ are the constant multiples of the coboundaries $\sigma \mapsto 1|\sigma - 1$. Thus $\dim \bar{H}^1 = \dim \bar{Z}^1 - 1$.

Let $S_{k,v}(G)$ be the space of cusp forms and let λ be the positive number such that $\begin{pmatrix} 1 & \lambda \\ 0 & 1 \end{pmatrix}$ generates the group of translations in G. Let $f \in S_{k,v}(G)$ and suppose that $f(z)$ has the Fourier expansion

$$f(z) = \sum_{n=1}^{\infty} a_n e^{\frac{2\pi i n z}{\lambda}}. \tag{2}$$

Denote by $f^*(z)$ the $(k-1)$-fold integral of $f(z)$ given by

$$f^*(z) = \sum_{n=1}^{\infty} a_n \left(\frac{2\pi i}{\lambda}\right)^{-k-1} e^{\frac{2\pi i n z}{\lambda}}. \tag{3}$$

Let $P_\sigma = f^* - f^*|_{2-k,v}\sigma$. Then $P_\sigma \in X(\mathbb{C})$ for all $\sigma \in G$ and so $P \in \bar{Z}^1(G,X(\mathbb{C}))$. Define homomorphisms $\delta_0 : S_{k,v}(G) \to \bar{Z}^1(G,X(\mathbb{C}))$ and $\delta : S_{k,v}(G) \to \bar{H}^1(G,X(\mathbb{C}))$ by

$$\delta_0 f = P \quad \text{and} \quad \delta f = \text{cohomology class of } \delta_0 f. \tag{4}$$

The maps δ_0 and δ are injective and the image of δ has dimension (over $\mathbb{C}$) equal to half that of $\bar{H}^1(G,X(\mathbb{C}))$. To get an isomorphism, Shimura ([9] and [10]) defines maps ρ_0 and ρ from $S_k(G)$ to $\bar{Z}^1(G,X(\mathbb{R}))$ and $\bar{H}^1(G,X(\mathbb{R}))$ respectively, by letting $\rho_0 f$ be the "real part" of the cocycle $\delta_0 f$. ("Real part" here applies only to coefficients of polynomials, not to the variable z.) Then ρf is the cohomology class of $\rho_0 f$ and the map ρ is an isomorphism. Furthermore, ρ commutes with the action of Hecke operators (double cosets) on $S_k(G)$ and $\bar{H}^1(G,X(K))$ and thus is an isomorphism of Hecke modules.

Let $\varepsilon = \begin{pmatrix} 1 & 0 \\ 0 & -1 \end{pmatrix}$ and suppose that ε normalizes G. Define the action of ε on functions f on the upper half plane by $(f|\varepsilon)(z) = \overline{f(-\bar{z})}$. (Note that ε is only $\mathbb{R}$-linear, not $\mathbb{C}$ linear.) If $f(z)$ has a Fourier expansion (2), then $f|\varepsilon = f$ if and only if the Fourier coefficients a_n are real and $f|\varepsilon = -f$ if and only if $(if)|\varepsilon = f$. The action of ε on a cocycle P is given by $(P|\varepsilon)_\sigma = P_{\varepsilon\sigma\varepsilon^{-1}}|\varepsilon$ and this action induces an automorphism of order two on $\bar{H}^1(G,X(\mathbb{R}))$. Denote the eigenspaces of ε corresponding to eigenvalues ± 1 by $\bar{Z}^1_\pm(G,X(\mathbb{R}))$ and $\bar{H}^1_\pm(G,X(\mathbb{R}))$ and let $S_k(G)(\mathbb{R})$ be the space of cusp forms with real coefficients. Since $\rho_0(f|\varepsilon) = (\rho_0 f)|\varepsilon$, ρ restricts to isomorphisms from $S_k(G)(R)$ to $\bar{H}^1_+(G,X(\mathbb{R}))$ and $iS_k(G)(\mathbb{R})$ to $\bar{H}^1_-(G,X(\mathbb{R}))$. If $f \in S_k(G)(\mathbb{R})$, then

$$P_\sigma = f^* - f^*|\sigma = P^+_\sigma + iP^-_\sigma \tag{5}$$

where $P^+_\sigma = \rho_0(f) \in \bar{Z}^1_+(G,X_{k-2,v}(\mathbb{R}))$ and $P^-_\sigma = \rho_0(-if) \in \bar{Z}^1(G,X_{k-2,v}(\mathbb{R}))$.

In the case of $\Gamma_0(N)$ we can take advantage of the fact that ρ is a Hecke-module isomorphism. The action of the Hecke operators T_p, U_q, W_q (see [1]) on $\Gamma_0(N)$ can be used to break $S_k(\Gamma_0(N))$ into spaces of newforms and oldforms and the space of newforms breaks up into one-dimensional eigenspaces with distinct families of eigenvalues for the T_p. This decomposition is carried over by ρ into $\bar{H}^1(G,X(\mathbb{R}))$. In fact it is carried over by ρ_0 into $\bar{Z}^1(G,X(\mathbb{R}))$, since the coboundaries lie in the same eigenspace for the Hecke algebra as an Eisenstein series for $\Gamma_0(N)$. Note that $\bar{Z}^1(G,X(\mathbb{R}))$ has a basis in $\bar{Z}^1(G,X(\mathbb{Q}))$ and this latter space is preserved by the Hecke operators. Finally, by the theory of Atkin-Lehner the newforms are actually cusp forms on $G = \Gamma_o^*(N)$, the group generated by $\Gamma_o(N)$ and the involutions W_q, $q|N$. Thus we get the following theorem.

Theorem 1: Let $f(z)$ be a newform (of Hauptypus) of weight k on $\Gamma_o(N)$. Suppose $f(z) = \sum_{n=1}^{\infty} a_n e^{2\pi inz}$, $a_1 = 1$ and $f^*(z) = (2\pi i)^{-k+1} \sum_{n=1}^{\infty} a_n n^{-k+1} e^{2\pi inz}$. Let K be the (totally real) field generated over $\mathbb{Q}$ by the a_n. There exist real numbers w^+ and w^- (depending only on f) such that for all $\sigma \in \Gamma_o^*(N)$, there are polynomials $A_\sigma^+(z)$ and $A_\sigma^-(z)$ of degree at most $k-2$ with coefficients in K such that

$$(f^* - f^*|_{2-k}\sigma)(z) = w^+A_\sigma^+(z) + iw^- A_\sigma^-(z). \tag{6}$$

3. Coefficients of the Eichler cocycles

In order to apply Theorem 1, we must relate the coefficients of the polynomials $A_\sigma^\pm(z)$ to f. The principal tool used is the Hecke correspondence between Fourier series and Dirichlet series via Mellin transform as described in the following Proposition.

Proposition 1. Let $\lambda > 0$, $f(z) = \sum_{n=0}^{\infty} a_n e^{2\pi inz/\lambda}$, $g(z) = \sum_{n=0}^{\infty} b_n e^{2\pi inz/\lambda}$, $\varphi(s) = \sum_{n=1}^{\infty} a_n n^{-s}$, $\psi(s) = \sum_{n=1}^{\infty} b_n n^{-s}$, $\Phi(s) = (2\pi/\lambda)^{-s} \Gamma(s)\varphi(s)$, $\Psi(s) = (2\pi/\lambda)^{-s}\Gamma(s)\psi(s)$. Assume that for some real number c, the complex numbers a_n, b_n satisfy $a_n, b_n = O(n^c)$. Let $\gamma, k \in \mathbb{C}$. The following are equivalent.

A. $\Phi(s) = \gamma\Psi(k-s)$ and there is a rational function $R(s)$ such that $\Phi(s) - R(s)$ is entire and bounded in vertical strips (EBV).

B. $f(z) = \gamma\left(\frac{z}{i}\right)^{-k} g\left(-\frac{1}{z}\right) + \sum \operatorname{Res} R(s)\left(\frac{z}{i}\right)^{-s}ds$, where the sum is over the poles of $R(s)$.

Next, observe that differentiation of $f(z)$ essentially corresponds to changing $\varphi(s)$ to $\varphi(s-1)$. In general this leads to a more complicated functional equation. However, let k be an integer, $k \geq 2$ and let $\varphi^*(s) = \varphi(s+k-1)$ and $\psi^*(s) = \psi(s+k-1)$. In this case, if $\Phi^*(s) = (2\pi/\lambda)^{-s}\Gamma(s)\varphi^*(s)$ and $\Psi^*(s) = (2\pi/\lambda)^{-s}\Gamma(s)\psi^*(s)$, and if Φ and Ψ satisfy condition A of Proposition 1, then

$$\Phi^*(s) = (-1)^{k-1}\gamma\Psi^*(2-k-s). \tag{7}$$

In addition, there is a rational function $R^*(s)$ such that $\Phi^*(s) - R^*(s)$ is EBV. A residue computation yields the following Theorem.

<u>Theorem 2</u>: Let $f(z) = \sum_{n=0}^{\infty} a_n e^{2\pi inz/\lambda}$, $g(z) = \sum_{n=0}^{\infty} b_n e^{2\pi inz/\lambda}$,

$$f^*(z) = \frac{a_0 z^{k-1}}{(k-1)!} + \left(\frac{2\pi i}{\lambda}\right)^{-(k-1)} \sum_{n=0}^{\infty} a_n n^{-k+1} e^{2\pi inz/\lambda},$$

$$g^*(z) = \frac{b_0 z^{k-1}}{(k-1)!} + \left(\frac{2\pi i}{\lambda}\right)^{-(k-1)} \sum_{n=0}^{\infty} b_n n^{-k+1} e^{2\pi inz/\lambda} \quad \text{and}$$

$\varphi(s) = \sum_{n=1}^{\infty} a_n n^{-s}$. If k is a positive integer and γ a complex number such that $f(z) = \gamma z^{-k} g\left(-\frac{1}{z}\right)$, then

$$f^*(z) = \gamma z^{k-2} g^*\left(-\frac{1}{z}\right) + \sum_{j=0}^{k-2} \frac{\varphi(k-1-j)}{j!} \left(\frac{2\pi i}{\lambda}\right)^{-(k-1-j)} z^j.$$

Let G be a subgroup of $SL_2(\mathbb{R})$, v a character of G and k a positive integer. Let $\sigma = \begin{pmatrix} a & b \\ c & d \end{pmatrix} \in G$ with $c \neq 0$. If f is any function on the upper half plane, define f_σ by

$$f_\sigma(z) = f\left(\frac{z}{|c|} - \frac{d}{c}\right).$$

The following Proposition is immediate.

Proposition 2: Let $\sigma \in G$, $\sigma = \begin{pmatrix} a & b \\ c & d \end{pmatrix}$. The following are equivalent:

A. $f|_{k,v}\sigma = f$.

B. $f_\sigma(z) = (\operatorname{sgn} c)^{-k}\, v(\sigma) z^{-k}\, f_{\sigma^{-1}}(-\frac{1}{z})$.

Thus, modulo some regularity conditions, f is an automorphic form on G if and only if the f_σ satisfy condition B of Proposition 2 for all $\sigma \in G$ — or even for a set of generators of G. By Proposition 1, this condition can be restated in terms of functional equations for the related Dirichlet series. We do not do so explicitly but simply combine Proposition 2 and Theorem 2:

Theorem 3: Let f be an automorphic form of weight k and multiplier v on a discrete subgroup G of $SL_2(\mathbb{R})$. Suppose G contains a (minimal) translation $\begin{pmatrix} 1 & \lambda \\ 0 & 1 \end{pmatrix}$, $\lambda > 0$, and that $f(z)$ has the Fourier expansion $f(z) = \sum_{n=0}^{\infty} a_n e^{2\pi i n z/\lambda}$. For each $\sigma \in G$, $\sigma = \begin{pmatrix} a & b \\ c & d \end{pmatrix}$, $c \neq 0$, let $\varphi_\sigma(s) = \sum_{n=1}^{\infty} e^{-\frac{2\pi i n d}{c\lambda}} a_n n^{-s}$. Let $f^*(z)$ be the $(k-1)$-fold integral of $f(z)$, $f^*(z) = \frac{a_0 z^{k-1}}{(k-1)!} + \left(\frac{2\pi i}{\lambda}\right)^{-(k-1)} \sum_{n=1}^{\infty} n^{-k+1} a_n e^{2\pi i n z/\lambda}$. Then

$$(f^* - f^*|_{2-k,v}\sigma)(z) = \left(\frac{2\pi i}{\lambda}\right)^{-(k-1)} \sum_{j=0}^{k=2} \frac{\varphi_\sigma(k-1-j)}{j!} \left(\frac{2\pi i}{\lambda}\right)^{j} (z+\frac{d}{c})^j. \quad (8)$$

Now let $f(z)$ be a newform of weight k for $\Gamma_0(N)$, $f(z) = \sum_{n=1}^{\infty} a_n e^{2\pi i n z}$, $a_1 = 1$. Define mixed Dirichlet-Fourier series

$\varphi(s,u) = \sum_{n=1}^{\infty} a_n e^{2\pi i n u} n^{-s}$; $\varphi^+(s,u) = \sum_{n=1}^{\infty} a_n (\cos 2\pi n u) n^{-s}$; $\varphi^-(s,u) =$

$\sum_{n=1}^{\infty} a_n (\sin 2\pi nu) n^{-s}$. Let K be the field generated over $\mathbb{Q}$ by the a_n and let w^+ and w^- be the real numbers defined by Theorem 1.

Theorem 4: Let c be a positive integer such that $((N,c), N/(N,c)) = 1$. If d is any integer such that $(d,c) = 1$, then $\varphi(s, -d/c)$ is an entire function and if j is any integer, $0 \le j \le k-2$, then

$$\pi^j \varphi^{\pm}(k-1-j, -d/c) \in \begin{cases} Kw^{\pm} & \text{if } j \text{ is even} \\ Kw^{\mp} & \text{if } j \text{ is odd.} \end{cases}$$

Proof: This is a direct consequence of Theorems 1 and 3 provided there is a matrix $\sigma \in \Gamma_0{}^*(N)$ such that $\varphi_\sigma(s) = \varphi(s, -d/c)$. If N divides c, there is such a σ in $\Gamma_0(N)$. The others occur in $\Gamma_0{}^*(N)$ because of the presence of the W_q for $q|N$.

Remark: Using the fact that $a_q = 0$ if q is a prime such that $q^2|N$, it is possible to remove the condition on c from the hypotheses of Theorem 4. However, in this case K may have to be replaced by the maximum real subfield of $K(e^{2\pi i/c})$.

§4. Eisenstein Series

Theorem 3 is valid for Eisenstein series as well as cusp forms. In [8], the period polynomials for the Eisenstein series of arbitrary level N are evaluated. We describe here one consequence of a special case of this computation. Let $G_k(z)$ be the Eisenstein series for the full modular group Γ:

$$G_k(z) = \sum_{c,d=-\infty}^{\infty}{}' (cz+d)^{-k} \quad (k > 2,\ k \text{ even}).$$

It was mentioned above (just before Theorem 1) that the coboundaries in $\bar{Z}^1$ are eigenvectors for the Hecke operators T_p. Let $\tilde{Z}^1(G,X(K))$ be the space of cocycles P such that $P_\sigma = 0$ if σ is a translation. Then $\delta_o(G_k) \in \tilde{Z}^1(\Gamma, X_{k-2}(\mathbb{C}))$ and the cocycle $\delta_o(G_k)$ is an eigenvector for the Hecke operators T_p with the same eigenvalues as the coboundaries. The eigenspace corresponding to these eigenvalues is just two dimensional. Thus the theory discussed in §2 predicts that the (k-1)-fold integral $G_k^*(z)$ of $G_k(z)$ should satisfy the condition

$$(G_K^*(z) - G_k^*|_{2-k}\sigma)(z) = Q_\sigma(z) + \alpha(1 - 1|_{2-k}\sigma), \quad \sigma \in \Gamma,$$

where the $Q_\sigma(z)$ are polynomials with rational coefficients and α is a constant.

In [8] it is shown that the coefficients of $Q_\sigma(z)$ are generalized Dedekind sums involving Bernoulli polynomials. But perhaps more surprisingly, the constant α turns out to be $(2\pi i)^{-(k-1)}\zeta(k-1)$, where $\zeta(s)$ is the Riemann zeta function!

References

1. A.O.L. Atkin and J. Lehner, Hecke operators on $\Gamma_o(m)$, Math. Ann., vol. 185(1970), 134-160.

2. J. Manin, Cyclotomic fields and modular curves, Russian Math. Surveys 26(1971), no. 6.

3. J. Manin, Parabolic points and zeta functions of modular curves, Math. USSR Izvestia, vol. 6(1972) No. 1, 19-64.

4. J. Manin, Periods of parabolic forms and p-adic Hecke series, Math. USSR Sbornik, vol. 21(1973), no. 3, 371-393.

5. B. Mazur and H.P.F. Swinnerton-Dyer, Arithmetic of Weil curves, Invent. Math. 25(1974), 1-61.

6. V. Miller, Ph.D. thesis, Harvard (1975).

7. M.J. Razar, Dirichlet series and Eichler cohomology, to appear.

8. M.J. Razar, Integrals and periods of Eisenstein series, to appear.

9. G. Shimura, Sur les integrales attachées aux formes automorphes, J. Math. Soc. Japan, vol. 11(1959), 291-311.

10. G. Shimura, Introduction to the Theory of Automorphic Functions, Princeton University Press, Princeton, N.J. (1971).

11. G. Shimura, The special values of the zeta functions associated with cusp forms, (preprint).

ANALYTIC PROPERTIES OF

EULER PRODUCTS OF AUTOMORPHIC REPRESENTATIONS

by C.J. Moreno

CONTENTS

0. INTRODUCTION.

This is a survey of the results that have been obtained by studying the analytic properties of the Euler products of automorphic representations on the group $GL_2(\mathbf{A})$ over the rationals. We shall use as far as possible the notation of Gelbart's monograph [2]. Also implicit in our presentation is the frequent use of the well known correspondence between automorphic representations and automorphic forms that are eigenfunctions of all the Hecke operators. Full details of our proofs will appear elsewhere.

1. THE RANKIN TRICK.

The significance of Rankin's convolution idea in the study of Euler products is now well understood. For convenience we examine some of its implications in the simplest possible situation. Let π and π' be auotmorphic representations which are unramified everywhere, that is, each local component is a class one representation; assume furthermore that the components at the infinite prime belong to the same holomorphic discrete series parametrized by the weight k. To these representations one associates in a natural way an Euler product $L(s,\pi \times \pi')$; the meromorphic continuation of this Euler product is provided by the following integral formula whose form clearly embodies the essence of Rankin's trick :

$$L(\tfrac{1+s}{2},\pi \times \pi') = \int_F f(z)\overline{g(z)}\, E(z,s)(\mathrm{Im}\, z)^k d\Omega,$$

where f and g are respectively primitive cusp forms associated to π

and π', F is a fundamental domain in the Poincaré upper half plane for the modular group Γ, $d\Omega$ is an SL_2-invariant measure and $E(z,s)$ is the Eisenstein series for Γ. The functional equation for $L(s,\pi\times\pi')$, and hence its meromorphic continuation, can be read off immediately from the well known ([4], p. 46) functional equation

$$E(z,s) = \frac{\Lambda(s)}{\Lambda(s+1)} E(z,-s),$$

where

$$(\mathrm{Im}\ z)^{\frac{1+s}{2}} + \frac{\Lambda(s)}{\Lambda(s+1)} (\mathrm{Im}\ z)^{\frac{1-s}{2}}$$

is the constant term of the Fourier expansion of $E(z,s)$; in fact $\Lambda(s)$ is non other than Riemann's zeta function completed with its local factor at infinity. The resulting functional equation for $L(s,\pi\times\pi')$ is

$$L(\tfrac{1+s}{2},\pi\times\pi') = \frac{\Lambda(s)}{\Lambda(s+1)} L(\tfrac{1-s}{2}, \pi\times\pi').$$

When the components of π and π' at the infinite prime are arbitrary and ramification is allowed at the finite primes the above functional equation has to be replaced by a vector equation and the scalar $\frac{\Lambda(s)}{\Lambda(s+1)}$ has to be replaced by the constant term matrix of suitably constructed Eisenstein series. In many cases one actually gets scalar functional equations. In the general situation the construction of the necessary Eisenstein series depends on the type of local components at infinity, the conductors and the central characters of the two representations π and π'; when this data coincides and the central character of π' is the complex conjugate of that of π the resulting Eisenstein series has a simple pole at $s = 1$. The residue of $L(s,\pi\times\pi')$ is a constant multiple of the Petersson inner product (f,g). This fact applied to an automorphic representation π and its contragredient leads to the interesting result, already known to Rankin in special cases, that the Fourier coefficients of the associated cusp form satisfy on the average the Petersson-Ramanujan conjecture. When π' is not the contragredient of π, the Euler product $L(s,\pi\times\pi')$ is free of poles or zeros to the right of the line $Re(s) = 1$. The appropriate generalization ([3], p. 127) of the ideas of Rankin leads to the following interesting result, which we view as the main consequence of Rankin's trick : let π be an automorphic representation of $GL_2(\mathbf{A})$ and $L(s,\pi)$ its associated Euler product; then $L(s,\pi)$ does not vanish outside the critical strip which is comprised between $0 \leqslant Re(s) \leqslant 1$.

Langlands [5] has constructed some very general Euler products and it is a problem of considerable interest, we believe, to isolate the precise location of the critical strips.

2. ZERO-FREE REGIONS.

Rankin had already realized that the further investigation of the location of the zeros of the Euler product $L(s,\pi)$ of an automorphic representation within the critical strip depends on the analytic properties of $L(s,\pi\times\pi')$; he was in fact able, at least in the case of the automorphic representation $\pi = \pi(\Delta)$ connected with Ramanujan's modular form of weight twelve, to extend to these Euler products the method that Hadamard and de la Vallée Poussin had successfully used in proving the non-vanishing of $\Lambda(s)$ on the line $Re(s) = 1$. A simple generalization of Rankin's method leads to the following result : if π is an automorphic representation of $GL_2(\mathbf{A})$ and $L(s,\pi)$ is the associated Euler product, then $L(s,\pi) \neq 0$ for $Re(s) = 1$. In fact we can strengthen this result to show that $L(s,\pi)$ does not vanish on regions inside the critical strip which are similar to those that occur in the theory of Dirichlet L-functions. Here the estimates for these zero-free regions depend in an essential way on three parameters :

a) the type of the local component corresponding to the infinite prime,

b) the conductor of the representation and

c) the size of the representation measured in terms of the Petersson inner product $\|\pi\|$.

The parameters a) and b) arise in a natural way from the consideration of Γ-factors and the presence of the conductor in the functional equation. The third parameter seems to be unnatural and enters into the computations when use is made of the fact that the Petersson-Ramanujan conjecture is true on the average. This seems to be necessary if one wants to get results of some generality, which apply for example to real analytic cusp forms. If one assumes that the local component for the infinite prime of the representation π belongs to the holomorphic discrete series, then one can use the theorem of Deligne on the size of the eigenvalues of the Hecke operators and obtain zero-free regions which can be effectively described in terms of the conductor and the local component at the infinite prime.

The corresponding problems for the Euler products $L(s,\pi\times\pi')$ are more difficult and somewhat incomplete at the present moment. Nevertheless it can be proved, again using another idea of Rankin, that except for the possible simple pole at $s = 1$ the Euler product $L(s,\pi\times\pi')$ is free of zeros or poles in a logarithmic region of exactly the same type as for the Euler product $L(s,\pi)$.

The best results that have been obtained thus far for the zero-free regions for the Euler products $L(s,\pi)$ is the following density estimate :

$$\{\rho = \beta + i\gamma : L(\rho,\pi) = 0, \beta \geqslant \alpha, \ |\gamma| \leqslant T\} \ll T^{c(1-\alpha)},$$

where the implied constants depend on the parameters a), b) and c) already **mentioned above** . The proof of this result for any automorphic representation π uses the techniques of the large sieve as developed by Gallagher [1]. We believe that further improvement of these techniques, say along the lines of the recent work of Montgomery and Selberg, will lead to improved estimates for the constants.

3. EXPLICIT FORMULAS.

The study of the analytic properties of the Euler products $L(s,\pi)$ owes its interest to the possibility of relating the behaviour of the characters of the local factors of the automorphic representation $\pi = \hat{\otimes}\, \pi_p$ to the distribution of the zeros of $L(s,\pi)$ as Riemann had already realized in the case of $\Lambda(s)$. A problem in analytic number theory which promises to touch on much fertile ground and whose implications have not yet been fully noted is the working out of Riemann's program [7] for the Euler products of automorphic representations of linear reductive groups. One of our earlier results, which is in fact relatively easy to prove, is the following von Mangoldt formula which for simplicity we only state for the Ramanujan automorphic representation $\pi = \pi(\Delta)$:

$$\sum_{p^n \leqslant X} (\lambda_p^n + \overline{\lambda}_p^n)\log p = -\sum_{\rho} \frac{X^\rho}{\rho} - \log(X-1) - \frac{1}{2}\cdot\frac{L^{(2)}(0,\pi)}{L^{(1)}(0,\pi)} ,$$

where $\tau(p) = \lambda_p + \overline{\lambda}_p$ is Ramanujan's arithmetical function, $L(s,\pi)$ is the well known Euler product for $\tau(p)$ normalized so that its critical strip lies in $\frac{11}{2} < \mathrm{Re}(s) < \frac{13}{2}$ and the sum $\sum_\rho$ runs over all zeros of $L(s,\pi)$ in the critical strip.

For arbitrary automorphic representations π of $GL_2(\mathbf{A})$ we have obtained similar explicit formulas which have a more complex appearance but which in principle, as is to be expected, relate the characters of the local factors π_p of π and the zeros of the Euler product $L(s,\pi)$.

This last remark suggests the useful role that explicit formulas will play in the arithmetic study of automorphic representations. In this connection it should be remarked that the above explicit formula is derived from truncated explicit formulas where only a finite number of zeros appear explicitely.

4. ZETA DISTRIBUTIONS.

In analytic number theory the explicit formulas that are most frequently used are the truncated ones. There is a generalized explicit formula, considered by Weil [14], which seems to have received little attention up to now but that in the long run may prove to be more useful. To avoid introducing much notation we describe this formula in a simple but already significant situation. Let $\pi = \hat{\otimes}\, \pi_p$ be an automorphic representation of $GL_2(\mathbf{A})$. For our convenience we make the restrictive assumption that it is unramified, i.e. each π_p is a class one representation. Let $\{g_p\}$ be the semisimple conjugacy class in $GL_2(\mathbf{C})$ associated to the zonal spherical function of π_p. Let $r = \mathrm{sym}^3$ be the third symmetric power of the standard 2-dimensional representation of $GL_2(\mathbf{C})$ and let χ be its complex character. For the infinite prime $p = \infty$ we consider the characteristic polynomial

$$\det(1-r(g_\infty)T) = (1-\lambda_1 T)(1-\lambda_2 T)(1-\lambda_3 T)(1-\lambda_4 T).$$

The Euler product

$$L(s,\pi,r) = \prod_{i=1}^{4} \pi^{-\frac{s-\lambda_i}{2}}\, \Gamma\left(\frac{s-\lambda_i}{2}\right).\prod_p \det(1-p^{-s}r(g_p))^{-1}$$

was introduced by Langlands [5]; its functional equation

$$L(s,\pi,r) = \varepsilon(\pi,r)L(1-s,\pi,\tilde{r}),$$

where $\tilde{r}$ is the contragredient of r and $|\varepsilon(\pi,r)| = 1$, has been established, using Langlands' theory of general Eisenstein series, by S. Shahidi [11].

Let L be the class of complex valued functions h on the real line satisfying the following three conditions :

i) there is a real number $a > 0$ such that the function $h(x)\exp(\frac{1}{2}+a)|x|$ is integrable on the real line;

ii) h is continuous and continuously differentiable everywhere except at a finite number of points $\{\alpha\}$ where h and its derivative h' have only a discontinuity of the first kind such that $h(\alpha) = \frac{1}{2}(h(\alpha+)+h(\alpha-))$;

iii) there is a real number $b > 0$ such that $h(x)\exp(\frac{1}{2}+b)|x|$ and $h'(x)\exp(\frac{1}{2}+b)|x|$ decay to zero as $|x| \to \infty$.

Define the Mellin transform of a function h in L by

$$\hat{h}(s) = \int_{-\infty}^{\infty} h(t)e^{(s-\frac{1}{2})t}\,dt.$$

For each finite prime p and h in L we define a local distribution by

$$W(\pi_p,r)h = -\sum_{n=1}^{\infty} p^{-\frac{n}{2}}\chi(g_p^n)h(\log p^n)\log p;$$

for the infinite prime we put

$$W(\pi_\infty,r)h = \lim_{m\to\infty}\left\{-\int_0^\infty \frac{1-e^{-(m+1)2x}}{1-e^{-2x}}\cdot e^{-\frac{x}{2}}\left(\sum_{i=1}^{4} e^{\lambda_i x}\right)h(x)dx + 2h(0)\log\frac{m}{\pi}\right\};$$

adding up the local terms leads to a global distribution

$$W(\pi,r) = \sum_p W(\pi_p,r)$$

where the sum is taken over all primes. The symmetry arising from the functional equation of $L(s,\pi,r)$ is also reflected in the explicit formula; to exhibit its presence we now introduce the concept of a <u>Weyl</u>

transform which applied to the distribution $W(\pi,r)$ defined above gives a new distribution $W^{\omega}(\pi,r)$ according to the rule

$$W^{\omega}(\pi,r)h(t) = W(\pi,\tilde{r})h(-t),$$

for any function h in L, where $\tilde{r}$ is the contragredient of r.

We now have the explicit formula : for any function h in L,

$$\sum_{\rho} \hat{h}(\rho) = (W(\pi,r) + W^{\omega}(\pi,r)).h,$$

where the sum $\sum_{\rho}$ runs over the zeros (and poles) of $L(s,\pi,r)$. The most commendable atribute of this formula is the remarkable resemblance, which indeed is not accidental, of the right hand side with the constant term of an Eisenstein series associated with a rank one parabolic subgroup of a reductive linear group. One can almost visualize the right hand side as an average of distributions parametrized by the elements of a Weyl group, which in our case contains only 2 elements. An elementary argument, already used by Weil [14], shows that the Euler product $L(s,\pi,r)$ is an entire function and that its zeros have real part $\frac{1}{2}$ if and only if the distribution $W + W^{\omega}$ is positive. Because of this last remark it is to be expected that the symmetric nature of the distribution $W + W^{\omega}$ is closely related to the Hermitian property of positive distributions ([8], p. 131). The next step in this direction will be to develop similar explicit formulas for Euler products in more than one complex variable.

Another significant problem that must be solved here is to determine the correct structure of the contribution to the distribution $W + W^{\omega}$ when π admits ramification. At present we are only able to obtain a term which exhibits no obvious Weyl symmetry. We have now undertaken a study of the Herbrand distribution attached to automorphic representations of $GL_2(\mathbf{A})$ similar to those that appear in [14].

5. ZEROS ON THE REAL LINE.

The principle of the argument applied to the Euler product $L(s,\pi)$ of an automorphic representation π $GL_2(\mathbf{A})$ together with simple estimates for the Γ-factors shows that the multiplicity of a real zero of $L(s,\pi)$, if it has any at all, is bounded from above by an absolute constant times $\log(100\,\lambda^2 f(\pi))$, where λ is the eigenvalue of the casimir operator that parametrizes the local factor π_∞ of π and $f(\pi)$ is the conductor of π. In the particular case when π_∞ belongs to the holomorphic discrete series Serre obtains, using the finer arguments of Stark and Odlyzko, that the multiplicity $m(\rho)$ of a real zero ρ of $L(s,\pi)$ is bounded by $m(\rho) \leqslant 2.4 + \frac{3}{4}\log(k^2 f(\pi))$, where k is the "weight" of π. Further improvements in this direction would complement the Birch and Swinnerton-Dyer conjecture about the relation between the rank of the Mordell-Weil group of an elliptic curve and the multiplicity of the real zero of the Hasse-Weil zeta function at the real point on the critical line.

The principle of the argument applied to the Hasse-Weil zeta functions of algebraic varieties defined over number fields ([10], p. 3) also leads to conjectural logarithmic estimates for the ranks of Picard groups of these varieties ([13], p. 104).

When the Hasse-Weil zeta function of an elliptic curve E defined over the rationals is the Euler product of an automorphic representation of $GL_2(\mathbf{A})$, then its Hasse-Weil zeta function $L(s,E,k_n)$ over the cyclotomic extension k_n containing the n-th roots of 1 can be written out explicitly . If R is a fixed rectangle in the critical strip one can apply the methods of Siegel ([12], p. 47) to study the asymptotic distribution of the zeros of $L(s,E,k_n)$ inside R as $n \to \infty$. It is of some interest to investigate more closely the nature of these asymptotic results as the rectangle R decreases in area.

6. INFINITE PRIME NUMBER THEOREMS.

Let Δ be the automorphic representation connected with the Ramanujan modular form. Let $\tau(p) = 2p^{\frac{11}{2}} \cos \theta_p$, $0 \leq \theta_p \leq \pi$ be Ramanujan's arithmetical function. As a natural generalization of the prime number theorem for arithmetic progressions, Čebotarev's density theorem and Hecke's theorem on the uniform distribution of the arguments of grossencharacters one is led to consider the problem of the distribution of the complex parameters s_p that index the spherical functions associated with almost all the local components of an automorphic representation π. An example of this type of question is the problem of the distribution of the angles θ_p. It is known ([9],I-25) in this situation that if all the Euler products $L(s,\Delta,r)$, as r runs through all the symmetric powers of the standard two dimensional representation of SU(2), do not vanish on the line $Re(s) = 1$, then the angles θ_p are uniformly distributed in $[0,\pi]$ with respect ot the measure $\frac{2}{\pi} \sin^2 \phi d\phi$. The arguments that lead to this type of prime number theorem involve the use of an infinite number of Euler products and it is not clear how to get error terms comparable to those that are possible in the classical prime number theorems without making heavy assumptions on the distribution of the zeros of all the Euler products $L(s,\Delta,r)$. The problem of improving the error terms in distribution results like the Sato-Tate conjecture leads one to consider the following formal identity. Let S be a subset of the space of conjugacy classes in SU(2). Let χ_S be the characteristic function of S, or a smooth approximation of it, and consider the Fourier expansion of χ_S according to the characters χ_r of the finite dimensional complex representations of SU(2) :

$$\chi_S(g) = \sum_r a_r(S)\chi_r(g).$$

Now using the notation of our §4 we have for π an automorphic representation of $GL_2(\mathbf{A})$, h a function of the class L and $W(\pi,r)$ the associated zeta distribution, formally the identity

$$\sum_r a_r(S) \sum_{\rho(r)} \hat{h}(\rho) = \sum_r a_r(S)(W(\pi,r) + W^{\omega}(\pi,r)).h,$$

where for a fixed representation r, the second sum $\sum_{\rho(r)}$ runs over all the zeros of $L(s,\pi,r)$. A rigorous derivation of this formula as well as truncated forms of it, at least for special types of functions h, will undoubtedly be of some significance to the problems of finding the distribution of the eigenvalues of Hecke operators. These problems have obvious generalizations to automorphic representations of reductive linear groups whose clear formulation will be done elsewhere.

7. EFFECTIVE COMPUTABILITY.

The known effective versions of the Čebotarev density theorem can be used to establish the following result. If σ and σ' are two continuous representations of $Gal(\overline{Q}/Q)$ of the same degree and of the same conductor $f(\sigma)$ with characters χ and χ', then there is an effectively computable constant c depending on the degree of the representation σ but not on $f(\sigma)$ such that if $\chi(g_p) = \chi'(g_p)$ for all the Frobenius conjugacy classes g_p associated to primes $p \leqslant f(\sigma)^c$ then σ and σ' are equivalent. Experimental evidence seems to suggest that the numerical values obtained for the constant c by making effective the known large sieve techniques do not give the truth. In fact if σ and σ' are of degree 2 and if their corresponding Artin L-functions are actually Euler products of automorphic representations $\pi(\sigma)$ and $\pi(\sigma')$ of $GL_2(\mathbf{A})$ whose infinity components belong to the holomorphic discrete series, then a simple dimensionality argument shows that the constant c can be taken to be smaller than 2. A simple geometric implication of the effective Čebotarev density theorem is that one can effectively decide whether two cubic surfaces, defined by equations with integer coefficients, have the same Hasse-Weil zeta function. Along these lines one can also make the following observation : let E and E' be elliptic curves defined by equations with integer coefficients and of the same conductor f; if their Hasse-Weil zeta functions are actually Euler products of automorphic representations and if the reductions modulo p of E and E' have the same congruence zeta function for all primes $p \leq f^2$, then E and E' are isogenous.

At the other extreme of the spectrum Casselman and Miyake have proved a strong version of the multiplicity one theorem; namely if π and π' are two automorphic representations of $GL_2(\mathbf{A})$ and if all, except a finite number, of their local components are equivalent, then the representations π and π' are themselves equivalent. Both proofs depend on the converse theorem of Jacquet-Langlands' version of the Hecke theory and are clearly ineffective since one must be able to twist by an infinite number of grossencharacters.

Over the rationals one can prove an effective version of the Casselman-Miyake theorem by simple dimensionality arguments. This has been done for automorphic representations that correspond to holomorphic cusp forms by Winnie Li (thesis) and in the real analytic case by persuing a simple idea of Mass. A weakened form of the conclusion is that if π and π' are two automorphic representations of $GL_2(\mathbf{A})$ with the same conductor f, then the unitary equivalence of the local components π_p and π'_p for all primes p less than some power of the conductor f, then π and π' are equivalent.

At the present time we are unable to prove an effective version of the Casselman-Miyake theorem for automorphic representations of $GL_2(\mathbf{A})$

over arbitrary number fields. Nevertheless a combination of the Rankin trick, our explicit formulas for the Euler products of automorphic representations of $GL_2(\mathbf{A})$ and the large sieve technique yields a proof of the following result. If π and π' are two automorphic representations of $GL_2(\mathbf{A})$ whose conductors have norms bounded by N, then there is a constant c depending on π_∞ and π'_∞ such that if the local components π_p and π'_p are equivalent for all primes $p \leq N^c$, then π and π' have the same Euler product $L(s,\pi) = L(s,\tilde{\pi}')$. It can be shown that the constant c is effectively computable if Satake's analogue of the Petersson-Ramanujan conjecture holds for automorphic representations of $GL_2(\mathbf{A})$. The reason why we are unable to show that π is equivalent to π' is again the necessity, inherent in the converse theorem to the Hecke theory, of having to twist by an infinite number of grossencharacters. Without taking into consideration this difficulty, we believe that a problem in the theory of automorphic representations, whose solution will be of great significance in diophantine analysis, is the establishment of an effective strong multiplicity one theorem : one should be able to tell whether two automorphic representations are equivalent by comparing only a finite number of their local components.

8. THE MONTGOMERY PHENOMENON.

Some conjectural investigations of H. Montgomery ([6], p. 184) concerning the distribution of the zeros of Riemann's Euler product $\Lambda(s)$ on the critical line have led him to some interesting speculations. In the framework of the theory of automorphic representations these observations of Montgomery can be extended, without much technical difficulty, to say that if π is an automorphic representation of $GL_2(\mathbf{A})$ with associated Euler product $L(s,\pi)$ and if the corresponding zeros are located on the line Re $s = \frac{1}{2}$, then the pair correlation function of the zeros of $L(s,\pi$ is identical with the pair correlation function of the eigenvalues of a random complex hermitian or unitary matrix of large order. It would be of some interest to understand Montgomery's phenomenon for automorphic representations from the point of view of harmonic analysis on $GL_2(\mathbf{A})$ and related groups.

9. THE VALUES OF EULER PRODUCTS AT INTEGER POINTS.

To end this circle of ideas we mention an elementary, though quite pretty, result that falls out immediately from Rankin's trick already stated in §1. To simplify our notation we assume that π and π' are two automorphic representations which are both unramified and have the same local component at infinity, say a member of the holomorphic discrete series; assume furthermore that their corresponding cusp forms on the Poincaré upper half plane are f and g. Now, if π and π' are not equivalent, then $L(s,\pi\times\pi')$ in the notation of §1, is regular at $s = 1$ and in fact its value is given by

$$L(1,\pi\times\pi') = -\frac{1}{2\pi}\int_F (\operatorname{Im} z)^k f(z)\overline{g(z)}\log(\operatorname{Im} z)^6 |\Delta(z)|)d\Omega,$$

where $\Delta(z)$ is Ramanujan's modular form. The proof of this result is an exercise in the use of Kronecker's limit formula to write the constant term in the Laurent expansion about $s = 1$ of the Eisenstein series $E(s,z)$. Similar results, which involve more complicated expressions are also possible for arbitrary automorphic representations.

REFERENCES

[1] GALLAGHER P.X., A large sieve density estimate near s = 1, Inventiones Math. 11 (1970), 329-339.

[2] GELBART S., Automorphic Forms on Adele Groups, **Annals of Mathematics Studies, Princeton University** Press, 1974.

[3] JACQUET H., Automorphic forms on GL(2). Part II. Springer Lecture Notes, Volume 278, 1972.

[4] KUBOTA T., Elementary Theory of Eisenstein Series, Kodansha Ltd. and John Wiley, 1973.

[5] LANGLANDS R.P., Euler Products, Yale University Press, 1971.

[6] MONTGOMERY H.L., The pair correlation of zeros of the zeta function, Proc. Sympos. Pure Math., vol. 24, Amer. Math. Soc., Providence, R.I., 1973, 181-193.

[7] RIEMANN B., Ueber die Anzahl der Primzahlen unter einer gegebenen Grosse, Collected Works, Dover Publications, Inc., New York, 1953, 145-155.

[8] SCHWARTZ L., Théorie des Distributions, vol. II, Hermann, Paris, 1959.

[9] SERRE J-P., Abelian L-Adic Representations and Elliptic Curves, W.A. Benjamin, Inc., New York, 1968.

[10] SERRE J-P., Facteurs locaux de fonctions zeta des variétés algébriques (définitions ét conjectures), Séminaire Delange-Pisot-Poitou, no. 19 (1970).

[11] SHAHIDI F., Functional equation satisfied by certain L-functions, preprint (1976).

[12] SIEGEL C.L., On the zeros of the Dirichlet L-functions, Ann. of Math. 46, (1945), 409-422.

[13] TATE J.T., Algebraic cycles and poles of zeta functions, Proc. Purdue Univ. Conf., 1963, 93-110.

[14] WEIL A., Sur les formules explicites de la théorie des nombres, Isv. Acad. Nauk 36 (1972), 3-18.

International Summer School on Modular Functions
BONN 1976

MODULAR FORMS OF WEIGHT 1/2

by

J-P. Serre and H.M. Stark

Contents

INTRODUCTION

In his Annals paper on modular forms of half integral weight [8], Shimura mentions several open questions. One of them is the following : <u>is every form of weight</u> 1/2 <u>a linear combination of theta series in one variable</u> ?

We show that the answer is positive. The precise statements are given in §2, Theorems A and B; they give an explicit basis of modular forms (and cusp forms) of weight 1/2 and given level. The proof uses the fact that, for weight 1/2, the formula defining the Hecke operator $T(p^2)$ introduces unbounded powers of p in the denominators of the coefficients - unless some remarkable cancellations take place (§5). But it is a familiar fact that coefficients of modular forms (on congruence subgroups) have bounded denominators. Hence the above cancellations do hold, and they give us the information we need, when combined with basic properties of "newforms" à la Atkin-Lehner-Li (§§ 3,4). The details are carried out in §§ 6,7. As an Appendix, we have included a letter from Deligne sketching an alternative method, using the "group-representation" point of view.

In the above proofs, arithmetic arguments play an essential role. It would be interesting to have a more analytic proof; a natural line of attack would be to adapt Shimura's Main Theorem ([8], §3) to weight 1/2, but we have not investigated this.

We mention a possible application of Theorems A and B : since the weights 1/2 and 3/2 occur together in dimension formulae and trace formulae ([9], §5), the explicit knowledge of forms of weight 1/2 gives a way of computing these dimensions and traces for weight 3/2.

§1. SOME NOTATION

1.1. Upper half-plane and modular groups.

We use standard notations, cf. [3], [7]. The letter H denotes the upper half-plane $\{z \mid \mathrm{Im}(z) > 0\}$. If $z \in H$, we put $q = e^{2\pi i z}$. Let $\mathbf{GL}_2(\mathbf{R})^+$ be the subgroup of $\mathbf{GL}_2(\mathbf{R})$ consisting of matrices $A = \begin{pmatrix} a & b \\ c & d \end{pmatrix}$ with $\det(A) > 0$; we make $\mathbf{GL}_2(\mathbf{R})^+$ act on H by

$$z \mapsto Az = (az+b)/(cz+d).$$

Let N be a positive integer divisible by 4. We denote by $\Gamma_0(N)$ and $\Gamma_1(N)$ the subgroups of $\mathbf{SL}_2(\mathbf{Z})$ defined by :

$$\begin{pmatrix} a & b \\ c & d \end{pmatrix} \in \Gamma_0(N) \iff c \equiv 0 \pmod{N}$$

$$\begin{pmatrix} a & b \\ c & d \end{pmatrix} \in \Gamma_1(N) \iff a \equiv d \equiv 1 \pmod{N} \text{ and } c \equiv 0 \pmod{N}.$$

The group $\Gamma_1(N)$ is a normal subgroup of $\Gamma_0(N)$, and the map $\begin{pmatrix} a & b \\ c & d \end{pmatrix} \mapsto d$ induces an isomorphism of $\Gamma_0(N)/\Gamma_1(N)$ onto $(\mathbf{Z}/N\mathbf{Z})^*$.

1.2. Characters.

If $t \in \mathbf{Z}$, we denote by χ_t the primitive character of order $\leqslant 2$ corresponding to the field extension $\mathbf{Q}(t^{1/2})/\mathbf{Q}$. If t is a square, we have $\chi_t = 1$. It t is not a square, and the discriminant of $\mathbf{Q}(t^{1/2})/\mathbf{Q}$ is D, then χ_t is a quadratic character of conductor $|D|$, and we have

$$\chi_t(m) = \left(\frac{D}{m}\right) \qquad \text{(Kronecker symbol)}.$$

In particular, $\chi_t(m) = 0$ if and only if $(m,D) \neq 1$. (Recall that, if $t = u^2 d$, with $u \in \mathbf{Z}$, and d is square-free, we have $D = d$ if $d \equiv 1 \pmod 4$, and $D = 4d$ otherwise.)

1.3. Theta multiplier.

Let $\theta(z) = \prod_{n=1}^{\infty} (1-q^{2n})(1+q^{2n-1})^2 = \sum_{-\infty}^{+\infty} q^{n^2} = 1 + 2q + 2q^4 + \ldots$ be the standard theta function. If $A = \begin{pmatrix} a & b \\ c & d \end{pmatrix}$ belongs to $\Gamma_0(4)$, we have

$$\theta(Az) = j(A,z)\theta(z),$$

where $j(A,z)$ is the "θ-multiplier" of A. Recall (cf. for instance [8]) that, if $c \neq 0$, we have

$$j(A,z) = \varepsilon_d^{-1} \chi_c(d)(cz+d)^{1/2},$$

where
$$\varepsilon_d = \begin{cases} 1 & \text{if } d \equiv 1 \pmod 4 \\ i & \text{if } d \equiv -1 \pmod 4), \end{cases}$$

and $(cz+d)^{1/2}$ is the "principal" determination of the square root of $cz + d$, i.e. the one whose real part is > 0 (more generally, all fractional powers in this paper have to be understood as principal values). If $c = 0$, we have $A = \pm 1$, and $j(A,z)$ is obviously equal to 1.

1.4. Modular forms of half integral weight.

Let $\chi : (\mathbf{Z}/N\mathbf{Z})^* \to \mathbf{C}^*$ be a character (mod N), and let κ be a positive odd integer. A function f on H is called a modular form of type $(\kappa/2,\chi)$ on $\Gamma_0(N)$ if :

a) $f(Az) = \chi(d)\, j(A,z)^{\kappa} f(z)$ for every $A = \begin{pmatrix} a & b \\ c & d \end{pmatrix}$ in $\Gamma_0(N)$; this makes sense since $4|N$;

b) f is holomorphic, both on H and at the cusps (see [8]).

One then calls $\kappa/2$ the <u>weight</u> of f, and χ its <u>character</u>. The space of such functions will be denoted by $M_0(N,\kappa/2,\chi)$; it is clear that $M_0(N,\kappa/2,\chi)$ consists only of 0 unless χ is <u>even</u>, i.e. $\chi(-1) = 1$. We put

$$M_1(N,\kappa/2) = \bigoplus_{\chi} M_0(N,\kappa/2,\chi),$$

where the sum is taken over all (even) characters of $(\mathbf{Z}/N\mathbf{Z})^*$; this space is the space of modular forms of weight $\kappa/2$ on $\Gamma_1(N)$.

A modular form which vanishes at all cusps is called a <u>cusp form</u>. The subspace of $M_0(N,\kappa/2,\chi)$ (resp. $M_1(N,\kappa/2)$) made up by cusp forms will be denoted by $S_0(N,\kappa/2,\chi)$ (resp. $S_1(N,\kappa/2)$).

<u>EXAMPLE</u> : <u>theta series with characters</u>.

Let ψ be an even primitive character of conductor $r = r(\psi)$. We put

$$\theta_\psi(z) = \sum_{-\infty}^{\infty} \psi(n)q^{n^2}.$$

When $\psi = 1$, θ_ψ is equal to θ. When $\psi \neq 1$, θ_ψ is equal to :

$$2\sum_{\substack{n\geqslant 1\\(n,r)=1}} \psi(n)q^{n^2} = 2(q + \psi(2)q^4 + \dots).$$

We have $\theta_\psi \in M_0(4r^2,1/2,\psi)$, cf. [8], p.457. This implies that, if t is an integer $\geqslant 1$, the series $\theta_{\psi,t}$ defined by

$$\theta_{\psi,t}(z) = \theta_\psi(tz) = \sum_{-\infty}^{\infty} \psi(n)q^{tn^2}$$

belongs to $M_0(4r^2t,1/2,\chi_t\psi)$, see for instance Lemma 2 below.

<u>Warning</u>. One should not confuse θ_ψ with the series $\sum \psi(n)^2q^{n^2}$ obtained by <u>twisting</u> θ with the character ψ, cf. §7.

1.5. <u>Petersson scalar product</u>.

If $z \in H$, we put $x = \mathrm{Re}(z)$, $y = \mathrm{Im}(z)$. The measure $dxdy/y^2$ is invariant by $\mathbf{GL}_2(\mathbf{R})^+$. If f,g belong to $M_1(N,\kappa/2)$, the function

$$F_{f,g}(z) = f(z)\overline{g(z)}y^{\kappa/2}$$

is invariant by $\Gamma_1(N)$. Hence $F_{f,g}(z)y^{-2}dxdy$ is invariant by $\Gamma_1(N)$ and defines a measure $\mu_{f,g}$ on $H/\Gamma_1(N)$. One checks immediately that $\mu_{f,g}$ is a <u>bounded measure</u> in each of the following two cases :

i) one of the forms f,g is a <u>cusp form</u>;

ii) $\kappa = 1$ (this was first noticed by Deligne).

In each case, the <u>Petersson scalar product</u> $<f,g>$ of f and g is defined as the (absolutely convergent) integral :

$$<f,g> = \frac{1}{c(N)}\int \mu_{f,g} = \frac{1}{c(N)}\int_{H/\Gamma_1(N)} f(z)\overline{g(z)}\, y^{\kappa/2-2}dxdy,$$

where c(N) is the index of $\Gamma_1(N)$ in $\mathbf{SL}_2(\mathbf{Z})$.

This is a hermitian scalar product. One has $<f,f> > 0$ if $<f,f>$ is defined and $f \neq 0$.

§2. STATEMENT OF RESULTS

2.1. Basis of modular forms of weight 1/2.

Our main result (Theorem A below) states that every modular form of weight 1/2 is a linear combination of theta series with characters. More precisely, let χ be an even character (mod N); let $\Omega(N,\chi)$ be the set of pairs (ψ,t), where t is an integer ≥ 1, and ψ is an even primitive character with conductor $r(\psi)$, such that :

(i) $4r(\psi)^2t$ divides N,

(ii) $\chi(n) = \psi(n)\chi_t(n)$ for all n prime to N.

Condition (ii) is equivalent to saying that ψ is the primitive character associated with $\chi\chi_t$; hence ψ is determined by t and χ. Conversely, t and ψ determine χ.

THEOREM A. The theta series $\theta_{\psi,t} = \sum_{-\infty}^{\infty} \psi(n)q^{tn^2}$, with $(\psi,t) \in \Omega(N,\chi)$, make up a basis of $M_0(N,1/2,\chi)$.

This will be proved in §6.

Call $\Omega(N)$ the set of pairs (ψ,t) satisfying condition (i) above; this set is the union of the $\Omega(N,\chi)$, for all even characters χ (mod N); hence Theorem A implies :

COROLLARY 1. The series $\theta_{\psi,t}$, with $(\psi,t) \in \Omega(N)$, make up a basis of the space $M_1(N,1/2)$ of modular forms of weight 1/2 on $\Gamma_1(N)$.

In particular :

COROLLARY 2. If $f = \sum_{n=0}^{\infty} a(n)q^n$ is a modular form of weight 1/2 on $\Gamma_1(N)$, then $a(n) = 0$ if n is not of the form tm^2, where t is a divisor of $N/4$, and $m \in \mathbf{Z}$.

COROLLARY 3. Let $f = \sum_{n=0}^{\infty} a(n)q^n$ be a formal power series with complex coefficients. The following properties are equivalent :

1) f is a modular form of weight 1/2 on some $\Gamma_1(N)$.

2) f is a linear combination of theta series

$$\theta_{n_0,r,t} = \sum_{\substack{n \equiv n_0 \pmod r \\ n \in \mathbf{Z}}} q^{tn^2}$$

3) For each square-free integer $t \geqslant 1$, there is a periodic function ε_t on $\mathbf{Z}$ such that :

3.1) $a(tn^2) = \varepsilon_t(n)$ for every $n \geqslant 1$;

3.2) each ε_t is even (i.e. $\varepsilon_t(n) = \varepsilon_t(-n)$ for all $n \in \mathbf{Z}$);

3.3) ε_t is 0 for all but finitely many t;

3.4) $a(0) = \frac{1}{2}\sum_t \varepsilon_t(0)$.

PROOF. The equivalence of 2) and 3) is elementary. The fact that a theta series is a modular form is well known (cf. for instance [8], §2); hence 2) implies 1). Corollary 2 above shows that 1) implies 3).

COROLLARY 4. *Let* $f = \sum_{n=0}^{\infty} a(n)q^n$ *be a non-zero modular form of weight* 1/2 *on some* $\Gamma_1(N)$. *Then* :

a) $|a(n)| = O(1)$;

b) *for every* $\rho \geqslant 0$, *there is a constant* $c_\rho > 0$ *such that*

$$\sum_{n \leqslant x} |a(n)|^\rho = c_\rho x^{1/2} + O(1) \text{ for } x \to \infty.$$

(If $\rho = 0$ and $a(n) = 0$, we put $|a(n)|^\rho = 0$.)

PROOF. This follows from Corollary 3.

REMARK. If f and g are modular forms of weight 1/2 on $\Gamma_1(N)$, their product $F = f.g$ is a modular form of weight 1. By Theorem A, F is a linear combination of series

$$\sum_{n,m} \alpha(n)\beta(m)\, q^{an^2+bm^2} ,$$

where α and β are characters. This shows that F is a linear combination of Eisenstein series and cusp forms of dihedral type associated with imaginary quadratic fields (cf. [3], §4). Hence, one cannot use products of forms of weight 1/2 to construct "exotic" modular forms of weight 1.

2.2. Cusp forms of weight 1/2.

If ψ is a character with conductor r, one may write ψ in a unique way as $\psi = \prod_{p|r} \psi_p$, where the conductor of ψ_p is the highest power of p dividing r; we call ψ_p the pth-*component* of ψ (in the Galois interpretation of characters, ψ_p is just the restriction of ψ to the inertia group at p). We say that ψ is totally even if all the ψ_p's are even, i.e. if $\psi_p(-1) = 1$ for all $p|r$; this is equivalent to saying that ψ is the square of a character (which can be chosen of conductor r, if r is odd, and of conductor 2r, if r is even).

Denote by $\Omega_e(N,\chi)$ the subset of $\Omega(N,\chi)$ (see above) made up of the (ψ,t) such that ψ is totally even, and put

$$\Omega_c(N,\chi) = \Omega(N,\chi) - \Omega_e(N,\chi).$$

Define similarly

$$\Omega_e(N) = \bigcup_\chi \Omega_e(N,\chi) \ , \ \Omega_c(N) = \bigcup_\chi \Omega_c(N,\chi).$$

THEOREM B. The series $\theta_{\psi,t}$, with $(\psi,t) \in \Omega_c(N,\chi)$, make up a basis of the space $S_0(N,1/2,\chi)$ of cusp forms of $M_0(N,1/2,\chi)$. The series $\theta_{\psi,t}$, with $(\psi,t) \in \Omega_e(N,\chi)$, make up a basis of the orthogonal complement of $S_0(N,1/2,\chi)$ in $M_0(N,1/2,\chi)$ for the Petersson scalar product.

This theorem will be proved in §7. It implies :

COROLLARY 1. The series $\theta_{\psi,t}$, with $(\psi,t) \in \Omega_c(N)$, make up a basis of the space $S_1(N,1/2)$ of cusp forms of weight 1/2 on $\Gamma_1(N)$.

COROLLARY 2. We have $S_1(N,1/2) \neq 0$ if and only if N is divisible by either $64p^2$ where p is an odd prime, or $4p^2p'^2$, where p and p' are distinct odd primes.

Indeed, Cor. 1 shows that $S_1(N,1/2)$ is non-zero if and only if there exists an even character ψ with conductor $r(\psi)$, which is not totally even, and which is such that $r(\psi)^2$ divides N/4. Since ψ is even, at least two p^{th}-components of ψ are odd; this shows that $r(\psi)$ is divisible by either 4p, where p is an odd prime, or by pp', where p and p' are distinct odd primes; hence N is divisible by either $4.(4p)^2 = 64p^2$ or $4(pp')^2 = 4p^2p'^2$. Conversely, if N is divisible by $64p^2$ (resp. by $4p^2p'^2$), one takes for ψ the product of an odd character of conductor p by an odd character of conductor 4 (resp. p'); it is clear that ψ has the required properties.

EXAMPLES. The above results allow an easy determination of the spaces of modular form of weight 1/2 on $\Gamma_0(N)$ and $\Gamma_1(N)$: all one has to do is to make a list of the divisors t of N/4, and, for each such t, determine the even characters ψ with conductor $r(\psi)$ such that $r(\psi)^2$ divides N/4t. The pairs (ψ,t) thus obtained make up the set $\Omega(N)$. We give two examples :

i) $N = 4p_1\ldots p_h$, where the p_i's are distinct primes. In this case t is a product of some of the p_i's, and $r(\psi)$ must be equal to 1, hence $\psi = 1$. Applying Cor. 1 to Th. A, we see that the series

$$\theta(tz) = \sum_{-\infty}^{\infty} q^{tn^2} \qquad \text{(where } t \text{ divides } p_1\ldots p_h)$$

make up a basis of $M_1(N,1/2)$. Moreover, we have $\theta(tz) \in M_0(N,1/2,\chi_t)$; since the χ_t's are pairwise distinct, each $M_0(N,1/2,\chi_t)$ is one-dimensional, and we have $M_0(N,1/2,\chi) = 0$ if χ is not equal to one of the χ_t's (in particular if χ is not real).

ii) Let us determine $S_1(N,1/2)$ for $N < 900$. If this space is $\neq 0$, Cor. 2 to Th. B shows that N is divisible by either $64p^2$ or $4p^2p'^2$ where p,p' are distinct odd primes; the first case is possible only if $N = 576 = 64\cdot 3^2$; the second one is impossible (since it implies $N \geqslant 4.3^2 5^2 = 900$, which contradicts the assumption made on N). Hence we have N = 576, and it is easy to see that the only element of $\Omega_c(N)$ is the pair (ψ,t) with t = 1 and $\psi = \chi_3$ (which has conductor 12). The corresponding theta series is

$$\theta_{\chi_3} = \sum_{n\equiv\pm 1 \pmod{12}} q^{n^2} - \sum_{n\equiv\pm 5 \pmod{12}} q^{n^2}$$
$$= 2(q - q^{25} - q^{49} + q^{121} + q^{169} + \ldots).$$

It follows from a classical result of Euler (cf. for instance [4], p. 931 or [8], p. 457) that $\frac{1}{2}\theta_{\chi_3}$ is equal to

$$\eta(24z) = q \prod_{n=1}^{\infty} (1-q^{24n}).$$

Up to a scalar factor, this series is thus the <u>only</u> cusp form of weight 1/2 and level $N < 900$.

§3. OPERATORS

3.1. Conventions on characters.

From now on, all characters are assumed to be <u>primitive</u>; this is necessary when dealing with different levels. We say that such a character χ is <u>definable</u> (mod m) when its conductor $r(\chi)$ divides m. The product $\chi\chi'$ of two characters χ and χ' is the primitive character associated with $n \mapsto \chi(n)\chi'(n)$; hence, we have

$$(\chi\chi')(n) = \chi(n)\chi'(n)$$

<u>if</u> n is prime to $r(\chi)r(\chi')$, but maybe not otherwise.

3.2. The group $\underline{G}$.

Following Shimura [8], we introduce the group extension $\underline{G}$ of $\mathbf{GL}_2(\mathbf{R})^+$ whose elements consist of pairs $\{M,\phi(z)\}$, where $M = \begin{pmatrix} r & s \\ t & u \end{pmatrix}$ belongs to $\mathbf{GL}_2(\mathbf{R})^+$ and $\phi(z)^2 = \alpha \det(M)^{-1/2}(tz+u)$, with $|\alpha| = 1$. The multiplication law in $\underline{G}$ is given by

$$\{M,\phi(z)\}\{N,\psi(z)\} = \{MN,\phi(Nz)\psi(z)\}.$$

When dealing with forms of weight $\kappa/2$ it is convenient to define the "slash operator" $f|_\kappa \xi = f|\xi$ by :

$$(f|\xi)(z) = \phi(z)^{-\kappa}f(Mz) \qquad \text{where } \xi = \{M,\phi\} \in \underline{G},$$

and, for $\xi_i \in \underline{G}$ and $c_i \in \mathbf{C}$:

$$f|(\sum c_i\xi_i) = \sum c_i(f|\xi_i).$$

If $A \in \Gamma_0(4)$, we define $A^* \in \underline{G}$ by $A^* = \{A,j(A,z)\}$, where $j(A,z)$ is the θ-multiplier of A, cf. §1. Thus, if $f \in M_0(N,\kappa/2,\chi)$ and $A = \begin{pmatrix} a & b \\ c & d \end{pmatrix} \in \Gamma_0(N)$, we have $f|A^* = \chi(d)f$.

It follows from the definition of j that

$$(1) \qquad A^*B^* = (AB)^* \quad \text{if } A,B \in \Gamma_0(4).$$

Computations in $\underline{G}$ are greatly aided by making use of (1) whenever possible.

3.3. Hecke operators.

For a prime p, with $p \nmid N$, we define $T(p^2)$ on $M_0(N,\kappa/2,\chi)$ as in Shimura [8] by :

$$T(p^2) = p^{\kappa/2-2}\Big[\sum_{j=0}^{p^2-1} \{\begin{pmatrix} 1 & j \\ 0 & p^2 \end{pmatrix}, p^{1/2}\} + \chi(p)\sum_{j=1}^{p-1} \{\begin{pmatrix} p & j \\ 0 & p \end{pmatrix}, \varepsilon_p^{-1}\chi_{-j}(p)\}$$

$$+ \chi(p^2)\{\begin{pmatrix} p^2 & 0 \\ 0 & 1 \end{pmatrix}, p^{-1/2}\}\Big]$$

where $\varepsilon_p = 1$ or i according as $p \equiv 1$ or 3 (mod 4), cf. §1. For a prime p with $p \mid N$ (for instance $p = 2$), we define $T(p^2)$ by

$$T(p^2) = p^{\kappa/2-2}\sum_{j=0}^{p^2-1} \{\begin{pmatrix} 1 & j \\ 0 & p^2 \end{pmatrix}, p^{1/2}\},$$

and, if $4p \mid N$, we define $T(p)$ by

$$T(p) = p^{\kappa/4-1}\sum_{j=0}^{p-1} \{\begin{pmatrix} 1 & j \\ 0 & p \end{pmatrix}, p^{1/4}\}.$$

LEMMA 1. *Let* $f = \sum_{n=0}^{\infty} a(n)q^n$ *be an element of* $M_0(N,\kappa/2,\chi)$, *and let* $f|T(p^2) = \sum_{n=0}^{\infty} b(n)q^n$. *Then* $f|T(p^2)$ *belongs to* $M_0(N,\kappa/2,\chi)$ *also, and* we have

$$b(n) = \begin{cases} a(np^2) & \text{if } p \mid N, \\ a(np^2) + p^{(\kappa-3)/2}\chi(p)\chi_{-4}(p)^{(\kappa-1)/2}(\frac{n}{p})a(n) + \\ \quad + p^{\kappa-2}\chi(p^2)a(n/p^2) & \text{if } p \nmid N, \end{cases}$$

where $(\frac{n}{p})$ is the Legendre symbol. If $4p \mid N$, then $f|T(p)$ belongs to $M_0(N,\kappa/2,\chi\chi_p)$ and is equal to $\sum_{n=0}^{\infty} a(np)q^n$. Any two such operators commute.

PROOF. The statements about $T(p^2)$ are proved in Shimura, loc. cit. Those about $T(p)$, when $4p \mid N$, are proved by a simple computation.

3.4. Other operators.

We need the shift $V(m) = m^{-\kappa/4}\ \{(\begin{smallmatrix} m & 0 \\ 0 & 1 \end{smallmatrix}), m^{-1/4}\}$ which acts by

$$[f|V(m)](z) = f(mz).$$

We need also the symmetry $W(N) = \{(\begin{smallmatrix} 0 & -1 \\ N & 0 \end{smallmatrix}), N^{1/4}(-iz)^{1/2}\}$, which acts by

$$[f|W(N)](z) = N^{-\kappa/4}(-iz)^{-\kappa/2}f(-1/Nz),$$

so that $[f|W(N)]|W(N) = f$ for all f.

The conjugation operator H is defined by :

$$(f|H)(z) = \overline{f(-\overline{z})} = \sum_{n=0}^{\infty} \overline{a(n)}q^n \quad \text{if} \quad f = \sum_{n=0}^{\infty} a(n)q^n.$$

LEMMA 2. The operators $V(m)$, $W(N)$ and H take $M_0(N,\kappa/2,\chi)$ to $M_0(Nm,\kappa/2,\chi\chi_m)$, $M_0(N,\kappa/2,\overline{\chi}\chi_N)$ and $M_0(N,\kappa/2,\overline{\chi})$ respectively. Further, if f belongs to $M_0(N,\kappa/2,\chi)$, we have :

$$[f|V(m)]|T(p^2) = [f|T(p^2)]|V(m) \quad \text{when } p \nmid m,$$

$$[f|H]|T(p^2) = [f|T(p^2)]|H,$$

$$[f|W(N)]|T(p^2) = \overline{\chi}(p^2)[f|T(p^2)]|W(N) \quad \text{when } p \nmid N.$$

PROOF. Again, the proof involves simple computations in G and is left to the reader. Care should be exercised in the commutativity results since the definition of $T(p^2)$ depends on the character appearing in the space containing the function to which $T(p^2)$ is applied.

The following operators will be used in §4 only. To define the first one, suppose the prime p_0 divides $N/4$, and write $\Gamma_0(N/p_0)$ as a disjoint union of cosets modulo $\Gamma_0(N)$:

$$\Gamma_0(N/p_0) = \coprod_{j=1}^{\mu} \Gamma_0(N)A_j, \text{ with } A_j = \begin{pmatrix} a_j & b_j \\ c_j & d_j \end{pmatrix}, \text{ and } \mu = (\Gamma_0(N/p_0) : \Gamma_0(N)).$$

We define the trace operator $S'(\chi) = S'(\chi,N,p_0)$ on $M_0(N,\kappa/2,\chi)$ by

$$S'(\chi) = \sum_{j=1}^{\mu} \chi(a_j)A_j^* = \sum_{j=1}^{\mu} \overline{\chi}(d_j)A_j^*.$$

It is easily seen that this operator does not depend on the choice of the A_j's. Moreover, if χ is definable (mod N/p_0), $S'(\chi)$ takes $M_0(N,\kappa/2,\chi)$ to $M_0(N/p_0,\kappa/2,\chi)$ and commutes with $T(p^2)$ for $p \nmid N$; if f belongs to $M_0(N/p_0,\kappa/2,\chi)$, we have

$$f|S'(\chi) = \mu f.$$

For our purposes, it is more important to find an operator which goes from level N to level N/p_0 and which undoes the action of the shift operator $V(p_0)$. To do this, we define $S(\chi) = S(\chi,N,p_0)$ on $M_0(N,\kappa/2,\chi)$ by :

$$S(\chi) = \frac{1}{\mu}\, p_0^{\kappa/4}\; W(N)\; S'(\overline{\chi}\chi_N)W(N/p_0).$$

LEMMA 3. Let p_0 be a prime such that $4p_0|N$, and $\chi\chi_{p_0}$ is definable (mod N/p_0). Then :

a) The operator $S(\chi,N,p_0)$ maps $M_0(N,\kappa/2,\chi)$ into $M_0(N/p_0,\kappa/2,\chi\chi_{p_0})$.

b) <u>If</u> m <u>is prime to</u> p_0, <u>and</u> f <u>belongs to</u> $M_0(N,\kappa/2,\chi)$, <u>then</u>

$$f|S(\chi,N,p_0) = f|S(\chi,Nm,p_0).$$

c) $S(\chi)$ <u>commutes with all</u> $T(p^2)$, <u>for</u> $p \nmid N$.

d) <u>If</u> $g \in M_0(N/p_0,\kappa/2,\chi\chi_{p_0})$, <u>then</u> $g|V(p_0) \in M_0(N,\kappa/2,\chi)$ <u>and</u>

$$[g|V(p_0)]|S(\chi,N,p_0) = g.$$

e) <u>Let</u> p <u>be a prime such that</u> $4p|N$, $p \neq p_0$, <u>and</u> $\chi\chi_p$ <u>is definable</u> (mod N/p). <u>If</u> $g \in M_0(N/p,\kappa/2,\chi\chi_p)$, <u>we have</u>

$$[g|V(p)]|S(\chi,N,p_0) = [g|S(\chi\chi_p,N/p,p_0)]|V(p).$$

<u>PROOF</u>. Assertion a) follows from Lemma 2 and from the fact that

$$\overline{\chi}\chi_N = \overline{\chi\chi_{p_0}}\chi_{N/p_0}$$

is definable (mod N/p_0).

If $\begin{pmatrix} a & b \\ c & d \end{pmatrix}$ belongs to $\Gamma_0(Nm/p_0)$, with $(m,p_0) = 1$, then

$$W(Nm)\begin{pmatrix} a & b \\ c & d \end{pmatrix}^* W(Nm/p_0) = \{m,1\}\ W(N)\begin{pmatrix} a & bm \\ c/m & d \end{pmatrix}^* W(N/p_0),$$

and b) follows, since $f|\{m,1\} = f$.

Assertion c) follows from the commutativity of the $T(p^2)$, $p \nmid N$, with $W(N)$, $S'(\overline{\chi}\chi_N)$ and $W(N/p_0)$.

As for d), we have

$$\{\begin{pmatrix} p_0 & 0 \\ 0 & 1 \end{pmatrix}, p_0^{-1/4}\}W(N) = \{p_0,1\}W(N/p_0),$$

hence

$$[g|V(p_0)]|W(N) = p_0^{-\kappa/4}\ g|W(N/p_0).$$

This is invariant by $\frac{1}{\mu}\ S'(\overline{\chi}\chi_N)$, and is sent to $p_0^{-\kappa/4}$ g by $W(N/p_0)$, whic proves d).

As for e), we have $4p_0p|N$, and $\chi\chi_{p_0}\chi_p$ is definable (mod N/pp_0). Further :

$$\{(\begin{smallmatrix} p & 0 \\ 0 & 1 \end{smallmatrix}), p^{-1/4}\} W(N) = \{p,1\} W(N/p),$$

$$W(N/p_0) = W(N/pp_0)\{(\begin{smallmatrix} p & 0 \\ 0 & 1 \end{smallmatrix}), p^{-1/4}\},$$

and $\overline{\chi}\chi_N = \overline{\chi\chi_p}\chi_{N/p}$. The formula

$$[g|V(p)]|S(\chi,N,p_0) = [g|S(\chi\chi_p,N/p,p_0)]|V(p)$$

follows from this, after a simple computation.

Let p be any prime. We shall need the operator

$$K(p) = 1 - T(p,Np)V(p),$$

where T(p,Np) is the Hecke operator T(p) relative to the level Np (see above).

LEMMA 4. *If* $f = \sum_{n=0}^{\infty} a(n)q^n$ *belongs to* $M_0(N,\kappa/2,\chi)$, *then* $f|K(p)$ *belongs to* $M_0(Np^2,\kappa/2,\chi)$ *and is equal to* $\sum_{(n,p)=1} a(n)q^n$. *Further*, *if* $p' \nmid Np$, *then* $T(p'^2)$ *and* $K(p)$ *commute*.

PROOF. This is immediate.

REMARK. All the above operators take cusp forms to cusp forms.

§4. NEWFORMS

4.1. *Definitions*.

Let $f \in M_0(N,\kappa/2,\chi)$ be an eigenform of all but finitely many $T(p^2)$. We say that f is an *oldform* (compare [1], [5]) if there exists a prime p dividing N/4 such that :

either χ is definable (mod N/p) and f belongs to $M_0(N/p,\kappa/2,\chi)$,

or $\chi\chi_p$ is definable (mod N/p) and $f = g|V(p)$, with $g \in M_0(N/p,\kappa/2,\chi\chi_p)$.

We denote by $M_0^{old}(N,\kappa/2,\chi)$ the subspace of $M_0(N,\kappa/2,\chi)$ spanned by old forms. If $f \in M_0(N,\kappa/2,\chi)$ is an eigenform of all but finitely many $T(p^2)$, and f does not belong to $M_0^{old}(N,\kappa/2,\chi)$, we say that f is a newform of level N.

LEMMA 5. The symmetry operator $W(N) : M_0(N,\kappa/2,\chi) \to M_0(N,\kappa/2,\bar{\chi}\chi_N)$ and the conjugation operator $H : M_0(N,\kappa/2,\chi) \to M_0(N,\kappa/2,\bar{\chi})$ take oldforms to oldforms and newforms to newforms.

PROOF. By Lemma 2, W(N) and H take eigenforms to eigenforms. If f is an oldform of the first type above, i.e. $f \in M_0(N/p,\kappa/2,\chi)$, then

$$f|W(N) = p^{\kappa/4}[f|W(N/p)]|V(p)$$

is an oldform of the second type. Conversely, if $f = g|V(p)$ is an oldform of the second type, then $f|W(N) = p^{-\kappa/4}\, g|W(N/p)$ is an oldform of the first type. Hence W(N) takes oldforms to oldforms; the same is obviously true for the conjugation operator H. That W(N) and H take newforms to newforms follows from this, and from the fact that their square is the identity.

LEMMA 6. Let $h \in M_0^{old}(N,\kappa/2,\chi)$ be a non-zero eigenform of all but finitely many $T(p^2)$. Then there is a divisor N_1 of N, with $N_1 < N$, a character ψ definable (mod N_1) and a newform g in $M_0(N_1,\kappa/2,\psi)$ such that h and g have the same eigenvalues for all but finitely many $T(p^2)$.

PROOF. We use induction on N. By construction, $M_0^{old}(N,\kappa/2,\chi)$ has a basis (f_i) consisting of forms of the type g, or $g|V(p)$, where g is an eigenform of all but finitely many $T(p^2)$, and is of lower level. Hence h is a linear combination with non-zero coefficients of some of the f_i's, and each f_i occurring in h has the same eigenvalue for $T(p^2)$ as h does. The Lemma then follows from the induction assumption.

LEMMA 7. Let p be a prime, and let $f = \sum_{n=0}^{\infty} a(n)q^n$ be a non-zero element of $M_0(N,\kappa/2,\chi)$ such that $a(n) = 0$ for all n not divisible by p. Then p divides $N/4$, $\chi\chi_p$ is definable (mod N/p) and $f = g|V(p)$ with $g \in M_0(N/p,\kappa/2,\chi\chi_p)$.

PROOF. Put

$$g(z) = f(z/p) = \sum_{n=0}^{\infty} a(pn)q^n = p^{\kappa/4} f|\{\left(\begin{smallmatrix}1 & 0\\ 0 & p\end{smallmatrix}\right), p^{1/4}\}.$$

Let $N' = N/p$ if $4p|N$ and $N' = N$ otherwise. Let $\Gamma_0(N',p)$ be the subgroup of $\Gamma_0(N')$ consisting of matrices $\left(\begin{smallmatrix}a & b\\ c & d\end{smallmatrix}\right)$ with $b \equiv 0 \pmod p$; if $A = \left(\begin{smallmatrix}a & b\\ c & d\end{smallmatrix}\right)$ is such a matrix, put $A_1 = \left(\begin{smallmatrix}a & b/p\\ pc & d\end{smallmatrix}\right)$. We have $A_1 \in \Gamma_0(N)$, and

$$\{\left(\begin{smallmatrix}1 & 0\\ 0 & p\end{smallmatrix}\right), p^{1/4}\}\ A^* = \{1,\chi_p(d)\}\ A_1^*\{\left(\begin{smallmatrix}1 & 0\\ 0 & p\end{smallmatrix}\right), p^{1/4}\},$$

hence

$$g|A^* = \chi_p(d)\chi(d)g.$$

Since d is relatively prime to both p and N, this can be rewritten as

$$(*) \qquad g|A^* = (\chi\chi_p)(d)g.$$

By hypothesis, g has a q-expansion in integral powers of q, hence (*) holds for $A = \left(\begin{smallmatrix}1 & 1\\ 0 & 1\end{smallmatrix}\right)$. Since $\Gamma_0(N')$ is generated by $\Gamma_0(N',p)$ and $\left(\begin{smallmatrix}1 & 1\\ 0 & 1\end{smallmatrix}\right)$, this shows that (*) holds for any $A \in \Gamma_0(N')$. Since g is non-zero, this implies that $\chi\chi_p$ is definable (mod N'); this is easily seen to be possible only if p divides N/4, in which case $N' = N/p$ and (*) shows that g belongs to $M_0(N/p,\kappa/2,\chi\chi_p)$.

REMARKS. (1) If f is a cusp form, it is clear that g is also a cusp form.

(2) The above Lemma gives a characterization of oldforms of the second type.

THEOREM 1. Let m be an integer ≥ 1, and let $f = \sum_{n=0}^{\infty} a(n)q^n$ be an element of $M_0(N,\kappa/2,\chi)$ such that $a(n) = 0$ for all n with $(n,m) = 1$. Then f can be written as

$$f = \sum_p f_p|V(p), \qquad \text{with } f_p \in M_0(N/p,\kappa/2,\chi\chi_p),$$

where p runs through the primes such that $p|m$, $4p|N$, and $\chi\chi_p$ is definable (mod N/p).

If f is a cusp form, the f_p can be chosen to be cusp forms. If f is an eigenform of all but finitely many $T(p'^2)$, then the f_p may be further chosen so that they, too, are eigenforms of all but finitely many $T(p'^2)$, and have the same eigenvalues as f.

(Compare with the integral weight case, in [1] or [5].)

PROOF. Clearly, we may assume that m is square-free. We proceed by induction on the number r of prime factors of m. If $r = 0$, then $m = 1$ and all $a(n)$ are zero by hypothesis; there is nothing to prove. Now suppose $r \geq 1$ and that Theorem 1 has been proved for all m's which are products of strictly less than r primes (and all levels). Let p_0 be a prime divisor of m. Put $m = p_0m_0$, and

$$h = \sum_{(n,m_0)=1} a(n)q^n = f|\prod_{p|m_0} K(p), \qquad \text{cf. §3.}$$

If $h = 0$, we may replace m by m_0, and Theorem 1 follows from the induction hypothesis. Hence, we may assume that $h \neq 0$. By Lemma 4, we have $h \in M_0(Nm_0^2,\kappa/2,\chi)$. If $(n,m_0) = 1$ and $a(n) \neq 0$, by hypothesis we have $(n,p_0) \neq 1$ and Lemma 7 shows that $4p_0|Nm_0^2$, $\chi\chi_{p_0}$ is definable (mod Nm_0^2/p_0) and $h = g_{p_0}|V(p_0)$ with $g_{p_0} \in M_0(Nm_0^2/p_0,\kappa/2,\chi\chi_{p_0})$. This implies that $4p_0|N$ and that $\chi\chi_{p_0}$ is definable (mod N/p_0). Moreover, we have

$$f - h = f - g_{p_0}|V(p_0) = \sum_{n=0}^{\infty} b(n)q^n,$$

with $b(n) = 0$ if $(n,m_0) = 1$. By the induction hypothesis (applied to m_0 and to the level Nm_0^2), this shows that $f - g_{p_0}|V(p_0)$ can be written as

$$f - g_{p_0}|V(p_0) = \sum_p g_p|V(p),$$

where p runs through the primes such that $p|m_0$ and $\chi\chi_p$ is definable (mod Nm_0^2/p), with $g_p \in M_0(Nm_0^2/p,\kappa/2,\chi\chi_p)$. We now apply the operator $S(\chi) = S(\chi,N,p_0)$ of §3 to f. Using Lemma 3, the above formula gives

$$f|S(\chi) - g_{p_0} = \sum_p [g_p|S(\chi\chi_p,Nm_0^2/p,p_0)]|V(p).$$

Let now f_{p_0} be $f|S(\chi)$. We have $f_{p_0} \in M_0(N/p_0,\kappa/2,\chi\chi_{p_0})$. Moreover the above formula shows that the nth coefficient of $f_0 = f - f_{p_0}|V(p_0)$ is 0 if $(n,m_0) = 1$; this allows us to apply the induction hypothesis to f_0 and m_0, and we get the required decomposition of f. As for the other assertions of Theorem 1, they follow from the inductive construction of the f_p's and from Lemma 3.

COROLLARY. *If the form* f *of Theorem 1 is an eigenform of all but finitely many* $T(p'^2)$, *then* f *belongs to* $M_0^{old}(N,\kappa/2,\chi)$.

§5. THE "BOUNDED DENOMINATORS" ARGUMENT

5.1. *Coefficients of modular forms of half integral weight.*

LEMMA 8. (a) *There is a basis of* $M_0(N,\kappa/2,\chi)$ *consisting of forms whose coefficients belong to a number field.*

(b) *If* $f = \sum a(n)q^n$ *belongs to* $M_0(N,\kappa/2,\chi)$ *and the* a(n) *are algebraic numbers, then the* a(n) *have bounded denominators* (i.e. there exists a non-zero integer D such that D.a(n) is an algebraic integer for all n).

<u>PROOF</u>. The analogous statement for modular forms of <u>integral weight</u> is well known (cf. for instance [7], Th. 3.5.2 or [3], Prop. 2.7). We shall reduce to that case by the familiar device of multiplying by a fixed form f_0. We choose for f_0 the form

$$\theta^{3\kappa} = (1 + 2q + 2q^4 + \dots)^{3\kappa} = 1 + 6\kappa q + \dots .$$

The map $\Phi : f \mapsto \theta^{3\kappa} f$ sends $M_0(N,\kappa/2,\chi)$ into the space $M_0(N,2\kappa,\chi)$ of modular forms of type $(2\kappa,\chi)$ on $\Gamma_0(N)$. By the results quoted above, it follows that, if the coefficients of f are algebraic, those of $\theta^{3\kappa} f$ have bounded denominators; dividing by $\theta^{3\kappa}$ does not increase denominators, hence b) follows. As for a), one has to check that the image $\mathrm{Im}(\Phi)$ of Φ can be defined by linear equations with algebraic coefficients. This is so because θ does not vanish on the upper half-plane (as its expansion shows), nor at any cusp except those congruent mod $\Gamma_0(4)$ to 1/2; hence a modular form F in $M_0(N,2\kappa,\chi)$ belongs to $\mathrm{Im}(\Phi)$ if and only if it vanishes (with prescribed multiplicities) at these cusps, i.e. if some of the coefficients of its expansions at these cusps are zero; since it is known that these coefficients are algebraic linear combinations of the coefficients of F at the cusp ∞, the result follows.

<u>REMARKS</u>. (1) A similar argument shows that $M_1(N,\kappa/2)$ has a basis made up of forms with coefficients in $\mathbf{Z}$, and that the action of $(\mathbf{Z}/N\mathbf{Z})^*$ is $\mathbf{Z}$-linear with respect to that basis. This implies that, if $f = \sum a(n)q^n$ belongs to $M_0(N,\kappa/2,\chi)$ and σ is any automorphism of $\mathbf{C}$, the series

$$f^\sigma = \sum \sigma(a(n))q^n$$

belongs to $M_0(N,\kappa/2,\chi^\sigma)$, just as in the integral weight case ([3], 2.7.4). We will not need these facts.

(2) On <u>noncongruence subgroups</u>, part (a) of Lemma 8 remains true, but part (b) does not, as was first noticed by Atkin and Swinnerton-Dyer [2]. A simple example is

$$f(z) = \theta(z)^{1/2}\,\theta(3z)^{1/2} = 1 + q - \frac{1}{2}q^2 + \frac{3}{2}q^3 + \frac{11}{8}q^4 - \ldots ,$$

which is a modular form of weight 1/2 on a subgroup of index 2 of $\Gamma_1(12)$, and whose coefficients have unbounded powers of 2 in denominator (if n is a power of 2, the 2-adic valuation of the nth coefficient of f is 1-n). Similar examples exist in higher weights, integral as well as half integral : take for instance

$$f_m(z) = \theta(z)^{1/2}\,\theta(3z)^{m/2}, \qquad \text{with } m \text{ odd} \geq 1,$$

which is of weight $(m+1)/4$.

5.2. Eigenvectors of the Hecke operators for weight 1/2.

From now on, we restrict ourselves to weight 1/2, i.e. we take $\kappa = 1$.

LEMMA 9. *Let* $f = \sum_{n=0}^{\infty} a(n)q^n$ *be a non-zero element of* $M_0(N,1/2,\chi)$ *and let* p *be a prime, with* $p \nmid N$. *Assume that* $f|T(p^2) = c_p f$, *with* $c_p \in \mathbf{C}$. *Let* $m \geq 1$ *be such that* $p^2 \nmid m$. *Then* :

(a) *we have* $a(mp^{2n}) = a(m)\chi(p)^n(\frac{m}{p})^n$ *for every* $n \geq 0$.

(b) *If* $a(m) \neq 0$, *then* $p \nmid m$ *and* $c_p = \chi(p)(\frac{m}{p})(1+p^{-1})$.

PROOF. Since $T(p^2)$ maps forms with algebraic coefficients into themselves (cf. Lemma 1), it follows from Lemma 8 that the eigenvalue c_p is algebraic, and that the corresponding eigenspace is generated by forms with algebraic coefficients. Hence we may assume that the coefficients a(n) of f are algebraic numbers. Consider the power series

$$A(T) = \sum_{n=0}^{\infty} a(mp^{2n})T^n,$$

where T is an indeterminate. By [8], p. 452, we have

$$A(T) = a(m)\,\frac{1 - \alpha T}{(1-\beta T)(1-\gamma T)} ,$$

with $\alpha = \chi(p)p^{-1}(\frac{m}{p})$ and $\beta + \gamma = c_p$, $\beta\gamma = \chi(p^2)p^{-1}$ (note the negative exponent of p, which comes from the fact that $\kappa = 1$). This already shows that $a(m) = 0$ implies $A(T) = 0$, i.e. $a(mp^{2n}) = 0$ for all $n \geq 0$. Hence we may assume that $a(m) \neq 0$, in which case A(T) is a non-zero rational function of T. If we view A(T) as a p-adic function of T (over a suitable finite extension of the p-adic field $\mathbf{Q}_p$), Lemma 8 (b) shows that A(T) converges in the p-adic unit disk U defined by $|T|_p < 1$; hence A(T) cannot have a pole in U. However, since $\beta\gamma = \chi(p^2)p^{-1}$, either β^{-1} or γ^{-1} belongs to U; assume it is β^{-1}. In order that A(T) be holomorphic at β^{-1}, it is necessary that the factors $1 - \beta T$ and $1 - \alpha T$ cancel each other. We then have $\alpha = \beta$ and

$$A(T) = a(m)/(1-\gamma T), \quad \text{so that} \quad a(mp^{2n}) = \gamma^n a(m).$$

Since $\beta\gamma \neq 0$ we have $\alpha \neq 0$, hence $p \nmid m$. Moreover,

$$\gamma = \beta\gamma/\alpha = \chi(p^2)p^{-1}/\chi(p)p^{-1}(\tfrac{m}{p}) = \chi(p)(\tfrac{m}{p}).$$

This shows that $a(mp^{2n}) = \gamma^n\ a(m) = a(m)\chi(p)^n(\frac{m}{p})^n$, which proves (a). As for the last assertion of (b), it follows from $c_p = \beta + \gamma = \alpha + \gamma$.

THEOREM 2. Let $f = \sum_{n=0}^{\infty} a(n)q^n$ be a non-zero element of $M_0(N,1/2,\chi)$ and let N' be a multiple of N. Assume that, for all $p \nmid N'$, we have $f|T(p^2) = c_p f$, with $c_p \in \mathbf{C}$. Then there exists a unique square-free integer $t \geq 1$ such that $a(n) = 0$ if n/t is not a square. Moreover :

(i) $t|N'$.

(ii) $c_p = \chi(p)(\frac{t}{p})(1+p^{-1})$ if $p \nmid N'$.

(iii) $a(nu^2) = a(n)\chi(u)(\frac{t}{u})$ if $(u,N') = 1$, $u \geq 1$.

PROOF. Let m and m' be two integers ≥ 1 such that $a(m) \neq 0$ and $a(m') \neq 0$. We show first that m'/m is a square. Let P be the set of primes p with $p \nmid N'mm'$. If $p \in P$, Lemma 9 shows that

$$\chi(p)(\frac{m}{p})(1+p^{-1}) = c_p = \chi(p)(\frac{m'}{p})(1+p^{-1}),$$

hence $$(\frac{m}{p}) = (\frac{m'}{p}) \text{ for all } p \in P.$$

It is well known that this implies that m'/m is a square. We may write m and m' as $m = tv^2$, $m' = tv'^2$, with $v,v' \geqslant 1$ and t square-free $\geqslant 1$. This proves the first part of the Theorem, i.e. the existence of t. Write now v as $p^n u$, with $p \nmid N'$ and $(p,u) = 1$, so that $m = tp^{2n}u^2$. By Lemma 9, applied to tu^2, we have $a(m) = \chi(p)^n(\frac{tu^2}{p})^n a(tu^2)$ hence $a(tu^2) \neq 0$ and Lemma 9 (b) shows that $p \nmid tu^2$, hence $p \nmid t$, and $c_p = \chi(p)(\frac{t}{p})(1+p^{-1})$. Hence every prime factor of t divides N'; since t is square-free, this shows that $t|N'$, and (i) and (ii) are proved. As for (iii), it is enough to check it when $u = p$ with $p \nmid N'$; in that case, one writes n as m_0p^{2a}, with $p^2 \nmid m_0$, and applies Lemma 9 (a).

<u>COROLLARY</u>. <u>If</u> $a(1) \neq 0$, <u>then</u> $t = 1$ <u>and</u> $c_p = \chi(p)(1+p^{-1})$ <u>for</u> $p \nmid N'$. (Note that, in this case, the c_p's determine the character χ.)

Let now $\sum_{n=1}^{\infty} a(n)n^{-s}$ be the <u>Dirichlet series</u> associated with f. Let ψ be the character $\chi\chi_t$, so that $\psi(p) = \chi(p)(\frac{t}{p})$ if $p \nmid N'$. Assertions (i) and (iii) of Theorem 2 can be reformulated as :

<u>THEOREM</u> 2'. <u>Under the assumptions of Theorem</u> 2, <u>we have</u>

$$\sum_{n=1}^{\infty} a(n)n^{-s} = t^{-s}\Big(\sum_{n|N'^{\infty}} a(tn^2)n^{-2s}\Big)\prod_{p\nmid N'}(1-\psi(p)p^{-2s})^{-1}.$$

(The notation $A|B^{\infty}$ means that A divides some power of B, i.e. that every prime factor of A is a factor of B.)

§6. PROOF OF THEOREM A

6.1. Structure of newforms of weight 1/2.

Let $f = \sum_{n=0}^{\infty} a(n)q^n$ be a newform of level N (cf. §4) belonging to $M_0(N,1/2,\chi)$. By Theorem 2, there is a unique square-free integer $t \geqslant 1$ such that $a(n) = 0$ if n/t is not a square.

LEMMA 10. We have $t = 1$ and $a(1) \neq 0$.

PROOF. The product expansion of $\sum_{n=1}^{\infty} a(n)n^{-s}$ given in Theorem 2' shows that, if $a(1) = 0$, we have $a(n) = 0$ for every n such that $(n,N') = 1$; the Corollary to Theorem 1 then shows that f belongs to $M_0^{\mathrm{old}}(N,1/2,\chi)$, contrary to the assumption that f is a newform. Hence $a(1) \neq 0$, and this implies $t = 1$, cf. the Corollary to Theorem 2.

This Lemma allows us to divide f by $a(1)$; hence we may assume that f is normalized, i.e. that $a(1) = 1$.

LEMMA 11. Let $g \in M_0(N,1/2,\chi)$ be an eigenform of all but finitely many $T(p^2)$, with the same eigenvalues as f. Then g is a scalar multiple of f.

PROOF. Let c be the coefficient of q in the q-expansion of g, and set

$$h = g - cf,$$

so that the coefficient of q in the q-expansion of h is 0. Suppose $h \neq 0$. By Lemma 10, h is not a newform; since it is an eigenform of all but finitely many $T(p^2)$, it belongs to $M_0^{\mathrm{old}}(N,1/2,\chi)$. Hence, by Lemma 6, there are $N_1|N$, with $N_1 < N$, a character ψ definable (mod N_1) and a normalized newform g_1 in $M_0(N_1,1/2,\psi)$ with the same eigenvalues c_p as f and h, for all but finitely many $T(p^2)$. Since the c_p's determine the character (cf. the Corollary to Theorem 2) we have $\chi = \psi$ and so g_1 belongs to $M_0^{old}(N,1/2,\chi)$. On the other hand, the coefficient of q in the q-expansion of $f - g_1$ is 0; the same argument as above then shows that $f - g_1$ belongs to $M_0^{old}(N,1/2,\chi)$. Hence $f = g_1 + (f-g_1)$ belongs to $M_0^{old}(N,1/2,\chi)$. This contradicts the assumption that f is a newform. Hence $h = 0$, i.e. $g = cf$.

LEMMA 12. The form f is an eigenform of every $T(p^2)$. If we put $f|T(p^2) = c_p f$, we have

$$(*) \qquad \sum_{n=1}^{\infty} a(n)n^{-s} = \prod_{p|N} (1 - c_p p^{-2s})^{-1} \prod_{p\nmid N} (1 - \chi(p)p^{-2s})^{-1} .$$

Further, if $4p|N$, then $c_p = 0$.

PROOF. If we apply Lemma 11 to $g = f|T(p^2)$, we see that g is a multiple of f. Hence f is an eigenform of every $T(p^2)$, and the Euler product (*) follows from this and Theorem 2' (applied with $N' = N$, $t = 1$, $\psi = \chi$).

If $4p|N$, then Lemma 1 shows that

$$f|T(p) = \sum_{n=0}^{\infty} a(np)q^n = \sum_{m=0}^{\infty} a(m^2p^2)q^{pm^2} = c_p f|V(p)$$

belongs to $M_0(N,1/2,\chi\chi_p)$. If $c_p \neq 0$, Lemma 7 applied to $f|T(p)$ and to the character $\chi\chi_p$ shows that χ is definable (mod N/p) and that $f|T(p) = g|V(p)$ with $g \in M_0(N/p,1/2,\chi)$. We have $c_p f|V(p) = g|V(p)$, hence $c_p f = g$; this shows that f belongs to $M_0(N/p,1/2,\chi)$ and contradicts the assumption that f is a newform. Hence $c_p = 0$.

LEMMA 13. The level N of the newform f is a square, and $f|W(N)$ is a multiple of $f|H$.

(Recall that $W(N)$ and H are respectively the symmetry and conjugation operators, cf. §3.)

PROOF. If $p \nmid N$, we have $f|T(p^2) = c_p f$ with $c_p = (1+p^{-1})\chi(p)$, and, by Lemma 2,

$$[f|W(N)]|T(p^2) = \overline{\chi}(p)^2 c_p \; f|W(N) = \overline{c}_p \; f|W(N),$$

$$[f|H]|T(p^2) = (c_p f)|H = \overline{c}_p \; f|H \qquad \text{since } H \text{ is anti-linear.}$$

But $f|W(N)$ and $f|H$ are newforms of level N and characters $\overline{\chi}\chi_N$ and $\overline{\chi}$ respectively, cf. Lemma 5. Since they have the same eigenvalues $\overline{c}_p$ for all $T(p^2)$, $p \nmid N$, and these eigenvalues determine the character (cf. the Corollary to Theorem 2), we have $\overline{\chi}\chi_N = \overline{\chi}$ and N is a square. The fact that $f|W(N)$ and $f|H$ are proportional follows from this and from Lemma 11.

THEOREM 3. If f is a normalized newform in $M_0(N,1/2,\chi)$, and r is the conductor of χ, then $N = 4r^2$ and $f = \frac{1}{2}\,\theta_\chi$.

PROOF. We write $f = \sum_{n=0}^{\infty} a(n)q^n$ as above, and put

$$F(s) = \sum_{n=1}^{\infty} a(n)n^{-s} = \prod_{p \mid N} (1 - c_p p^{-2s})^{-1} \prod_{p \nmid N} (1 - \chi(p)p^{-2s})^{-1},$$

$$\overline{F}(s) = \sum_{n=1}^{\infty} \overline{a(n)}n^{-s}.$$

The Dirichlet series F and $\overline{F}$ converge for Re(s) large enough. Using Mellin transforms, and Lemma 13, we obtain by a standard argument the analytic continuation of F and $\overline{F}$ as entire functions of s (except for a simple pole at $s = 1/2$ if $a(0) \neq 0$), and the functional equation

$$(2\pi)^{-s}\,\Gamma(s)F(s) = C_1\left(\frac{2\pi}{N}\right)^{-(1/2-s)} \Gamma(\tfrac{1}{2}-s)\overline{F}(\tfrac{1}{2}-s),$$

where C_1 (and C_2, C_3, C_4 below) is a non-zero constant.

On the other hand, we know that the functions

$$G(s) = L(2s,\chi) = \sum_{n=1}^{\infty} \chi(n)n^{-2s} = \prod_{p \nmid r} (1 - \chi(p)p^{-2s})^{-1}$$

$$\overline{G}(s) = L(2s,\overline{\chi})$$

satisfy the functional equation

$$(2\pi)^{-s}\,\Gamma(s)G(s) = C_2\left(\frac{2\pi}{4r^2}\right)^{-(1/2-s)}\,\Gamma(\tfrac{1}{2}-s)\overline{G}(\tfrac{1}{2}-s).$$

Dividing these equations, we find

$$(*)\qquad \prod_{p|m}\left(\frac{1-c_p p^{-2s}}{1-\chi(p)p^{-2s}}\right) = C_3\left(\frac{N}{4r^2}\right)^{-(1/2-s)}\prod_{p|m}\left(\frac{1-\overline{c}_p p^{2s-1}}{1-\overline{\chi}(p)p^{2s-1}}\right),$$

where m is the product of the prime divisors p of N such that $c_p \neq \chi(p)$.

If, for some $p|m$, we have $\chi(p) \neq 0$, then the left side of (*) has an infinity of poles on the line $\mathrm{Re}(s) = 0$, only finitely many of which can appear on the right side. This shows that $p|m$ implies $\chi(p) = 0$, (i.e. $p|r$) and $c_p \neq 0$ since $c_p \neq \chi(p)$. We may now rewrite (*) as :

$$\prod_{p|m}(1-c_p p^{-2s}) = C_4\left(\frac{Nm^2}{4r^2}\right)^{s}\prod_{p|m}(1-c'_p p^{-2s}),$$

where $c'_p = p/\overline{c}_p$. The same argument as above (using zeros instead of poles) shows that, for every $p|m$, we have $c_p = c'_p$, i.e. $|c_p|^2 = p$; the above equation then gives $C_4 = 1$ and $Nm^2 = 4r^2$. But, by Lemma 12, we have $c_p = 0$ when $4p|N$. This shows that $m = 1$ or 2, and that $m = 2$ can occur only when $8 \nmid N$ and $\chi(2) = 0$; in the last case, r is divisible by 4 and the equation $Nm^2 = 4r^2$ shows that N is divisible by 16, which contradicts $8 \nmid N$. Hence only the case $m = 1$ is possible, and we have $N = 4r^2$, $F(s) = G(s)$. This shows that, for every $n \geq 1$, the coefficients of q^n in f and in $\frac{1}{2}\theta_\chi$ are the same. Hence $f - \frac{1}{2}\theta_\chi$ is a constant, and, since it is a modular form of weight 1/2, it is 0. This concludes the proof.

6.2. Alternative arguments.

(1) To show that the constant term of f and $\frac{1}{2}\theta_\chi$ agree, we could have used the well-known fact that they are equal to $-F(0)$ and $-G(0)$

respectively.

(2) Another way to rule out $|c_p|^2 = p$ is to prove a priori that $|c_p| \leqslant 1$. This may be done as follows. Choose $D \geqslant 1$ such that p is inert in $\mathbf{Q}(\sqrt{-D})$, and consider the modular form of weight 1 :

$$g(z) = f(z)\theta(Dz) = \left(\sum_{u=0}^{\infty} a(u)q^u\right)\left(\sum_{-\infty}^{\infty} q^{Dv^2}\right) = \sum_{u,v} a(u^2)q^{u^2+Dv^2}.$$

The p^{2n}-th coefficient of g is $a(p^{2n}) = (c_p)^n$. By [3], Cor. 9.2, this coefficient is $O(p^{2n\delta})$ for every $\delta > 0$. This obviously implies $|c_p| \leqslant 1$.

Theorem 3 has a converse :

THEOREM 4. If χ is an even character of conductor r, then $\frac{1}{2}\theta_\chi$ is a normalized newform in $M_0(4r^2,1/2,\chi)$.

(Recall that all characters are assumed to be primitive.)

PROOF. Let $N = 4r^2$. We know that θ_χ belongs to $M_0(N,1/2,\chi)$ and it is easily checked that it is an eigenform of all $T(p^2)$, with eigenvalue

$$c_p = (1+p^{-1})\chi(p) \quad \text{if } p \nmid N \qquad \text{(cf. Lemma 1).}$$

Thus, if θ_χ is not a newform, Lemma 6 shows that there are a divisor N_1 of N, with $N_1 < N$, a character ψ definable (mod N_1) and a newform f in $M_0(N_1,1/2,\psi)$ such that f and θ_χ have the same eigenvalues for all but finitely many $T(p^2)$. We thus have

$$(1+p^{-1})\psi(p) = c_p = (1+p^{-1})\chi(p) \qquad \text{for almost all } p,$$

and this implies $\psi = \chi$, hence $N_1 = 4r^2$ by Theorem 3. This contradicts $N_1 < N$. Hence θ_χ is a newform, and $\frac{1}{2}\theta_\chi$ is obviously normalized.

6.3. Proof of Theorem A.

Let χ be an even character definable (mod N). With the notations of

§2, we want to prove that the theta series $\theta_{\psi,t} = \theta_\psi|V(t)$, with $(\psi,t) \in \Omega(N,\chi)$, make a basis of $M_0(N,1/2,\chi)$. The proof splits into two parts :

a) Linear independence of the $\theta_{\psi,t}$.

Since t and χ determine ψ, every t occurs as the second entry of at most one (ψ,t) in $\Omega(N,\chi)$. Suppose then that we have

$$\lambda_1 \theta_{\psi_1,t_1} + \ldots + \lambda_m \theta_{\psi_m,t_m} = 0,$$

with $t_1 < t_2 < \ldots < t_m$ and $\lambda_i \neq 0$ for all i. The coefficient of q^{t_1} in θ_{ψ_1,t_1} is equal to 2; in θ_{ψ_j,t_j}, $j \geq 2$, it is equal to 0. This shows that $2\lambda_1 = 0$, hence $\lambda_1 = 0$. This contradiction proves the linear independence of the $\theta_{\psi,t}$.

b) The $\theta_{\psi,t}$ with $(\psi,t) \in \Omega(N,\chi)$, generate $M_0(N,1/2,\chi)$.

We need :

LEMMA 14. There is a basis of $M_0(N,1/2,\chi)$ consisting of eigenforms for all the $T(p^2)$, $p \nmid N$.

PROOF. Put on $M_0(N,1/2,\chi)$ the Petersson scalar product $\langle f,g \rangle$, cf. §1. A standard computation shows that, if $p \nmid N$, we have

$$\langle f|T(p^2),g \rangle = \chi(p^2)\langle f,g|T(p^2) \rangle,$$

hence $\overline{\chi}(p)T(p^2)$ is hermitian. The Lemma follows from this, and from the fact that the $T(p^2)$ commute.

We can now prove assertion b), using induction on N. By Lemma 14, it is enough to show that any eigenform f of all $T(p^2)$, $p \nmid N$, is a linear combination of the $\theta_{\psi,t}$ with $(\psi,t) \in \Omega(N,\chi)$. If f is a newform, this follows from Theorem 3. If not, we may assume f is an oldform of one of the two types of §4 :

either χ is definable (mod N/p) and f belongs to $M_0(N/p,1/2,\chi)$,

<u>or</u> $\chi\chi_p$ is definable (mod N/p) and $f = g|V(p)$ with $g \in M_0(N,N/p,1/2,\chi\chi_p)$. In the first case, the induction assumption shows that f is a linear combination of the $\theta_{\psi,t}$ with $(\psi,t) \in \Omega(N/p,\chi)$ and <u>a fortiori</u> with $(\psi,t) \in \Omega(N,\chi)$. In the second case, g is a linear combination of the $\theta_{\psi,t}$, with $(\psi,t) \in \Omega(N/p,\chi\chi_p)$, and hence f is a linear combination of the $\theta_{\psi,tp}$, with $(\psi,tp) \in \Omega(N,\chi)$.

<u>REMARK</u>. It is possible to prove Lemma 14 without using Petersson products. Indeed, assume that some $T(p^2)$, $p \nmid N$, is not diagonalizable. Then there exists an eigenvalue c_p of $T(p^2)$ and a non-zero element g of $M_0(N,1/2,\chi)$ such that

$$g|U \neq 0 \quad \text{and} \quad g|U^2 = 0, \quad \text{where } U = T(p^2) - c_p.$$

Using Lemma 8, one may further assume that the coefficients of g are algebraic numbers. A computation similar to that of Lemma 9 then shows that these coefficients have <u>unbounded powers of</u> p in denominators, and this contradicts Lemma 8. Hence, each $T(p^2)$ is diagonalizable. Since these operators commute, Lemma 14 follows.

§7. PROOF OF THEOREM B

7.1. Twists.

Let $f = \sum_{n=0}^{\infty} a(n)q^n$ be a modular form of weight $k = \kappa/2$ on some $\Gamma_1(N)$. Let M be an integer $\geqslant 1$, and ε a function on $\mathbf{Z}$ with period M (i.e. a function on $\mathbf{Z}/M\mathbf{Z}$). We put

$$f * \varepsilon = \sum_{n=0}^{\infty} a(n)\varepsilon(n)q^n.$$

Let $\hat{\varepsilon}$ be the Fourier transform of ε on $\mathbf{Z}/M\mathbf{Z}$, defined by :

$$\hat{\varepsilon}(m) = \frac{1}{M} \sum_{n \in \mathbf{Z}/M\mathbf{Z}} \varepsilon(n) \exp(-2\pi i n m/M).$$

We then have

$$\varepsilon(n) = \sum_{m \in \mathbf{Z}/M\mathbf{Z}} \hat{\varepsilon}(m) \exp(2\pi i n m/M),$$

hence

$$(f * \varepsilon)(z) = \sum_{m \in \mathbf{Z}/M\mathbf{Z}} \hat{\varepsilon}(m) f(z+\tfrac{m}{M}).$$

From this, one deduces easily that $f * \varepsilon$ is a modular form of weight k on $\Gamma_1(NM^2)$.

7.2. Characterization of cusp forms.

We keep the above notation, and we put

$$\phi_f(s) = \sum_{n=1}^{\infty} a(n) n^{-s}.$$

THEOREM 5. *The following properties are equivalent* :

i) f *vanishes at all cusps* m/M, *with* $m \in \mathbf{Z}$;

ii) *for every function* ε *on* $\mathbf{Z}$, *with period* M, *the function* $\phi_{f*\varepsilon}(s) = \sum_{n=1}^{\infty} a(n)\varepsilon(n) n^{-s}$ *is holomorphic at* $s = k$.

(This is also true when k is an integer, instead of a half integer; the proof is the same.)

PROOF. Consider first the case where $M = 1$. Assertion i) then means that f vanishes at the cusp 0, and assertion ii) that $\phi_f(s)$ is holomorphic at $s = k$. If we put

$$g = f|W(N) = \sum_{n=0}^{\infty} b(n) q^n,$$

then i) is equivalent to :

i') g *vanishes at the cusp* ∞, *i.e.* $b(0)$ *is* 0,

while the functional equation relating $\phi_f(s)$ and $\phi_g(k-s)$ shows that ii) is equivalent to :

ii') $(2\pi)^{-s}\Gamma(s)\phi_g(s)$ is holomorphic at $s = 0$, i.e. $\phi_g(0) = 0$.

The equivalence of i') and ii') then follows from the known relation

$$b(0) = -\phi_g(0).$$

Consider now the general case. By applying the above to $f * \varepsilon$ (with N replaced by NM^2), we see that ii) is equivalent to :

iii) for every function ε on $\mathbf{Z}$, with period M, the modular form $f * \varepsilon$ vanishes at the cusp 0.

Using the above formulae, this is in turn equivalent to :

iv) for every $m \in \mathbf{Z}/M\mathbf{Z}$, the modular form $f(z+\frac{m}{M})$ vanishes at the cusp 0, and it is clear that iv) is equivalent to i).

COROLLARY. The following properties are equivalent :

a) f is a cusp form;

b) for every periodic function ε on $\mathbf{Z}$, the function $\phi_{f*\varepsilon}(s)$ is holomorphic at $s = k$.

Indeed, Theorem 5 shows that b) is equivalent to the fact that f vanishes at all cusps $\neq \infty$; since ∞ is $\Gamma_1(N)$-equivalent to 1/N, this means that f is a cusp form.

REMARK. When f belongs to some $M_0(N,\kappa/2,\chi)$, it is enough to check property b) for functions ε with period N. Indeed, by Theorem 5, this implies the vanishing of f at all cusps m/N, with $m \in \mathbf{Z}$, and it is known that every cusp is $\Gamma_0(N)$-equivalent to one of these.

We now go back to the case $\kappa = 1$, $k = 1/2$:

LEMMA 15. Let ψ be an even character which is not totally even (cf. §2). Then θ_ψ is a cusp form.

PROOF. Let ε be a periodic function on $\mathbf{Z}$. By the Corollary to Theorem

5, it is enough to prove that the Dirichlet series

$$F_\varepsilon(s) = 2 \sum_{n=1}^{\infty} \varepsilon(n^2)\psi(n)n^{-2s}$$

is holomorphic at s = 1/2. Let M ⩾ 1 be a period of ε, which we may assume to be a multiple of the conductor r(ψ) of ψ. We have

$$F_\varepsilon(s) = 2 \sum_{m \in \mathbf{Z}/M\mathbf{Z}} \varepsilon(m^2)\psi(m)F_{m,M}(2s),$$

where

$$F_{m,M}(s) = \sum_{\substack{n \equiv m \pmod M \\ m \geqslant 1}} n^{-s}.$$

It is an elementary fact that $F_{m,M}(s)$ has a simple pole at s = 1 with residue 1/M. Hence $F_\varepsilon(s)$ has at most a simple pole at s = 1/2, with residue R(ε,ψ)/M, where

$$R(\varepsilon,\psi) = \sum_{m \in \mathbf{Z}/M\mathbf{Z}} \varepsilon(m^2)\psi(m),$$

and we have to prove that R(ε,ψ) = 0. By assumption, there is a prime ℓ dividing r(ψ) such that the ℓth component ψ_ℓ of ψ is odd. Let us write M as $\ell^a M'$, with $(\ell,M') = 1$, so that the ring **Z**/M**Z** splits as $\mathbf{Z}/\ell^a\mathbf{Z} \times \mathbf{Z}/M'\mathbf{Z}$. Let x_ℓ be the element of **Z**/M**Z** whose first component (in the above decomposition) is -1, and the second component is 1. The fact that ψ_ℓ is odd means that $\psi(x_\ell) = -1$. Since x_ℓ is invertible in **Z**/M**Z**, we have

$$R(\varepsilon,\psi) = \sum_{m \in \mathbf{Z}/M\mathbf{Z}} \varepsilon((x_\ell m)^2)\psi(x_\ell m) = \sum_{m \in \mathbf{Z}/M\mathbf{Z}} \varepsilon(m^2)\psi(x_\ell m)$$

$$= - \sum_{m \in \mathbf{Z}/M\mathbf{Z}} \varepsilon(m^2)\psi(m) = -R(\varepsilon,\psi)$$

which shows that R(ε,ψ) = 0, as wanted.

<u>LEMMA</u> 16. <u>Let</u> ψ <u>be a totally even character</u>, <u>and</u> T <u>a finite set of integers</u> ⩾ 1. <u>If the modular form</u> $f = \sum_{t \in T} c_t\theta_{\psi,t}$ $(c_t \in \mathbf{C})$

is a cusp form, then all c_t *are* 0.

PROOF. Assume the c_t are not all 0, and let t_0 be the smallest $t \in T$ such that $c_t \neq 0$. Choose an integer $M \geqslant 1$ which is divisible by $2r(\psi)$ and by all $t \in T$. The first divisibility condition, together with the assumption that ψ is totally even, implies that there is a character α definable (mod M) such that $\alpha^2 = \psi$. Define now a periodic function ε on $\mathbf{Z}$ by

$$\varepsilon(n) = \begin{cases} \overline{\alpha}(n/t_0) & \text{if } t_0 \mid n \text{ and } n/t_0 \text{ is prime to } M \\ 0 & \text{otherwise.} \end{cases}$$

We have

$$\varepsilon(t_0 n^2) = \begin{cases} \overline{\psi}(n) & \text{if } (n,M) = 1 \\ 0 & \text{if } (n,M) \neq 1 \end{cases}$$

and

$$\varepsilon(tn^2) = 0 \text{ if } t \in T,\ t > t_0 \text{ (since } (tn^2,M) \geqslant t > t_0).$$

Using the minimality of t_0, this shows that the Dirichlet series $\Phi_{f * \varepsilon}(s)$ is equal to

$$2c_{t_0} \sum_{\substack{(n,M)=1 \\ n\geqslant 1}} \overline{\psi}(n)\psi(n)(t_0 n^2)^{-s} = 2c_{t_0} t_0^{-s} \sum_{\substack{(n,M)=1 \\ n\geqslant 1}} n^{-2s}.$$

The same argument as in the proof of Lemma 15 shows that the residue of this function at $s = 1/2$ is equal to

$$c_{t_0} t_0^{-1/2}\ \phi(M)/M = c_{t_0} t_0^{-1/2} \prod_{p \mid M} (1-\frac{1}{p}),$$

which is $\neq 0$. By Theorem 5, we thus see that f is not a cusp form.

7.3. Proof of Theorem B.

Let $N, \chi, \Omega_c(N,\chi), \Omega_e(N,\chi)$ be as defined in §2. We have three assertions to prove :

a) The $\theta_{\psi,t}$, with $(\psi,t) \in \Omega_c(N,\chi)$, are cusp forms.
Indeed, Lemma 15 shows that θ_ψ is a cusp form, and this obviously implies the same property for $\theta_{\psi,t}$.

b) No linear combination (except 0) of the $\theta_{\psi,t}$, with $(\psi,t) \in \Omega_e(N,\chi)$, is a cusp form.
Let V be the space of the linear combinations of the $\theta_{\psi,t}$, with $(\psi,t) \in \Omega_e(N,\chi)$, which are cusp forms. It is clear that V is stable under the $T(p^2)$, $p \nmid N$. Hence, if V is non-zero, it contains a common eigenform f of the $T(p^2)$, $p \nmid N$. Since the eigenvalue of $\theta_{\psi,t}$ is $(1+p^{-1})\psi(p)$, the form f has to be a linear combination of the $\theta_{\psi,t}$ for a fixed character ψ, and this contradicts Lemma 16.

c) If $(\psi,t) \in \Omega_c(N,\chi)$ and $(\psi',t') \in \Omega_e(N,\chi)$, then $\theta_{\psi,t}$ and $\theta_{\psi',t'}$ are orthogonal for the Petersson scalar product.
Indeed, since $\psi \neq \psi'$, there is a $p \nmid N$ such that $\psi(p) \neq \psi'(p)$. Hence, $\theta_{\psi,t}$ and $\theta_{\psi',t'}$ are eigenforms of $T(p^2)$ corresponding to different eigenvalues. Since $\overline{\chi}(p)T(p^2)$ is hermitian (cf. the proof of Lemma 14, §6) this implies that these two functions are orthogonal.

7.4. The space $E_1(N,1/2)$.

Let $E_0(N,1/2,\chi)$ be the space of linear combinations of the $\theta_{\psi,t}$ with $(\psi,t) \in \Omega_e(N,\chi)$. By Theorem B, we have the orthogonal decomposition

$$M_0(N,1/2,\chi) = E_0(N,1/2,\chi) \oplus S_0(N,1/2,\chi),$$

where $S_0(N,1/2,\chi)$ is the space of cusp forms. Similarly, if we put

$E_1(N,1/2) = \oplus E_0(N,1/2,\chi)$, we have

$$M_1(N,1/2) = E_1(N,1/2) \oplus S_1(N,1/2).$$

The elements of $E_1(N,1/2)$ can be characterized as follows :

THEOREM 6. Let f be an element of $M_1(N,1/2)$. The following properties are equivalent :

i) f belongs to $E_1(N,1/2)$.

ii) f is a linear combination of $\theta(az+b)$, with $a \in \mathbf{Z}$, $a \geq 1$, and $b \in \mathbf{Q}$.

iii) f is orthogonal to all cusp forms of all levels.

PROOF. Clearly ii) implies iii) since θ is in $E_1(M,1/2)$ for every M, and so is orthogonal to all cusp forms; the same is then true of $\theta(az+b)$ for any a and b. We have already shown that iii) implies i). Finally, if ψ is a totally even character, we may write ψ as α^2 where the character α is ramified at the same primes as ψ; we have $\theta_\psi = \theta * \alpha$, hence θ_ψ is a linear combination of the $\theta(z+b)$, with $b \in \mathbf{Q}$; this shows that θ_ψ has property ii), hence that i) implies ii).

REMARK. Maass [6] has shown that $\theta(z)$ can be defined as an "Eisenstein series", by analytic continuation à la Hecke. The same is true for all the $\theta(az+b)$, hence for all the elements of $E_1(N,1/2)$.

APPENDIX

Free translation of a letter from Pierre DELIGNE,

dated March 1, 1976

... Using the same trick as in my Antwerp's paper (vol. II, p.90, proof of 2.5.6), one can deduce directly from your Theorem 2 the structure of the modular forms of weight 1/2 (on congruence subgroups of $\mathbf{SL}_2(\mathbf{Z})$). The final result is :

THEOREM. The q-expansions of the modular forms of weight 1/2 are

$$(1) \qquad \sum_t \sum_{u \in \mathbf{Z}} \phi_t(u) q^{tu^2},$$

where t runs through a finite subset of $\mathbf{Q}^{*+}$, and, for each t, ϕ_t is a periodic function on $\mathbf{Z}$ (i.e. the restriction of a locally constant function on $\hat{\mathbf{Z}}$).

PROOF. Let H be the space of modular forms of weight 1/2, and Θ the subspace of H consisting of the theta series (1). We put on H the Petersson scalar product (which always converges). The metaplectic 2-covering $\widetilde{\mathbf{SL}}_2(\mathbf{A}_f)$ of $\mathbf{SL}_2(\mathbf{A}_f)$ acts on H, preserves the scalar product, and leaves Θ stable. Under this action, H decomposes into a direct sum of irreducible representations. Let H_i be one of them. We want to prove that H_i is contained in Θ.

One checks immediately that, if N and χ are suitably chosen, H_i has a non-zero intersection with $M_0(N,1/2,\chi)$. The Hecke operators $T(p^2)$ associated with all primes p (including those dividing N) come from the action of (the group ring of) $\widetilde{\mathbf{SL}}_2(\mathbf{A}_f)$, and commute with each other. Hence they have a non-zero common eigenvector f in $H_i \cap M_0(N,1/2,\chi)$. By your Theorem 2, one has

$$f = \sum_{u \in \mathbf{Z}} a(tu^2) q^{tu^2} \qquad (t \text{ square-free, } t|N),$$

and

$$a(mu^2) = a(m)\psi(u) \quad \text{if } (u,N)=1, \ \psi \text{ being some character (mod 2N)},$$
$$a(mp^2) = \lambda_p a(m) \quad \text{if } p|N \quad \text{(cf. Shimura [8], 1.7).}$$

Consider now

$$g = \sum_{(u,N)=1} a(tu^2) q^{tu^2}.$$

It is clear that g is a non-zero element of Θ. On the other hand, g is (up to a scalar factor) the transform of f by $\prod_{p|N} L_p$, where L_p is the operator which transforms $h(z)$ into $h(z) - \lambda_p h(p^2 z)$. Since L_p can be defined by the element $1 - \lambda_p \begin{pmatrix} p & 0 \\ 0 & p^{-1} \end{pmatrix}$ of the group ring of $\widetilde{\mathbf{SL}}_2(\mathbf{Q}_p)$, this shows that g belongs to H_i, hence $H_i \cap \Theta \neq 0$. Since H_i is irreducible, this implies $H_i \subset \Theta$, q.e.d.

Yours,

P. Deligne

PS. These arguments should extend to any totally real number field.

BIBLIOGRAPHY

[1] A.O.L. ATKIN and J. LEHNER, Hecke operators on $\Gamma_0(m)$, Math. Ann. 185 (1970), p. 134-160.

[2] A.O.L. ATKIN and H.P.F. SWINNERTON-DYER, Modular forms on noncongruence subgroups, Proc. Symp. Pure Math. XIX, p. 1-25, Amer. Math. Soc., 1971

[3] P. DELIGNE and J-P. SERRE, Formes modulaires de poids 1, Ann. Sci. E.N.S. (4) 7 (1974), p. 507-530.

[4] E. HECKE, Mathematische Werke (zw. Aufl.) Vandenhoeck und Ruprecht, Göttingen, 1970.

[5] W. LI, Newforms and Functional Equations, Math. Ann. 212 (1975), p. 285-315.

[6] H. MAASS, Konstruktion ganzer Modulformen halbzahliger Dimension mit θ-Multiplikatoren in einer und zwei Variablen, Abh. Math. Sem. Univ. Hamburg 12 (1937), p. 133-162.

[7] G. SHIMURA, Introduction to the arithmetic theory of automorphic functions, Publ. Math. Soc. Japan, 11, Princeton Univ. Press, 1971.

[8] G. SHIMURA, On modular forms of half integral weight, Ann. of Math. 97 (1973), p. 440-481.

[9] G. SHIMURA, On the trace formula for Hecke operators, Acta Math. 132 (1974), p. 245-281.

International Summer School on Modular Functions
BONN 1976

DIMENSIONS DES ESPACES DE FORMES MODULAIRES

par H. COHEN et J. OESTERLÉ

I) Introduction.

La connaissance explicite des dimensions des espaces de formes modulaires est nécessaire dans de nombreux problèmes. Les formules qui les donnent sont connues de beaucoup de gens et il existe plusieurs méthodes permettant de les obtenir (théorème de Riemann-Roch, application des formules de trace données par Shimura dans [3]). Néanmoins on ne les trouve pas dans la littérature courante ; cet exposé, tout en s'abstenant de fournir les démonstrations, se propose donc de combler cette lacune. En outre, une table donnant $\dim S_k(\Gamma_0(N),\chi)$ et $\dim(M_k(\Gamma_0(N),\chi)$ pour $k\in\frac{1}{2}+\mathbb{Z}$, $N\leq 200$, et tout caractère χ, figure à la fin de l'article.

II) Notations.

Soit $\mathfrak{H}=\{z\in\mathbb{C}/\mathrm{Im}\ z>0\}$ le demi-plan de Poincaré sur lequel agit $SL_2(\mathbb{R})$ par : $\begin{pmatrix} a & b \\ c & d \end{pmatrix}.z=\frac{az+b}{cz+d}$.

Soit N un entier naturel non nul et notons $\Gamma_0(N)$ le sous-groupe de $SL_2(\mathbb{Z})$ formé des matrices $\begin{pmatrix} a & b \\ c & d \end{pmatrix}$ telles que $c\equiv 0(N)$.

Soit χ un caractère multiplicatif modulo N, i.e. un homomorphisme de $(\mathbb{Z}/N\mathbb{Z})^*$ dans $\mathbb{C}^*$. Soit f le conducteur de χ, c'est-à-dire le plus petit diviseur de N tel que χ se factorise à travers

$(\mathbb{Z}/f\mathbb{Z})^*$.

Soit $k \in \frac{1}{2}\mathbb{Z}$ un entier ou demi-entier.

Nous ferons les hypothèses suivantes :

- Si k est entier, $\chi(-1) = (-1)^k$.

- Si $k \in \frac{1}{2} + \mathbb{Z}$, N est multiple de 4 et $\chi(-1) = 1$.

Nous définirons le symbole $(\frac{c}{d})$ pour $c \in \mathbb{Z}$, et $d \in \mathbb{Z}^*$ par les conditions suivantes :

$d \to (\frac{c}{d})$ est complètement multiplicative

$(\frac{c}{-1}) = -1$ si $c < 0$ et $(\frac{c}{-1}) = 1$ si $c \geqslant 0$

$(\frac{c}{2}) = (-1)^{(c^2-1)/8}$ si c est impair

$(\frac{c}{p})$ est le symbole de Legendre si p est premier impair et c premier à p

$(\frac{c}{d}) = 0$ si $\text{pgcd}(c,d) \neq 1$.

La notation a^b , pour $a \in \mathbb{C}^*$ et $b \in \mathbb{C}$ désignera $\exp(b(\log|a| + i \operatorname{Arg} a))$, avec $-\pi < \operatorname{Arg} a \leqslant \pi$.

Nous noterons $M_k(\Gamma_o(N),\chi)$ (resp. $S_k(\Gamma_o(N),\chi)$) et nous appellerons espace des formes modulaires entières (resp. paraboliques) de poids k , de niveau N et de caractère χ l'espace des fonctions f définies sur $\mathfrak{H}$ ayant les propriétés suivantes :

a) f est holomorphe sur $\mathfrak{H}$.

b) Si $k \in \mathbb{Z}$, $f(\frac{az+b}{cz+d}) = \chi(d)(cz+d)^k f(z)$ pour toute matrice $\begin{pmatrix} a & b \\ c & d \end{pmatrix} \in \Gamma_o(N)$ et tout $z \in \mathfrak{H}$.

b') Si $k \in \frac{1}{2} + \mathbb{Z}$, $f(\frac{az+b}{cz+d}) = \chi(d)(\frac{c}{d})(\frac{-1}{d})^{-k}(cz+d)^k f(z)$ pour toute matrice $\begin{pmatrix} a & b \\ c & d \end{pmatrix} \in \Gamma_o(N)$ et tout $z \in \mathfrak{H}$.

c) f est holomorphe (resp. s'annule) aux pointes (cf. [2] pour la signification de cet énoncé).

On démontre que les espaces $M_k(\Gamma_o(N),\chi)$ et $S_k(\Gamma_o(N),\chi)$ sont de dimension finie. On a même les résultats suivants :

(i) $\dim S_k(\Gamma_o(N),\chi) = 0$ si $k \leq 0$,

(ii) $\dim M_k(\Gamma_o(N),\chi) = 0$ si $k < 0$, ou bien si $k = 0$ et que χ n'est pas le caractère trivial χ_o,

(iii) $\dim M_o(\Gamma_o(N),\chi_o) = 1$.

III) Les résultats.

Dans les théorèmes 1 et 2, nous allons donner la valeur de $\dim S_k(\Gamma_o(N),\chi) - \dim M_{2-k}(\Gamma_o(N),\chi)$, valable pour toute valeur de k dans $\frac{1}{2}\mathbb{Z}$, le théorème 1 traitant le cas où k est dans $\mathbb{Z}$, et le théorème 2 le cas où $k \in \frac{1}{2}+\mathbb{Z}$.

Compte tenu de (i), (ii), (iii), les formules ainsi obtenues permettent de calculer la valeur de $\dim S_k(\Gamma_o(N),\chi)$ pour $k \geq 2$, et aussi celle de $\dim M_k(\Gamma_o(N),\chi)$ pour $k \geq 2$ (faire $k = 2-k$ dans la formule).

Pour $k = 1/2$ et $k = 3/2$, les formules donnent la valeur de $\dim S_{1/2}(\Gamma_o(N),\chi) - \dim M_{3/2}(\Gamma_o(N),\chi)$ et de $\dim S_{3/2}(\Gamma_o(N),\chi) - \dim M_{1/2}(\Gamma_o(N),\chi)$.

Pour un N et un χ fixé, Serre et Stark exhibent dans [1] une base explicite de $S_{1/2}(\Gamma_o(N),\chi)$ et $M_{1/2}(\Gamma_o(N),\chi)$. Les formules précédentes permettent alors d'obtenir la dimension de $M_{3/2}(\Gamma_o(N),\chi)$ et $S_{3/2}(\Gamma_o(N),\chi)$.

Théorème 1. Soit $k \in \mathbb{Z}$ (et donc $\chi(-1) = (-1)^k$). On a :

$$\dim S_k(\Gamma_o(N),\chi) - \dim M_{2-k}(\Gamma_o(N),\chi) =$$

$$\frac{k-1}{12} N \prod_{p|N} \left(1 + \frac{1}{p}\right) - \frac{1}{2} \prod_{p|N} \lambda(r_p, s_p, p) + \varepsilon_k \sum_{\substack{x \bmod N \\ x^2+1 \equiv 0(N)}} \chi(x)$$

$$+ \mu_k \sum_{\substack{x \bmod N \\ x^2+x+1 \equiv 0(N)}} \chi(x)$$

avec les notations suivantes :

si $p|N$, r_p (resp. s_p) désigne l'exposant de p dans la décomposition en facteurs premiers de N (resp. de f)

. $\lambda(r_p,s_p,p)$ vaut : $\begin{cases} p^{r'}+p^{r'-1} & \text{si } 2s_p \leqslant r_p = 2r' \\ 2p^{r'} & \text{si } 2s_p \leqslant r_p = 2r'+1 \\ 2p^{r_p-s_p} & \text{si } 2s_p > r_p \end{cases}$

. ε_k vaut : $\begin{cases} 0 & \text{si } k \text{ impair} \\ -\frac{1}{4} & \text{si } k \equiv 2 \pmod 4 \\ \frac{1}{4} & \text{si } k \equiv 0 \pmod 4 \end{cases}$

. μ_k vaut : $\begin{cases} 0 & \text{si } k \equiv 1 \pmod 3 \\ -\frac{1}{3} & \text{si } k \equiv 2 \pmod 3 \\ \frac{1}{3} & \text{si } k \equiv 0 \pmod 3 \end{cases}$

Remarques concernant l'énoncé du théorème :

1°) L'expression $\prod_{p|N} \lambda(r_p,s_p,p)$ est en fait égale à la somme $\sum_{\substack{c|N \\ (c,\frac{N}{c})|\frac{N}{f}}} \varphi((c,\frac{N}{c}))$ où φ est la fonction d'Euler. Mais c'est le produit qui est le plus maniable en pratique.

2°) Si $\chi = \chi_0$, caractère trivial, la somme $\sum_{\substack{x \bmod N \\ x^2+1 \equiv 0(N)}} \chi(x)$ (resp. $\sum_{\substack{x \bmod N \\ x^2+x+1 \equiv 0(N)}} \chi(x)$) est égale au nombre de racines de l'équation $x^2+1=0$ (resp. $x^2+x+1=0$) dans $\mathbb{Z}/N\mathbb{Z}$. Ce nombre vaut :

$$\begin{cases} 0 & \text{si } 4|N \\ \prod_{p|N} (1+(\frac{-4}{p})) & \text{si } 4\nmid N \end{cases} \qquad \left(\text{resp.} \begin{cases} 0 & \text{si } 9|N \\ \prod_{p|N} (1+(\frac{-3}{p})) & \text{si } 9\nmid N \end{cases}\right)$$

Théorème 2. Soit $k \in \frac{1}{2}+\mathbb{Z}$ (et donc $4|N$ et $\chi(-1)=1$). On a :

$$\dim S_k(\Gamma_0(N),\chi) - \dim M_{2-k}(\Gamma_0(N),\chi) = \frac{k-1}{12}N \prod_{p|N} (1+\frac{1}{p}) - \frac{\zeta}{2} \prod_{\substack{p|N \\ p\neq 2}} \lambda(r_p,s_p,p)$$

où λ, r_p et s_p ont même signification qu'au théorème 1, et où ζ est défini par le diagramme suivant :

$r_2 \geqslant 4$				$\zeta = \lambda(r_2, s_2, 2)$
$r_2 = 3$				$\zeta = 3$
$r_2 = 2$	Condition (C)			$\zeta = 2$
	non (C)	$k - \frac{1}{2} \in \mathbb{Z}$	$s_2 = 0$	$\zeta = 3/2$
			$s_2 = 2$	$\zeta = 5/2$
		$k - \frac{3}{2} \in \mathbb{Z}$	$s_2 = 0$	$\zeta = 5/2$
			$s_2 = 2$	$\zeta = 3/2$

<u>La condition</u> (C) <u>étant la condition suivante</u> :

$$(C) \Longleftrightarrow \exists p \text{ premier},\ p \equiv 3(4)\ ,\ p \mid N\ ,\ r_p \text{ impair ou } 0 < r_p < 2s_p\ .$$

<u>Par suite</u> :

$$\text{non } (C) \Longleftrightarrow (\forall p \text{ premier}),\ (p \equiv 3(4) \quad \text{et} \quad p \mid N \Longrightarrow r_p \ \text{ pair et } \ r_p \geqslant 2s_p).$$

TABLES : DIMENSIONS EN POIDS DEMI-ENTIER

I) Rappel des notations.

$k \in \frac{1}{2} + \mathbb{Z}$, N est un entier naturel multiple de 4 et χ un caractère multiplicatif modulo N de conducteur f tel que $\chi(-1) = 1$; on note $M_k(\Gamma_o(N),\chi)$, $S_k(\Gamma_o(N),\chi)$, $E_k(\Gamma_o(N),\chi)$ l'espace des formes modulaires entières, paraboliques, d'Eisenstein respectivement, relativement au groupe $\Gamma_o(N)$ et au caractère χ , et de poids k .

χ étant astreint à être pair, f ne peut prendre que des valeurs distinctes de 3 et 4 et non congrues à 2 modulo 4.

II) Présentation des tables .

Les tables qui suivent permettent de calculer facilement les valeurs des dimensions des espaces $M_k(\Gamma_o(N),\chi)$, $S_k(\Gamma_o(N),\chi)$, $E_k(\Gamma_o(N),\chi)$ pour tout $k \in \frac{1}{2} + \mathbb{Z}$, et tout caractère χ modulo N , pourvu que le niveau N soit inférieur ou égal à 200.

Une première table donne les valeurs d'une certaine quantité a(N) en fonction de N .

Une deuxième table, ordonnée en quatre colonnes, donne les valeurs de certaines quantités b(N,f) et c(N,f) qui ne dépendent que du niveau N, et du conducteur f de χ . Dans la première colonne figurent les valeurs de N . Dans la seconde colonne figurent les valeurs de f . Parfois, cette colonne est vide ; cela signifie que pour le niveau N correspondant, b(N,f) et c(N,f) sont en fait indépendants de f . Parfois, on trouvera sur une même ligne de cette seconde colonne, plusieurs valeurs de f distinctes, séparées par une virgule ; cela signifie que $b(N,f)$ et $c(N,f)$ sont les mêmes pour toutes ces valeurs de f.

Une troisième table donne la dimension, que nous noterons $d(N,\chi)$, de l'espace $M_{1/2}(\Gamma_o(N),\chi)$. Des règles générales permettant de trouver $d(N,\chi)$ y figurent, et elles sont suivies de la liste des exceptions.

III) Résultats.

Avec les notations précédentes on a :

si $k < 0$ $\quad \dim S_k(\Gamma_o(N),\chi) = \dim M_k(\Gamma_o(N),\chi) = 0$

si $k = \frac{1}{2}$ $\quad \dim S_{1/2}(\Gamma_o(N),\chi) = 0$ (attention : ceci n'est plus vrai si $N > 200$)

$\dim M_{1/2}(\Gamma_o(N),\chi) = d(N,\chi)$

si $k = \frac{3}{2}$ $\quad \dim S_{3/2}(\Gamma_o(N),\chi) = d(N,\chi) + b(N,f) - c(N,f)$

$\dim M_{3/2}(\Gamma_o(N),\chi) = b(N,f)$

si $k = 2k' + \frac{1}{2}$ avec $k' \geqslant 1$ $\quad \dim S_k(\Gamma_o(N),\chi) = a(N)k' - b(N,f)$

$\dim M_k(\Gamma_o(N),\chi) = a(N)k' - b(N,f) + c(N,f)$

$\dim E_k(\Gamma_o(N),\chi) = c(N,f)$

si $k = 2k' + \frac{3}{2}$ avec $k' \geqslant 1$ $\quad \dim S_k(\Gamma_o(N),\chi) = a(N)k' + b(N,f) - c(N,f)$

$\dim M_k(\Gamma_o(N),\chi) = a(N)k' + b(N,f)$

$\dim E_k(\Gamma_o(N),\chi) = c(N,f)$

1ère table :

N	4	8	12	16	20	24	28	32	36	40	44	48	52	56	60	64	68
a(N)	1	2	4	4	6	8	8	8	12	12	12	16	14	16	24	16	18

N	72	76	80	84	88	92	96	100	104	108	112	116	120	124	128	132	136
a(N)	24	20	24	32	24	24	32	30	28	36	32	30	48	32	32	48	36

N	140	144	148	152	156	160	164	168	172	176	180	184	188	192	196	200
a(N)	48	48	38	40	56	48	42	64	44	48	72	48	48	64	56	60

2ème table :

N	f	c(N,f)	b(N,f)
4		2	1
8		3	2
12		4	3
16	1	6	4
	8	4	3
	16	2	2
20	1,5	4	3
	20	4	4
24		6	5
28		4	4
32	1,8	8	6
	16	4	4
	32	2	3
36	1	8	6
	12	8	8
	9,36	4	5
40		6	6
44		4	5
48	1,12	12	10
	8,24	8	8
	16,48	4	6
52	1,13	4	5
	52	4	6
56		6	7
60		8	10
64	1,8	12	10
	16	8	8
	32	4	6
	64	2	5
68	1,17	4	6
	68	4	7
72	1,8,12,24	12	12
	9,36,72	6	9
76		4	7
80	1,5,20	12	12
	8,40	8	10
	16,80	4	8
84		8	12
88		6	9
92		4	8
96	1,8,12,24	16	16
	16,48	8	12
	32,96	4	10
100	1,5	12	12
	20	12	15
	25	4	9
	100	4	10
104		6	10
108	1,9,12,36	12	15
	27,108	4	11

N	f	c(N,f)	b(N,f)
112	1,7,28	12	14
	8,56	8	12
	16,112	4	10
116	1,29	4	9
	116	4	10
120		12	18
124		4	10
128	1,8,16	16	16
	32	8	12
	64	4	10
	128	2	9
132		8	16
136		6	12
140		8	16
144	1,12	24	24
	8,24	16	20
	9,36	12	18
	16,48,72	8	16
	144	4	14
148	1,37	4	11
	148	4	12
152		6	13
156		8	18
160	1,5,8,20,40	16	20
	16,80	8	16
	32,160	4	14
164	1,41	4	12
	164	4	13
168		12	22
172		4	13
176	1,11,44	12	18
	8,88	8	16
	16,176	4	14
180	1,5,15	16	24
	12,20,60	16	28
	9,36,45,180	8	22
184		6	15
188		4	14
192	1,12,8,24	24	28
	16,48	16	24
	32,96	8	20
	64,192	4	18
196	1,7	16	20
	28	16	24
	49,196	4	16
200	1,5,8,20,40	18	24
	25,100,200	6	18

<u>3ème table</u> : Valeurs de $d(N,\chi) = \dim M_{\frac{1}{2}}(\Gamma_0(N),\chi)$ pour $N < 200$.

- <u>Premier cas</u> : χ n'est pas le caractère unité et n'est pas d'ordre 2.

Dans ce cas on a $d(N,\chi) = 0$ sauf dans le cas suivant :

Si $N = 196$ et que χ est l'un des deux caractères d'ordre 3 et de conducteur 7, alors $d(N,\chi) = 1$.

- <u>Deuxième cas</u> : χ est le caractère unité, i.e : $\chi = \chi_0$.

Dans ce cas, on a la table suivante :

N	4	8	12	16	20	24	28	32	36	40	44	48	52	56	60	64	68
$d(N,\chi_0)$	1	1	1	2	1	1	1	2	2	1	1	2	1	1	1	3	1

N	72	76	80	84	88	92	96	100	104	108	112	116	120	124	128	132	136
$d(N,\chi_0)$	2	1	2	1	1	1	2	2	1	2	2	1	1	1	3	1	1

N	140	144	148	152	156	160	164	168	172	176	180	184	188	192	196	200
$d(N,\chi_0)$	1	4	1	1	1	2	1	1	1	2	2	1	1	3	2	2

- <u>Troisième cas</u> : χ est d'ordre 2.

Dans ce cas $d(N,\chi) = 1$, sauf pour la table d'exceptions que nous présentons ci-dessous avec en première colonne la valeur de N, en seconde colonne l'entier f (il est à remarquer que si χ quadratique pair, alors $\chi = (\frac{f}{\cdot})$ où f est son conducteur) ; dans la troisième colonne on trouve $d(N,\chi)$.

N	f	d(N,χ)
32	8	2
48	12	2
64	8	2
72	8	2
80	5	2
96	8	2
	12	2
	24	2
100	5	2

N	f	d(N,χ)
108	12	2
112	28	2
128	8	3
144	8	2
	12	2
160	8	2
	5	2
	40	2
176	44	2

N	f	d(N,χ)
180	5	2
192	8	2
	12	3
	24	2
200	8	2
	5	2
	40	2

IV) Exemples.

1°) Calcul de $\dim[S_{3/2}(\Gamma_o(108),(\frac{12}{\cdot}))]$. Remarque : $(\frac{12}{\cdot})$ est le caractère d'ordre 2 associé à l'extension $\mathbb{Q}(\sqrt{3})/\mathbb{Q}$; son conducteur f est égal à 12.

On a : $d(108,(\frac{12}{\cdot})) = 2$ (cf 3ème table, 3e cas, exceptions)

$b(108,12) = 15$ (2ème table)

$c(108,12) = 12$ (2ème table)

Donc : $\dim S_{3/2}(\Gamma_o(108),(\frac{12}{\cdot})) = 2+15-12 = 5$.

2°) Calcul de $\dim[M_{9/2}(\Gamma_o(156),\chi)]$ où χ est un caractère d'ordre 19 et de conducteur 39.

On a : $\frac{9}{2} = 2k' + \frac{1}{2}$ avec $k' = 2$

$a(156) = 56$ (1ère table)

$b(156,39) = 18$ (2ème table)

$c(156,39) = 8$ (2ème table)

Donc la dimension cherchée est $56 \times 2 - 18 + 8 = 102$.

-:-:-:-:-:-:-

Bibliographie

[1] J.-P. SERRE et H. M. STARK : Modular forms of weight ½, (publié dans ce même volume).

[2] G. SHIMURA : Introduction to the theory of automorphic forms. Princeton University Press, (1971).

[3] G. SHIMURA : On the trace formula for Hecke operators. Acta Mathematica, vol. 132, (1974).

International Summer School on Modular Functions
BONN 1976

FACTEURS GAMMA ET ÉQUATIONS FONCTIONNELLES

par M.-F. VIGNÉRAS

Introduction.

On considère des séries de Dirichlet $\varphi(s)$ de la forme

(1) $$\sum_{n\geqslant 1} a_n n^{-s} \quad , \quad s\in\mathbb{C} \ , \ a_n\in\mathbb{C}$$

vérifiant une équation fonctionnelle :

(2) $$\Phi(s) = A^{-s}\Psi(-s) \quad , \quad A>0$$

avec les conditions suivantes :

1) $\varphi(s)$ converge pour $\mathrm{Re}(s)$ assez grand vers une fonction non identiquement nulle.

2) $\psi(s)$ est une autre série de Dirichlet vérifiant 1), de la forme (1), $\psi(s)=\sum_{n\geqslant 1} b_n n^{-s}$, $s\in\mathbb{C}$, $b_n\in\mathbb{C}$.

3) $\Delta_1(s)$, $\Delta_2(s)$ sont des facteurs gamma

(3) $$\Delta_1(s) = \prod_{f=1}^{G} \Gamma(p_f s + c_f)$$
$$\Delta_2(s) = \prod_{g=G+1}^{H} \Gamma(p_g s + c_g)$$

où $G\geqslant 1$, $H-G\geqslant 1$, les constantes c_f, c_g sont des nombres complexes et les constantes p_f, p_g des nombres rationnels strictement positifs. On pose

$$p = p_1 + \ldots + p_H$$

$$P = p_1^{p_1} \times \ldots \times p_H^{p_H} .$$

4) Les fonctions

$$\Phi(s) = (2P\pi)^{-s/2} \Delta_1(s) \varphi(s) \tag{4}$$

$$\Psi(s) = (2P\pi)^{-s/2} \Delta_2(s) \psi(s)$$

admettent des prolongements holomorphes à l'extérieur d'un ensemble borné, tels que pour tout σ_1 , σ_2 finis

$$\lim_{|t|\to\infty} \Phi(\sigma+it) = \lim_{|t|\to\infty} \Psi(\sigma+it) = 0 \quad \text{uniformément pour } \sigma_1 \leq \sigma \leq \sigma_2 .$$

On dira que $(A, \psi(s), \Delta_1(s), \Delta_2(s))$ sont les données de l'équation fonctionnelle vérifiée par $\varphi(s)$. On a utilisé les notations de Bochner [1]. On remarquera que l'équation fonctionnelle avec $k-s$, au lieu de $-s$, $k \in \mathbb{C}$

$$\Phi(s) = A^{-s} \Psi(k-s)$$

se ramène à une équation fonctionnelle du type précédent avec les données $(A, (2P\pi)^{-k/2} \psi(k+s), \Delta_1(s), \Delta_2(k+s))$.

Bochner étudie dans [1] les équations fonctionnelles des séries de Dirichlet générales, c'est-à-dire les équations fonctionnelles (2), vérifiant 1) à 4) mais où $\varphi(s)$, $\psi(s)$ ne sont plus astreintes à la condition d'être de la forme (1) et sont des séries de Dirichlet générales, de la forme :

$$(1') \quad \begin{aligned} \varphi(s) &= \sum_{n\geq 1} a_n \lambda_n^{-s} , \quad a_n \in \mathbb{C} , \ 0 < \lambda_1 < \lambda_2 \ldots < \lambda_n \to \infty \\ \psi(s) &= \sum_{n\geq 1} b_n \mu_n^{-s} , \quad b_n \in \mathbb{C} , \ 0 < \mu_1 < \mu_2 \ldots < \mu_n \to \infty \end{aligned}$$

Nous démontrerons dans le paragraphe 1 que les résultats de Bochner [1] ont pour conséquence les théorèmes 1 et 2 suivants :

<u>Théorème</u> 1. <u>Il n'existe pas d'équation fonctionnelle vérifiant 1) à 4) si</u> $\Sigma p_f \neq \Sigma p_g$ <u>ou si</u> $\Sigma p_f < \frac{1}{2}$.

Si l'on suppose :

5) $\Sigma p_f = \Sigma p_g = \frac{1}{2}$

les solutions des équations fonctionnelles s'obtiennent à partir des fonctions zêta partielles

$$\zeta_{a,A}(s) = \sum_{\substack{n \in \mathbb{Z} \\ n \equiv a \pmod A \\ n \neq 0}} |n|^{-s} \qquad \zeta^*_{a,A}(s) = \sum_{\substack{n \in \mathbb{Z} \\ n \equiv a \pmod A \\ n \neq 0}} n|n|^{-s} \quad , \quad a,A \in \mathbb{N}.$$

Plus précisément :

<u>Théorème</u> 2. <u>Soit</u> $\varphi(s)$ <u>une série de Dirichlet de la forme</u> (1) <u>vérifiant une équation fonctionnelle</u> (2) <u>satisfaisant</u> 1) <u>à</u> 5). <u>Alors</u>,

- <u>les facteurs gamma de l'équation fonctionnelle sont de la forme</u> :

$$\Delta_1(s) = \Gamma\left(\frac{s+c_1}{2}\right) f_1(s) \qquad \Delta_2(s) = \Gamma\left(\frac{s+c_2}{2}\right) f_2(s)$$

<u>où</u> $f_1(s)$ <u>et</u> $f_2(s)$ <u>sont deux fonctions entières d'ordre</u> 1 <u>et où</u> c_1 <u>et</u> c_2 <u>sont deux nombres complexes vérifiant</u> :

$$c_1 + c_2 = 1 \text{ \underline{ou} } 3$$

- $A \in \mathbb{N}$
- <u>Si</u> $c_1+c_2 = 1$, $\varphi(s)$ <u>est une combinaison linéaire quelconque des fonctions</u> :

$$\zeta_{a,A}(s+c_1) \quad , \quad a = 0,\dots,A-1 .$$

- <u>Si</u> $c_1+c_2 = 3$, $\varphi(s)$ <u>est une combinaison linéaire quelconque des fonctions</u> :

$$\zeta^*_{a,A}(s+c_1) \quad , \quad a = 0,\dots,A-1 .$$

Ce théorème admet le corollaire suivant qui peut s'interpréter comme une généralisation de la caractérisation de la fonction zêta de Riemann par son équation fonctionnelle, donnée par Hamburger [4] :

Corollaire 1. Les séries de Dirichlet de la forme (1) vérifiant une équation fonctionnelle (2) satisfaisant 1) à 5) sont les combinaisons linéaires des fonctions $L(s,\chi)$ de Dirichlet, où les caractères χ sont soit tous pairs, soit tous impairs.

On rappelle que la fonction $L(s,\chi)$ de Dirichlet est définie par

(5) $$L(s,\chi) = \sum_{n\geq 1} \chi(n)\, n^{-s} \quad , \ \mathrm{Re}(s) > 1 \ .$$

Pour son équation fonctionnelle voir [10].

La transformation de Mellin permet d'appliquer les résultats de Bochner [1] aux formes modulaires. Il y a lieu de considérer les séries thêta :

$$\theta_{a,A}(z) = \sum_{\substack{n\in\mathbb{Z}\\ n\equiv a \,(\mathrm{mod}\, A)}} \exp(2i\pi n^2 z) \qquad \theta^*_{a,A}(z) = \sum_{\substack{n\in\mathbb{Z}\\ n\equiv a\,(\mathrm{mod}\, A)}} n \exp(2i\pi n^2 z) \qquad a, A \in \mathbb{N}.$$

Théorème 3. Soit $f(z) = \sum_{n\geq 0} a_n \exp(2i\pi nz)$ une forme modulaire de poids $k/2$ (k entier) pour un groupe de congruence. Supposons qu'il existe un ensemble D fini d'entiers positifs, sans facteurs carrés, deux à deux distincts tel que $a_n = 0$ si n n'est pas de la forme dm^2, avec $d \in D$, $m \in \mathbb{N}$. Alors,

- $k/2 = 1/2$ ou $3/2$.

- Si $k/2 = 1/2$, $f(z)$ est une combinaison linéaire quelconque des séries thêta

$$\theta_{a,A}(dz) \quad , \quad d \in D \quad , \quad a = 0,\dots,A-1 \quad , \quad A \in \mathbb{N} \ .$$

- Si $k/2 = 3/2$, $f(z)$ est une combinaison linéaire quelconque des séries thêta

$$\theta^*_{a,A}(dz) \quad , \quad d \in D \quad , \quad a = 0,\dots,A-1 \quad , \quad A \in \mathbb{N} \ .$$

Les séries de Dirichlet peuvent vérifier plusieurs équations fonctionnelles provenant des relations entre facteurs gamma, exclusivement (théorème 4 du n° 4), paragraphe 1). Ces relations sont données dans l'appendice écrit par J.-P. Serre.

L'auteur remercie Jean-Pierre Serre qui énonça comme probables ces théorèmes dans son cours de 1976-1977 au Collège de France, et qui l'aida considérablement par ses suggestions.

§1. Démonstration des théorèmes 1 et 2.

1) Un théorème de Bochner [1].

En étendant une idée de Siegel [8] pour démontrer la caractérisation de la fonction zêta de Riemann par son équation fonctionnelle, Bochner démontre le théorème suivant ([1], th. 2.6, th. 3.2, th. 3.4), concernant les équations fonctionnelles des séries de Dirichlet générales avec $p = A = 1$.

Théorème (Bochner [1]). 1) Soit $p = A = 1$, il n'existe pas d'équations fonctionnelles de séries de Dirichlet générales si $\Sigma p_f \neq \Sigma p_g$ ou si la densité de Polya de l'une des suites (λ_n) ou (μ_n) est nulle.

2) Si $A = 1$ et $\Sigma p_f = \Sigma p_g = \frac{1}{2}$, pour tout $r \in \mathbb{N}$ assez grand, l'équation fonctionnelle (2) des séries de Dirichlet générales $\varphi(s) = \Sigma\, a_n \lambda_n^{-s}$, $\psi(s) = \Sigma\, b_n \mu_n^{-s}$ implique une relation de la forme

$$(6) \qquad \sum_{n\geq 1} b_n \mu_n^r e^{-2\pi\mu_n z} = K_r(z) + \sum_{n\geq 1} a_n \lambda_n^{-r} \Phi_r(z/\lambda_n)$$

vérifiée pour $z \in \mathbb{C}$, $\mathrm{Re}(z) > 0$, où $K_r(z)$ est une fonction résiduelle holomorphe de $\log z$ et $\Phi_r(z)$, holomorphe pour $\mathrm{Re}(z) > 0$, admet pour seules singularités sur l'axe imaginaire les points $\mp i$ et l'on a :

au voisinage de i $\quad \Phi_r(z) \sim C(iz+1)^{-\sigma-r}$

au voisinage de $-i$ $\quad \Phi_r(z) \sim Ce^{-2\pi id}\,(-1)^{H+G+r}(-iz+1)^{-\sigma-r}$

où l'on a posé

$$\sigma_h = c_h - \frac{1}{2}\ , \quad h = 1,\dots,H$$

$$\sigma = \sum_{f=1}^{G} \sigma_f - \sum_{g=G+1}^{H} \sigma_g + \frac{1}{2}\ ; \quad d = \sum_{g=G+1}^{H} c_g$$

C <u>est une constante dépendant des données</u> A , $\varphi(s)$, $\psi(s)$, $\Delta_1(s)$, $\Delta_2(s)$ <u>et de</u> r .

La démonstration de Bochner est suffisamment claire et élégante pour qu'il ne soit pas utile de donner plus d'indications sur ce théorème, quoiqu'il ne soit pas énoncé exactement ainsi dans [1]. Les idées principales sont les suivantes : par transformation de Mellin, on obtient la relation (6), puis on applique un théorème de Polya [5, p.89] et la théorie classique de Frobenius-Fuchs à une certaine équation différentielle linéaire vérifiée par $\Phi_r(z)$.

La remarque suivante sera essentielle : <u>le membre de droite de la relation</u> (6) <u>admet sur l'axe imaginaire pour seules singularités non nulles les points</u> $\mp i\lambda_n$ <u>tels que</u> $a_n \neq 0$. <u>Au voisinage d'un des deux points</u> $\mp i\lambda_{n_o}$ <u>tels que</u> $a_{n_o} \neq 0$, <u>il est équivalent à</u>

$$a_{n_o}\lambda_{n_o}^{-r}\Phi_r(z/\lambda_{n_o}).$$

En effet, soit $\lambda \in \mathbb{R}$ et considérons la somme restreinte aux n tels que $\lambda_n \neq \mp\lambda$:

$$\sum_{n\geqslant 1}{}' a_n\lambda_n^{-r}\Phi_r(i\lambda/\lambda_n).$$

Il existe $N(\lambda) > 0$, tel que pour $n \geqslant N(\lambda)$, on ait $|\lambda/\lambda_n| < \frac{1}{2}$; et une constante $M(r) > 0$, telle que pour $|y| < \frac{1}{2}$, $|\Phi_r(iy)| < M(r)$. On suppose que r est choisi assez grand pour appartenir au domaine de convergence absolue de la série $\sum_{n\geqslant 1} a_n\lambda_n^{-s}$. On en déduit que la somme précédente est convergente, d'où la remarque précédente.

On rappelle qu'une suite (μ_n) de nombres réels strictement croissante a pour <u>densité de Polya</u> δ si

$$\lim_{n\to\infty} (\mu_{n+1} - \mu_n) > 0 \quad \text{et si} \quad \delta = \lim_{n\to\infty} \frac{n}{\mu_n}$$

2) <u>Conséquences du théorème de Bochner</u>.

a) Soit une équation fonctionnelle satisfaisant 1) à 4) avec les données $(A,\varphi(s),\psi(s),\Delta_1(s),\Delta_2(s))$ où $\varphi(s)$ et $\psi(s)$ sont deux séries de Dirichlet de la forme (1) :

$$\varphi(s) = \sum_{n\geq 1} a_n n^{-s} \qquad \psi(s) = \sum_{n\geq 1} b_n n^{-s} .$$

On voit facilement que l'équation fonctionnelle écrite en $\frac{s}{p}$ est une équation fonctionnelle avec les hypothèses de Bochner pour les données $(1,\tilde{\varphi}(s),\tilde{\psi}(s),\Delta_1(\frac{s}{p}),\Delta_2(\frac{s}{p}))$, où :

$$\tilde{\varphi}(s) = \sum_{n\geq 1} a_n \lambda_n^{-s} , \quad \tilde{\psi}(s) = \sum_{n\geq 1} b_n \mu_n^{-s}$$

$$\lambda_n = n^{1/p} , \qquad \mu_n = \frac{p}{2\pi} \left(\frac{2\pi n}{A}\right)^{1/p} .$$

Les densités de Polya des suites (λ_n) et (μ_n) sont bien définies et égales, selon les valeurs de p, à :

0 si $p<1$

1 et A respectivement, si $p=1$

∞ si $p>1$.

Le théorème 1 résulte immédiatement du théorème de Bochner.

Examinons maintenant le cas d'une équation fonctionnelle (2) satisfaisant 1) à 5). On se ramène aux hypothèses de la deuxième partie du théorème de Bochner en posant :

$$\lambda_n = n \qquad \mu_n = \frac{n}{A} .$$

Le premier membre de la relation (6) est alors une fonction <u>périodique</u> de période iA. Cette information jointe à la connaissance des <u>singularités</u> du deuxième membre sur l'axe imaginaire permettra d'obtenir le théorème 2.

On se ramène à $\sigma + r \in \mathbb{Z}$, en remplaçant $\varphi(s)$ par $\varphi(s+c)$ où c est une constante convenable. En effet l'équation fonctionnelle (2) au point $s+c$, est équivalente à

$$(2P\pi)^{-s/2}\ \Delta_1(s+c)\ \varphi(s+c) = A^{-s}(2P\pi)^{s/2}\ \Delta_2(-s-c)\ (2P\pi/A)^c\psi(-s-c)$$

qui est une équation fonctionnelle du même type, associée à $\tilde{A} = A$, et à:

$$\tilde{\Delta}_1(s) = \Delta_1(s+c) \quad , \quad \tilde{\Delta}_2(s) = \Delta_2(s-c) \quad ,$$

$$\tilde{\varphi}(s) = \varphi(s+c) \quad , \quad \tilde{\psi}(s) = (2P\pi/A)^c\psi(s-c) \ ;$$

dont la constante $\tilde{\sigma}$ correspondante est égale à $\sigma+c$.

La périodicité, de période iA , et le fait que les pôles sur l'axe imaginaire sont tous de la forme $\mp in$, $n \in \mathbb{N}$, montrent que A est un <u>entier</u>.

Au voisinage d'un point in , tel que $a_n \neq 0$, le second membre de (6) est équivalent à

$$C\ a_n\ n^\sigma\ (iz+n)^{-\sigma-r}$$

et au voisinage de $-in$, à

$$C\ a_n\ n^\sigma\ (-1)^{H+G+\sigma}\ e^{-2i\pi d}\ (iz-n)^{-\sigma-r} \ .$$

La périodicité montre que si $n \equiv b \pmod{A}$, $0 \leqslant b \leqslant A-1$, on a :

$$a_n n^\sigma = a_b b^\sigma = a_{A-b}(A-b)^\sigma(-1)^{H+G+\sigma}\ e^{-2i\pi d} \ . \tag{7}$$

En appliquant à (6) l'opérateur $z \to \overline{f(\bar{z})}$, on obtient une relation (6') ayant les mêmes propriétés essentielles que (6)

$$\sum_{n\geqslant 1} \bar{b}_n \left(\frac{n}{A}\right)^r e^{-2\pi\frac{n}{A}z} = \overline{K_r(\bar{z})} + \sum_{n\geqslant 1} \bar{a}_n\ n^{-r}\ \overline{\Phi_r(\bar{z}/n)}. \tag{6'}$$

Le second membre de (6') est équivalent au voisinage d'un point in , tel que $a_n \neq 0$ à

$$\bar{C}\ \bar{a}_n\ n^\sigma\ (-1)^{H+G+r} e^{2i\pi d}\ (iz+n)^{-\sigma-r}$$

et au voisinage de $-in$, à

$$\bar{C}\ \bar{a}_n\ n^\sigma\ (-1)^{\sigma+r}\ (iz-n)^{-\sigma-r} \ .$$

La périodicité, de période iA , du premier membre, montre que

$$\bar{a}_b\ b^\sigma\ (-1)^{H+G+r}\ e^{2i\pi d} = \bar{a}_{A-b}(A-b)^\sigma\ (-1)^{\sigma+r} \ . \tag{8}$$

Les relations (7) et (8) montrent que $2d \in \mathbb{Z}$ et que l'on a , pour tout

$n \equiv b \pmod{A}$

$$a_n\, n^{\sigma} = a_b\, b^{\sigma} = (-1)^{H+G+\sigma+2d}\, a_{A-b}(A-b)^{\sigma} .$$

Nous démontrons ainsi qu'il existe une constante c_1 telle que $\varphi(s)$ soit une combinaison linéaire, soit des fonctions zêta partielles $\zeta_{a,A}(s+c_1)$, soit des fonctions $\zeta^*_{a,A}(s+c_1)$.

3) Rappels sur les fonctions zêta d'Hurwitz.

Le fait que les fonctions zêta partielles $\zeta_{a,A}(s)$ et $\zeta^*_{a,A}(s)$ vérifient une équation fonctionnelle avec les facteurs

$\Delta_1(s) = \Gamma(\frac{s}{2})$ $\quad \Delta_2(s) = \Gamma(\frac{s+1}{2})$ et $\Gamma(\frac{s+3}{2})$ respectivement

est dû à Hurwitz ([10]). On démontre que les fonctions zêta, indexées par $\alpha \in]0,1]$, définies sur le demi-plan $\mathrm{Re}(s) > 1$ par les séries

$$Z(s,\alpha) = \sum_{n\geq 0} (n+\alpha)^{-s}$$

admettent un prolongement analytique sur tout le plan, avec un pôle simple en $s = 1$ et vérifient une équation fonctionnelle

$$Z(s,\alpha) = \frac{2\Gamma(1-s)}{(2\pi)^{1-s}} \sum_{m=1}^{\infty} \frac{\sin \pi(\frac{s}{2}+2m\alpha)}{m^{1-s}} , \quad \mathrm{Re}(s) < 0$$

qui peut s'écrire sous la forme

$$Z(s,\alpha) = \frac{\pi^{s/2}\Gamma(\frac{1-s}{2})}{\Gamma(\frac{s}{2})\pi^{\frac{1-s}{2}}} \sum_{m=1}^{\infty} \frac{\cos 2\pi m\alpha}{m^{1-s}} + \frac{\pi^{s/2}\Gamma(\frac{1-s+1}{2})}{\Gamma(\frac{s+1}{2})\pi^{\frac{1-s}{2}}} \sum_{m=1}^{\infty} \frac{\sin 2\pi m\alpha}{m^{1-s}} .$$

On en déduit l'équation fonctionnelle des séries de Dirichlet

$$\zeta_{a,A}(s) = A^s(Z(s,\tfrac{a}{A}) + Z(s,1-\tfrac{a}{A}))$$

$$\zeta^*_{a,A}(s) = A^s(Z(s-1,\tfrac{a}{A}) - Z(s-1,1-\tfrac{a}{A})) .$$

Il reste à déterminer les facteurs gamma pour achever la démonstration du théorème 2.

4) Facteurs gamma des équations fonctionnelles.

Une série de Dirichlet $\varphi(s)$ peut vérifier plusieurs équations fonctionnelles provenant des relations entre facteurs gamma. On note F l'ensemble des données des équations fonctionnelles satisfaisant 1) à 4) vérifiées par $\varphi(s)$. On suppose $F \neq \emptyset$, et on se donne $(A,\psi(s),\Delta_1(s),\Delta_2(s)) \in F$.

Afin d'éviter les changements triviaux d'équation fonctionnelle, provenant de :

$$(A,\psi(s) = \sum_{n\geq 1} b_n n^{-s}) \to (AB, \sum_{n\geq 1} b_n (nB)^{-s}) \quad , \quad B \text{ entier, } B \geq 2$$

on supposera que :

6) Le p.g.c.d. des n tels que $b_n \neq 0$ est égal à 1.

On notera $\tilde{A}$ un nombre réel positif, $\tilde{\psi}(s)$ une série de Dirichlet satisfaisant 2) et 6), $\tilde{\Delta}_1(s)$, $\tilde{\Delta}_2(s)$ des facteurs gamma satisfaisant 3)*. On note E le groupe des fonctions de la forme e^{as+b} , $a,b \in \mathbb{C}$ (voir l'appendice (Serre)).

<u>Théorème</u> 4. <u>Pour que</u> $(\tilde{A},\tilde{\psi}(s),\tilde{\Delta}_1(s),\tilde{\Delta}_2(s)) \in F$, <u>il faut et il suffit que</u> $\tilde{A} = A$ <u>et qu'il existe</u> $f(s) \in E$, <u>telle que</u> $\tilde{\psi}(s) = f(0)\, \psi(s)$ <u>et</u> :

$$\text{(9)} \qquad \frac{\tilde{\Delta}_1(s)}{\Delta_1(s)} = f(s)\, \frac{\tilde{\Delta}_2(-s)}{\Delta_2(-s)} \,.$$

<u>On a alors</u> :

$$f(s) = f(0)\, \left(\frac{\tilde{p}}{p}\right)^s$$

$$\tilde{p} = p$$

$$\sum c_{\tilde{f}} - \sum c_{\tilde{g}} = \sum c_f - \sum c_g \,.$$

<u>Démonstration</u> : On suppose que $(\tilde{A},\tilde{\psi}(s),\tilde{\Delta}_1(s),\tilde{\Delta}_2(s)) \in F$. On doit avoir

$$\text{(10)} \qquad \frac{\tilde{\Delta}_1(s)}{\Delta_1(s)} = \left(\frac{\tilde{p}A}{p\tilde{A}}\right)^s \frac{\tilde{\psi}(-s)}{\psi(-s)}\, \frac{\tilde{\Delta}_2(-s)}{\Delta_2(-s)} \,.$$

On peut déterminer M , tel que pour $\mathrm{Re}(s) < -M$, le second membre de (10) n'ait ni pôle, ni zéro . Le nombre de pôles de $\Delta_1(s)$ dans $\mathrm{Re}(s) < -R$ est équivalent à $R \sum p_f$ quand R tend vers l'infini, et $\Delta_1(s)$ ne s'annule pas. On en déduit que $\sum p_f = \sum p_{\tilde{f}}$ et d'après le théorème 1, $p = \tilde{p}$. On pose alors

---

* $\tilde{x}$ se définit comme x .

$$\lambda_n = n^{1/p} \quad , \quad \tilde{\mu}_n = \frac{p}{2\pi}\left(\frac{2\pi n}{\tilde{A}}\right)^{1/p} = \mu_n\left(\frac{A}{\tilde{A}}\right)^{1/p}$$

de façon à se ramener aux hypothèses de Bochner. On a alors pour r assez grand :

$$(11) \qquad \sum_{n\geq 1} a_n \lambda_n^r e^{-2\pi\lambda_n z} = \tilde{L}_r(z) + \sum_{n\geq 1} \tilde{b}_n \tilde{\mu}_n^{-r} \tilde{\Psi}_r(z/\tilde{\mu}_n) .$$

Cela résulte de la relation (6) appliquée au couple $(\tilde{\psi}(s),\varphi(s))$ au lieu du couple $(\varphi(s),\psi(s))$. Les fonctions $\tilde{L}_r(z)$, $\tilde{\Psi}_r(z)$ ont respectivement les mêmes propriétés que $K_r(z)$, $\Phi_r(z)$. Le premier membre de (11) est déterminé par $\varphi(s)$. En particulier son comportement sur l'axe imaginaire est connu :

a) ses singularités : $\mp i\,\tilde{\mu}_n$ lorsque $\tilde{b}_n \neq 0$

b) leur ordre : $\tilde{\sigma}+r$

c) le coefficient : $\tilde{D}\,\tilde{b}_n \tilde{\mu}_n^{\tilde{\sigma}}$, où $\tilde{D}$ est une constante indépendante de n.

On déduit de a) que lorsque $\tilde{b}_n \neq 0$, il existe $m\geq 1$, entier, tel que $b_m \neq 0$ et $\tilde{\mu}_n = \mu_m$; c'est équivalent à dire que $n = m(\frac{\tilde{A}}{A})^{1/p}$. La condition 6) implique $\tilde{A} = A$.

On déduit de b) que $\Sigma c_{\tilde{f}} - \Sigma c_{\tilde{g}} = \Sigma c_f - \Sigma c_g$ (se reporter à la définition de σ, théorème de Bochner, paragraphe 1).

On déduit de a), b) et c) que $\tilde{b}_n = C\, b_n$ où C est une constante indépendante de n. Comme $\tilde{\mu}_n = \mu_n$, on a $\tilde{\psi}(s) = C\,\psi(s)$. La relation (10) s'écrit alors :

$$\frac{\tilde{\Delta}_1(s)}{\Delta_1(s)} = C\left(\frac{\tilde{p}}{p}\right)^s \frac{\tilde{\Delta}_2(-s)}{\Delta_2(-s)} .$$

La réciproque est immédiate.

<u>Corollaire</u>. <u>A chaque couple</u> $(\tilde{\Delta}_1(s),\tilde{\Delta}_2(s))$ <u>de facteurs gamma satisfaisant</u> 2) <u>et</u> (9), <u>correspond une et une seule équation fonctionnelle de</u> $\varphi(s)$ <u>satisfaisant</u> 1) <u>à</u> 4) <u>et</u> 6).

On introduit de nouvelles notations : soit $a(z)$, $z \in \mathbb{C}$ des entiers presque tous nuls, on pose pour $s \in \mathbb{C}$:

$$G_a(s) = \prod_{z \in \mathbb{C}} \Gamma(s+z)^{a(z)} \quad , \quad H_a(s) = G_a(-s) \ .$$

Un nombre rationnel $q \neq 0$ est dit <u>admissible</u> pour $(\Delta_1(s), \Delta_2(s))$ s'il existe $f_1(s)$, $f_2(s) \in E$ et deux familles d'entiers $a(z) \geqslant 0$, $b(z) \geqslant 0$, presque tous nuls, tels que :

$$\Delta_1(s) = G_a(qs)\, f_1(s) \quad , \quad \Delta_2(s) = G_b(qs)\, f_2(s) \ .$$

Les familles $a(z)$, $b(z)$ ainsi que les fonctions $f_1(s)$, $f_2(s)$ sont uniques* (cf. Appendice, lemme 2). Les formules de multiplication de Gauss $\Gamma(ns) = \prod_{k=0}^{n-1} \Gamma(s+\frac{k}{n})\, f(s)$, $f(s) \in E$, montrent que l'ensemble des nombres admissibles est infini.

Soit $q \in \mathbb{Q}$, un nombre admissible pour $(\Delta_1(s), \Delta_2(s))$ et $(\widetilde{\Delta}_1(s), \widetilde{\Delta}_2(s))$, la relation (9) est équivalente à :

$$G_{\widetilde{a}-a}(s) = f(s)\, H_{\widetilde{b}-b}(s) \ .$$

Serre a démontré (appendice) :

<u>Théorème</u> (Serre). <u>Pour qu'un couple</u> $(\widetilde{\Delta}_1(s), \widetilde{\Delta}_2(s))$ <u>de facteurs gamma satisfaisant</u> 2), <u>vérifie la relation</u> (9), <u>il faut et il suffit qu'il existe un nombre</u> $q \in \mathbb{Q}$ <u>admissible pour</u> $(\Delta_1(s), \Delta_2(s))$ <u>et</u> $(\widetilde{\Delta}_1(s), \widetilde{\Delta}_2(s))$ <u>tel que l'on ait les deux conditions suivantes</u> :

(*) <u>pour tout</u> $z_0 \in \mathbb{C}$, $\displaystyle\sum_{z \equiv z_0 (\mathrm{mod}\ \mathbb{Z})} \widetilde{a}(z) - a(z) = 0$

(**) <u>pour tout</u> $z \in \mathbb{C}$, $\widetilde{a}(z) - a(z) + \widetilde{b}(1-z) - b(1-z) = 0$.

On dit que 2 couples de facteurs gamma $(\widetilde{\Delta}_1(s), \widetilde{\Delta}_2(s))$ et $(\Delta_1(s), \Delta_2(s))$ sont <u>équivalents</u> mod E , s'il existe $f_1(s)$, $f_2(s) \in E$ tels que

$$\widetilde{\Delta}_1(s) = \Delta_1(s) f_1(s) \qquad \widetilde{\Delta}_2(s) = \Delta_2(s) f_2(s) \ .$$

Il est clair qu'il faut et qu'il suffit qu'il existe un nombre $q \in \mathbb{Q}$ admissible commun tel que pour tout $z \in \mathbb{C}$, on ait

---

* Elles dépendent de q , bien entendu.

$$\tilde{a}(z) = a(z) \quad \text{et} \quad \tilde{b}(z) = b(z) .$$

Soit Λ les classes d'équivalence des couples $(\tilde{\Delta}_1(s),\tilde{\Delta}_2(s))$ satisfaisant 2) et (9).

<u>Corollaire. L'ordre de</u> Λ <u>est égal à</u> 1 <u>dans les cas suivants</u> :

a) <u>Il existe</u> $q \in \mathbb{Q}$ <u>admissible pour</u> $(\Delta_1(s),\Delta_2(s))$ <u>tel que pour tout</u> $z_o \in \mathbb{C}$, <u>tous les</u> $a(z)$ <u>ou tous les</u> $b(-z)$, <u>tels que</u> $z \equiv z_o$ (mod $\mathbb{Z}$) <u>soient nuls</u>.

b) <u>Il existe</u> $r_1, r_2 \in \mathbb{N}$ <u>tel que</u> $\Delta_1(s) = \Gamma(\frac{s}{2})^{r_1}\Gamma(\frac{s+1}{2})^{r_2}$ <u>et</u> $\Delta_2(s) = \Delta_1(s+1)$.

<u>L'ordre de</u> Λ <u>est égal à</u> $2^{|k-1|}$ <u>s'il existe</u> $k \in \mathbb{Z}$ <u>tel que</u> $\Delta_1(s) = \Gamma(s)$, $\Delta_2(s) = \Gamma(s+k)$.

Le cas a) implique à équivalence près mod E, l'unicité des facteurs gamma des fonctions zêta partielles $\zeta_{a,A}(s)$ et $\zeta^*_{a,A}(s)$ ainsi que des séries de Dirichlet associées aux formes modulaires de poids demi-entier. Le cas b) montre l'unicité des facteurs gamma des fonctions L d'Artin généralisant les fonctions L de Dirichlet. Le cas c) montre que les séries de Dirichlet associées aux formes modulaires de poids entier k possèdent, à équivalence près mod E, 2^{k-1} équations fonctionnelles.

<u>Démonstration du corollaire</u> :

a) Si tous les $a(z)$ (resp. tous les $b(z)$) tels que $z \equiv z_o$ (mod $\mathbb{Z}$) sont nuls, la condition (*) implique qu'il en est de même pour tous les $\tilde{a}(z)$ (resp. tous les $\tilde{b}(z)$) et la condition (**) implique que $\tilde{b}(1-z_o) = b(1-z_o)$ (resp. $\tilde{a}(1-z_o) = a(1-z_o)$). On détermine ainsi tous les $\tilde{b}(-z)$ et $\tilde{a}(z)$ (resp. tous les $\tilde{a}(-z)$ et $\tilde{b}(z)$) tels que $z \equiv z_o$ (mod $\mathbb{Z}$).

b) Soit $(\tilde{\Delta}_1(s),\tilde{\Delta}_2(s))$ un autre couple de facteurs gamma. On peut déterminer un entier $n \geq 1$ tel que $\frac{1}{2n}$ soit admissible pour ce couple. Pour $\frac{1}{2n}$, les familles d'entiers $a(z)$ et $b(z)$ sont définies par :

$$k = 0,1,\ldots n-1 \quad a(\frac{2k}{2n}) = b(\frac{2k+1}{2n}) = r_1 \;, \quad a(\frac{2k+1}{2n}) = b(\frac{2k+2}{2n}) = r_2$$

$a(z) = 0$ et $b(z) = 0$ pour les autres z.

On démontre que $(\widetilde{\Delta}_1(s), \widetilde{\Delta}_2(s))$ est équivalent modulo E à $(\Gamma(\frac{s}{2})^{r_1}\Gamma(\frac{s+1}{2})^{r_2}, \Gamma(\frac{s+1}{2})^{r_1}\Gamma(\frac{s+2}{2})^{r_2})$ en vérifiant que les relations (*) et (**) admettent pour seule solution $\tilde{a}(z) = a(z)$, $\tilde{b}(z) = b(z)$, pour tout $z \in \mathbb{C}$. En effet, on voit facilement que pour $z \neq \frac{2k}{n}$, $\frac{2k+1}{n}$, on a $a(z) = b(1-z) = 0$; on en conclut que $\tilde{a}(z) = \tilde{b}(1-z) = 0$ grâce à (**). Si $z = \frac{2k}{n}$ ou $\frac{2k+1}{n}$, la relation (*) permet de voir que $\tilde{a}(z) = a(z)$ et $\tilde{b}(1-z) = b(1-z)$.

c) Le raisonnement est analogue : on montre que pour $\underbrace{\frac{1}{n} < \frac{1}{|k|+2}}_{\text{un nombre admissible}}$ $(n \in \mathbb{N})$, les relations (*) et (**) admettent $2^{|k-1|}$ solutions. On note X (resp. Y) l'ensemble des entiers compris entre 0 et n-1 (resp. 1-k et n-k). Pour le nombre admissible $\frac{1}{n}$, les familles $a(z)$ et $b(z)$ sont définies par :

$$a(z) = \begin{cases} 0 & z \neq j/n \;,\; j \in X \\ 1 & z = j/n \;,\; j \in X \end{cases} \qquad b(1-z) = \begin{cases} 0 & z \neq j/n \;,\; j \in Y \\ 1 & z = j/n \;,\; j \in Y \end{cases}$$

On en conclut grâce à (**) que pour $z \neq \frac{j}{n}$, $j \in X \cup Y$, on a $\tilde{a}(z) = \tilde{b}(1-z) = 0$. Pour $z_0 = \frac{j}{n}$, $j \in X \cap Y$, on vérifie facilement que si $z \equiv z_0 \pmod{\mathbb{Z}}$ alors $z \neq \frac{j}{n}$, $j \in X \cup Y$, on en conclut grâce à (*) que $\tilde{a}(z_0) = \tilde{b}(1-z_0) = 1$.

On a un bijection de $Y-(X\cap Y)$ sur $X-(Y\cap X)$ en posant $j' = j + (\operatorname{sign} k)n$ $j \in Y$, $j \notin X$. Pour $z = \frac{j}{n}$ et $z' = \frac{j'}{n}$, on écrit avec (*) les égalités

$$\tilde{a}(z) + \tilde{a}(z') = 1 \qquad \tilde{b}(1-z) + \tilde{b}(1-z') = 1$$

et avec (**) les égalités

$$\tilde{a}(z) + \tilde{b}(1-z) = 1 \qquad \tilde{a}(z') + \tilde{b}(1-z') = 1 \;.$$

Ce sont les seules relations que doivent vérifier ces quatre entiers naturels. Elles sont équivalentes à :

$$\tilde{a}(z) = \tilde{b}(1-z') \qquad \tilde{a}(z') = \tilde{b}(1-z) \qquad \tilde{a}(z) + \tilde{a}(z') = 1 .$$

Elles admettent <u>deux</u> solutions. Le nombre d'éléments de $Y-(X \cap Y)$ est $k-1$. Le corollaire est démontré.

Toutefois, si l'on cherche à déterminer les couples de facteurs gamma <u>admettant 1 comme nombre admissible</u>, on s'aperçoit qu'il n'y a que deux couples modulo E , représentés par $(\Gamma(s), \Gamma(k+s))$ et $(\Gamma(s+1-k), \Gamma(s+1))$. Ils proviennent de l'égalité $\Gamma(s)\,\Gamma(1-s) = \pi/\sin \pi s$ d'où on tire la relation

$$\Gamma(s)\,\Gamma(1-s) = (-1)^k\,\Gamma(s+k)\,\Gamma(1-k-s) .$$

Weil a utilisé ces deux équations fonctionnelles pour retrouver des relations vérifiés par les intégrales d'Eichler [9].

<u>Problème</u> : Existe-t-il des équations fonctionnelles avec $\frac{1}{2} < \Sigma\, p_f < 1$, ou plus généralement avec $\Sigma\, p_f \neq n/2$, $n \in \mathbb{N}^*$?

§2. <u>Le th. 3 sur les formes modulaires</u>.

La démonstration a lieu en deux temps. On suppose d'abord que D a un seul élément. En utilisant la méthode de Siegel [8] pour la démonstration du théorème de Hamburger (caractérisation de la fonction zêta de Riemann par son équation fonctionnelle), ou plutôt la généralisation de cette méthode de Bochner et Chandrasekharan [2], on obtient une relation du type (6), d'où on déduit le théorème. Puis on raisonne par récurrence sur le nombre d'éléments de D ; on utilise alors, de façon essentielle que la "tordue" d'une forme modulaire est une forme modulaire.

<u>Première partie</u> : $D = \{d\}$.

On suppose que $f(z) = \sum_{n \geq 0} a_n \exp(2i\pi n^2 dz)$ est une forme modulaire de poids $k/2$ (k entier) pour un groupe de congruence, de niveau N (pour une définition détaillée, se reporter à Shimura [7] ou à

Serre-Stark [6]).

Il existe une forme modulaire $g(z) = \sum_{n\geq 0} b_n^1 \exp(2i\pi nz)$ de même poids et de même niveau, telle que

$$g(z) = z^{-k/2} f(-1/Nz) .$$

Cette relation écrite au point $z = ix/2$, en posant $b_n = b_n^1 (i/2)^{k/2}$, donne :

$$\sum_{n\geq 0} b_n \exp(-\pi nx) = x^{-k/2} \sum_{n\geq 0} a_n \exp(-4\pi n^2 d/Nx) . \tag{12}$$

1) La transformation de Siegel-Bochner-Chandrasekharan.

Elle permet de transformer la relation précédente en une relation du type (6), c'est-à-dire avec un membre périodique et l'autre avec des pôles connus. L'idée est d'utiliser les deux formules élémentaires suivantes :

$$t \int_0^\infty \exp(-\pi t^2 x - \pi a^2/x)\, x^{\frac{1}{2}} \frac{dx}{x} = \exp(-2\pi at) \quad , \quad a > 0 \tag{13}$$

$$\int_0^\infty \exp(-\pi ax)\, x^\alpha \frac{dx}{x} = \Gamma(\alpha)\, (\pi a)^{-\alpha} \quad , \quad a > 0 \ , \ \alpha \in \mathbb{C} . \tag{14}$$

On considère la fonction $H(t,x) = t \exp(-\pi t^2 x)$ et on multiplie chaque terme de la somme du deuxième membre de la relation modulaire (12) par $x^{\alpha-1} H_0(t,x)$, en posant $\alpha = \frac{k+1}{2}$. En intégrant de 0 à ∞ en x , on obtient :

$$t \int_0^\infty a_n \exp(-\pi t^2 x - 4\pi n^2 d/Nx)\, x^{\frac{1}{2}} \frac{dx}{x} = a_n \exp(-2\pi nt \sqrt{\tfrac{4d}{N}}). \tag{15}$$

La série de terme général le second membre de (15) est absolument convergente pour $t > 0$ et définit par prolongement analytique une fonction holomorphe pour $t \in \mathbb{C}$, $\mathrm{Re}\, t > 0$.

En effectuant la même opération sur chaque terme de la somme du premier membre de (12), on obtient :

$$tb_n \int_0^\infty \exp(-\pi(t^2+n)x)\, x^\alpha \frac{dx}{x} = tb_n\, \Gamma(\alpha)(\pi(t^2+n))^{-\alpha} . \tag{16}$$

Si la série de Dirichlet $\sum_{n\geq 1} b_n n^{-s}$ converge absolument pour $s = \alpha$, la série de terme général le second membre de (16) définit une fonction holomorphe pour $t \in \mathbb{C}$, $\mathrm{Re}\, t > 0$. Sinon, on utilise l'idée de Bochner et de Chandrasekharan [2], qui consiste à remplacer la fonction auxiliaire $H_o(t,x)$ par sa dérivée en t d'ordre assez élevé : on pose

$$H_x(t,x) = \frac{d^{2r}}{dt^{2r}} H_o(t,x)$$

et on effectue les opérations juste décrites avec $H_r(t,x)$ au lieu de $H_o(t,x)$. Le premier membre de (12) donne des termes de la forme

$$(16_r) \qquad \Gamma(\alpha)\pi^{-\alpha}\, b_n \frac{d^{2r}}{dt^{2r}} (t(t^2+n)^{-\alpha}) .$$

Si r est assez grand, la série de terme général (16_r) converge absolument, et définit sur tout compact ne contenant pas les pôles $\mp i\sqrt{n}$, si $b_n \neq 0$, une fonction holomorphe. Le second membre de (12) donne des termes de la forme

$$(15_r) \qquad (-2\pi \sqrt{\tfrac{4d}{N}})^{2r}\, a_n\, n^{2r} \exp(-2\pi nt \sqrt{\tfrac{4d}{N}})$$

qui déterminent une série absolument convergente pour $t \in \mathbb{C}$, $\mathrm{Re}\, t > 0$. Nous avons obtenu la proposition suivante :

<u>Proposition</u> 1. <u>La relation</u> (12) <u>implique la relation suivante, pour tout</u> $r \in \mathbb{N}$ <u>assez grand</u> :

$$(17) \qquad \sum_{n\geq 0} b_n \frac{d^{2r}}{dt^{2r}} (t(t^2+n)^{-\alpha}) = \frac{\pi^\alpha}{\Gamma(\alpha)} (-2\pi \sqrt{\tfrac{4d}{N}})^{2r} \sum_{n\geq 0} a_n\, n^{2r}\, e^{-2\pi nt\sqrt{\frac{4d}{N}}} .$$

Une connaissance plus précise du comportement du premier membre aux pôles nous sera utile. On a par un calcul élémentaire le résultat suivant:

<u>Lemme</u> 1. <u>Si</u> $u = t-i\sqrt{n}$ <u>tend vers</u> 0, <u>on a l'équivalence suivante</u> :

$$(18) \qquad \frac{d^{2r}}{dt^{2r}} (t(t^2+n)^{-\alpha}) \sim \tfrac{1}{2}(2i\sqrt{n})^{1-\alpha} \Big[u^{-\alpha-2r}(-\alpha)(-\alpha-1)\dots(-\alpha-2r+1) + u^{1-\alpha-2r} \frac{(2-\alpha)(1-\alpha)(-\alpha)\dots(-\alpha-2r-2)}{2i\sqrt{n}}\Big].$$

2) <u>La démonstration du théorème</u> 3.

On applique le même raisonnement que dans le paragraphe 1, 2). Le second membre de (17) étant périodique, de période $i\sqrt{\frac{N}{4d}}$, on en conclut que si $i\sqrt{n}$ est un pôle c'est-à-dire $b_n \neq 0$, alors $i(\sqrt{n}+\sqrt{\frac{N}{4d}})$ est aussi un pôle, donc est égal à $i\sqrt{m}$, $m \in \mathbb{N}$.

On doit avoir simultanément, d'après (18) :

$$\left[\frac{b_n}{2}(2i\sqrt{n})^{1-\alpha} - \frac{b_m}{2}(2i\sqrt{m})^{1-\alpha}\right](-\alpha)(-\alpha-1)\ldots(-\alpha-2r+1) = 0$$

$$\left[\frac{b_n}{2}(2i\sqrt{n})^{-\alpha} - \frac{b_m}{2}(2i\sqrt{m})^{-\alpha}\right](2-\alpha)(1-\alpha)(-\alpha)\ldots(-\alpha-2r-2) = 0 .$$

La première égalité donne, car $\alpha > 0$, $b_n n^{\frac{1-\alpha}{2}} = b_m m^{\frac{1-\alpha}{2}}$, et la seconde $(2-\alpha)(1-\alpha) = 0$, c'est-à-dire $\alpha = 1$ ou 2 , ou encore $k = 1$ ou 3 . On a :

- si le poids $k/2 = 1/2$, $b_n = b_m$
- si le poids $k/2 = 3/2$, $b_n/\sqrt{n} = b_m/\sqrt{m}$.

L'égalité $\sqrt{m} = \sqrt{n} + \sqrt{\frac{N}{4d}}$ implique que $N/4d$ est un entier et que $nN/4d$ est un entier. Après avoir écrit $N = 4dd'A^2$, écriture unique si l'on suppose d et d' sans facteurs carrés, on en déduit que $n = d'n_1^2$ où $n_1 \in \mathbb{N}$. On modifie alors les notations en écrivant b_j ce que l'on avait noté b_n , si $n = d'j^2$. On conclut que :

- $g(z)$ est de poids $k/2 = 1/2$ ou $3/2$
- $g(z) = \sum_{n \geq 0} b_n \exp(2i\pi n^2 d'z)$
- si $k/2 = 1/2$, on a $b_n = b_{n+A}$ et si $k/2 = 3/2$, on a $b_n = nb'_n$ avec $b'_n = b'_{n+A}$.

Autrement dit, $g(z)$ est une combinaison linéaire soit des $\theta_{a,A}(d'z)$, soit des $\theta^*_{a,A}(d'z)$. Le théorème s'obtient en utilisant la symétrie en $f(z)$ et en $g(z)$ de (12).

<u>Deuxième partie</u> : <u>récurrence sur le nombre d'éléments de</u> D .

1) <u>Trois lemmes</u>.

<u>Lemme</u> 2. <u>Etant donnés deux nombres entiers</u> $d \geq 1$, $d' \geq 1$, <u>distincts et sans facteurs carrés et deux signes</u> $\varepsilon = \pm 1$, $\varepsilon' = \pm 1$, <u>il existe une infinité de nombres premiers</u> p <u>tels que</u> $(\frac{d}{p}) = \varepsilon$ <u>et</u> $(\frac{d'}{p}) = \varepsilon'$.

Référence : Borevitch-Chafarevitch [3] p. 383 exercice 4.

Pour tout entier $N \geqslant 1$, on note :

$$\Gamma(N) = \left\{ \begin{pmatrix} a & b \\ c & d \end{pmatrix} \in SL_2(\mathbb{Z}) \ , \ \begin{pmatrix} a & b \\ c & d \end{pmatrix} \equiv \begin{pmatrix} 1 & 0 \\ 0 & 1 \end{pmatrix} \bmod N \right\}$$

$$\Gamma_1(N) = \left\{ \begin{pmatrix} a & b \\ c & d \end{pmatrix} \in SL_2(\mathbb{Z}) \ , \ \begin{pmatrix} a & b \\ c & d \end{pmatrix} \equiv \begin{pmatrix} 1 & * \\ 0 & 1 \end{pmatrix} \bmod N \right\}$$

<u>Lemme</u> 3. <u>Si</u> $f(z) = \sum_{n \geqslant 0} a_n \exp(2i\pi nAz)$ <u>est une forme modulaire de poids</u> $k/2$ <u>pour</u> $\Gamma(N)$, <u>alors</u> $g(z) = f(z/A)$ <u>est modulaire pour</u> $\Gamma_1(N/(N,A))$ <u>si le poids est entier</u>, <u>et pour</u> $\Gamma_1(N/(\frac{N}{4},A))$, <u>si le poids est demi-entier</u>.

<u>Démonstration</u> : On a

- $g(z) = g(z+1)$
- si $\alpha = \begin{pmatrix} A & 0 \\ 0 & 1 \end{pmatrix}$, alors $SL_2(\mathbb{Z}) \cap \alpha^{-1}\Gamma(N)\alpha = \left\{ \begin{pmatrix} a & b \\ c & d \end{pmatrix} \in SL_2(\mathbb{Z}) \right.$,
  $\left. b \equiv 0 \bmod NA \ , \ c \equiv 0 \bmod N/(N,A) \right\}$.

Le groupe engendré par $\alpha^{-1}\Gamma(N)\alpha$ et $\begin{pmatrix} 1 & 1 \\ 0 & 1 \end{pmatrix}$ est le groupe $\Gamma_1(N/(N,A))$. On en déduit le lemme si le poids de $f(z)$ est entier. Si le poids est demi-entier, soit $\theta(z) = \sum_{n \geqslant 0} \exp(2i\pi n^2 z)$ la série thêta usuelle de poids $\frac{1}{2}$. Le lemme est vrai pour la forme modulaire $\theta(Az)f(z)$ de poids entier $\frac{k+1}{2}$, de niveau $4AN/(4A,N)$, donc $\theta(z)f(z/A)$ est modulaire pour $\Gamma_1(N/(\frac{N}{4},A))$. On en déduit le lemme.

<u>Lemme</u> 4. <u>Si</u> $f(z) = \sum_{n \geqslant 0} a_n \exp(2i\pi nz)$ <u>une forme modulaire de poids</u> $k/2$, <u>de niveau</u> N , <u>et si</u> χ <u>est un caractère de conducteur</u> M <u>premier à</u> N , <u>alors</u> $f{*}\chi(z) = \sum_{n \geqslant 0} a_n \chi(n) \exp(2i\pi nz)$ <u>est une forme modulaire de poids</u> $k/2$, <u>de niveau</u> NM^2 .

Références : [6, §7] et [7, §5].

2) <u>La démonstration du théorème</u> 3.

On suppose que les nombres n^2d tels que $a_{n^2d} \neq 0$ sont <u>premiers entre eux</u>. Si ce n'est pas le cas, on remplace $f(z)$ par $f(z/A)$ où A est le p.g.c.d. de ces nombres (lemme 3). L'ensemble D est remplacé par $\{d/\delta \ , \ d \in D \ , \ \delta = \text{p.g.c.d.}\{d, d \in D\}\}$, mais son nombre d'éléments reste le même.

Soient d et d' deux éléments distincts de D .

Il existe (lemme 2) une <u>infinité</u> de nombres premiers p tels que $(\frac{d'}{p}) = 1$, $(\frac{d''}{p}) = -1$. On choisit un tel nombre p , assujetti en outre à être premier au niveau N , et à tous les éléments de l'ensemble fini D . L'application $d \to (\frac{d}{p})$ définit une partition de D en deux sous-ensembles D_1 et D_2 non vides. Les deux fonctions :

$$f_1(z) = 2 \sum_{d \in D_1} \sum_{(n,p)=1} a_{n^2 d} \exp(2i\pi n^2 dz)$$

$$f_2(z) = 2 \sum_{d \in D_2} \sum_{(n,p)=1} a_{n^2 d} \exp(2i\pi n^2 dz)$$

respectivement la somme et la différence des formes modulaires (lemme 4) $F*\chi_0(z)$ et $F*(\frac{\cdot}{p})(z)$, où χ_0 est le caractère trivial modulo p , sont des formes modulaires de poids k/2 , de niveau Np^2 . Elles ne peuvent pas être toutes les deux nulles, sinon tous les nombres n tels que $a_n \neq 0$ seraient divisibles par p , ce qui est contraire à l'hypothèse que nous avons faite. Par récurrence, on en déduit que le <u>poids</u> est k/2 = 1/2 ou 3/2 .

Pour chaque d , tel que l'un des $a_{n^2 d}$, avec (n,p) = 1 , est non nul, on a $4d \mid Np^2$, d'où $4d \mid N$ puisque $p \nmid 4d$. La grande liberté de choix des nombres p nous permet de choisir pour tout $n^2 d$ tel que $a_{n^2 d} \neq 0$, un nombre p tel que (n,p) = 1 . Ceci montre que les éléments 4d , $d \in D$ <u>divisent</u> N .

Soit $d \in D$, on pose $N = 4dd'A^2$, où $d' \in \mathbb{N}$ est sans facteurs carrés, et on note a_j ce que l'on avait noté a_n , si $n = j^2 d$. Soit n , tel que $a_n \neq 0$; on choisit deux nombres premiers p et q distincts, premiers à n et n+A , avec les propriétés précédentes. Enfin soit $\lambda, \mu \in \mathbb{Z}$ tels que $\lambda p + \mu q = 1$

- si k/2 = 1/2 , on a par récurrence, $a_{n+\lambda Ap} = a_n$ (car (n,p) = 1) , mais on a aussi par récurrence, $a_{n+\lambda Ap} = a_{n+A}$ (car (n+A,q) = 1 et $\lambda Ap + \mu Aq = A$) donc $a_n = a_{n+A}$;
- si k/2 = 3/2 , on raisonne de même avec $a'_n = a_n/n$.

On en conclut le théorème 3.

Appendice (J.-P. Serre)

Relations entre facteurs gamma

1. Relations entre sinus.

La variable est notée s . Je désigne par E le groupe des fonctions de la forme e^{as+b} , $a,b \in \mathbb{C}$, autrement dit le groupe des fonctions entières d'ordre 1 sans zéro.

Considérons un produit

$$\prod_{z\in\mathbb{C}} \sin \pi(s+z)^{a(z)} = F_a(s) ,$$

où les a(z) sont des entiers, nuls sauf un nombre fini d'entre eux.

Lemme 1. *Pour que $F_a(s)$ appartienne à E , il faut et il suffit que la condition suivante soit satisfaite :*

(∗) *pour tout* $z_0 \in \mathbb{C}$, *on a* $\sum_{z\equiv z_0 (\mathrm{mod}\ \mathbb{Z})} a(z) = 0$.

On a alors $F_a(s) = \pm 1$.

Il est clair que $F_a(s)$ appartient à E si et seulement si $F_a(s)$ n'a ni zéro ni pôle. Or l'ordre de $F_a(s)$ au point $-z_0$ est égal à la somme des a(z) pour $z \equiv z_o \pmod{\mathbb{Z}}$: on trouve bien la condition (∗). Si cette condition est satisfaite, F_a est de la forme e^{as+b} , et, comme $F_a(s)$ est périodique de période 2 , le nombre a est de la forme πiN , avec $N\in\mathbb{Z}$. Si l'on prend s de la forme it , avec t réel $\to +\infty$, il s'ensuit que $F_a(it) = e^{b-\pi tN}$. Mais d'autre part, on a $\sin \pi(s+z) \sim e^{\pi t}e^{-i\pi z}/(-2i)$ pour $s = it$, comme on le voit facilement. On en conclut que $F_a(it)$ est équivalent à $(e^{\pi t}/(-2i))^S e^{-i\pi A}$, où $S = \sum_{z\in\mathbb{C}} a(z)$, $A = \sum_{z\in\mathbb{C}} za(z)$. Vu (∗), on a $S = 0$ (c'est clair), et $A\in\mathbb{Z}$ (regrouper les termes za(z) correspondant à une classe donnée z_0 mod. $\mathbb{Z}$, et remarquer que la somme partielle en question est égale à $\Sigma(z-z_0)a(z)$, qui est un entier). On obtient finalement $F_a(s) = (-1)^A$, d'où le résultat cherché (avec en prime la détermination du signe).

<u>Variante</u>. On aurait pu aussi tout exprimer en terme de $T = e^{\pi is}$ de façon à avoir une fonction rationnelle de T.

<u>Reformulation du lemme</u> 1. Considérons le groupe abélien formé par les $a = (a(z))_{z\in\mathbb{C}}$ tels que $F_a(s) \in E$. Le lemme 1 est équivalent à dire que <u>ce groupe est engendré par les</u> a du type suivant : $a(z) = 0$ pour $z \neq z_0$, z_0+1, $a(z_0) = 1$, $a(z_0+1) = -1$. En d'autres termes (plus parlants et moins précis) toute "relation $F_a(s) \in E$" est conséquence des "<u>relations élémentaires</u>"

$$\sin \pi(s+z_0)/\sin \pi(s+z_0+1) = -1 .$$

2. <u>Relations entre les</u> $\Gamma(s+z)$.

Il n'y en a pas :

<u>Lemme</u> 2. <u>Soit</u> $a(z)$ <u>une famille d'entiers presque tous nuls</u>. <u>Pour que le produit</u>

$$G_a(s) = \prod_{z\in\mathbb{C}} \Gamma(s+z)^{a(z)}$$

<u>appartienne à</u> E, <u>il faut et il suffit que tous les</u> $a(z)$ <u>soient nuls</u> (auquel cas le produit en question n'a aucun mérite à être égal à 1 ..).

Supposons que les $a(z)$ ne soient pas tous nuls, et soit X l'ensemble (fini) des z tels que $a(z) \neq 0$. Soit $z_0 \in X$ un élément tel que $\mathrm{Re}(z_0) \leqslant \mathrm{Re}(z)$ pour tout $z \in X$. On ne peut avoir

$$-z_0 = -z-n \quad , \text{ avec } z \in X \text{ , } n \text{ entier} \geqslant 0 \text{ ,}$$

que si $z = z_0$, $n = 0$. On en conclut que le point $-z_0$ ne peut être un pôle d'aucune des fonctions $\Gamma(s+z)$, $z \in X - \{z_0\}$. Il en résulte que l'ordre de $G_a(s)$ en $-z_0$ est égal à $-a(z_0) \neq 0$, ce qui montre bien que $G_a(s)$ n'appartient pas à E.

3. <u>Relations entre les</u> $\Gamma(s+z)$ <u>et les</u> $\Gamma(-s+z)$.

On va considérer deux familles $a(z)$, $b(z)$ comme ci-dessus, et s'intéresser aux fonctions

$$G_a(s) = \prod_{z\in\mathbb{C}} \Gamma(s+z)^{a(z)} \quad , \quad H_b(s) = \prod_{z\in\mathbb{C}} \Gamma(-s+z)^{b(z)} \quad .$$

<u>Théorème</u>. <u>Pour qu'il existe</u> $f(s) \in E$ <u>tel que</u>

$$G_a(s) = f(s).H_b(s) \ ,$$

<u>il faut et il suffit que les deux conditions suivantes soient satisfaites</u> :

(*) <u>pour tout</u> $z_0 \in \mathbb{C}$, <u>on a</u> $\sum_{z\equiv z_0 (\mathrm{mod}\ \mathbb{Z})} a(z) = 0$;

(**) $a(z) + b(1-z) = 0$ <u>pour tout</u> $z \in \mathbb{C}$.

<u>On a alors</u> $f(s) = \pm 1$.

On va, bien sûr, se servir de la relation $\Gamma(s)\Gamma(1-s) = \pi/\sin \pi s$. En remplaçant s par $s+z$, elle donne

$$\Gamma(s+z)\Gamma(-s+1-z) = \pi/\sin \pi(s+z) \ .$$

Convenons de noter a' la fonction définie par

$$a'(z) = a(1-z)$$

et définissons de même b'. L'identité ci-dessus donne :

$$G_a(s).H_{a'}(s) = \pi^A/F_a(s) \quad \text{où} \quad A = \sum_{z\in\mathbb{C}} a(z) \ .$$

Ceci étant, supposons que l'on ait $G_a(s) = f(s).H_b(s)$, avec $f(s) \in E$. En multipliant par $H_{a'}$, on en tire :

$$\pi^A/F_a(s) = f(s).H_{a'}(s).H_b(s) \ .$$

Mais le membre de droite n'a pas de zéros et pas de pôles pour $\mathrm{Re}(s) \leqslant M$, avec M convenable, alors que le membre de gauche est périodique de période 2. Il en résulte que les deux membres n'ont ni zéro ni pôle. On en conclut qu'ils appartiennent à E . En particulier, d'après le lemme 1, on a la condition (*), $A=0$, et $F_a = \pm 1$. Et, d'après le lemme 2, on a $a'+b=0$, ce qui est justement la condition (**). On en conclut que $f(s) = F_a(s) = \pm 1$ (et on sait déterminer le signe).

Inversement, supposons (*) et (**) vérifiées. On a $G_a . H_{a'} = 1/F_a = \pm 1$ d'après le lemme 1. Comme $1/H_{a'} = H_b$, on voit bien que $G_a = \pm H_b$.

<u>Reformulation</u>. Ici encore, on peut reformuler le théorème en disant que les relations $G_a/H_b \in E$ sont <u>conséquences</u> des relations élémentaires

$$\Gamma(s+z_o)/\Gamma(s+z_o+1) = -\Gamma(-s-z_o)/\Gamma(-s+1-z_o) .$$

Bibliographie

[1] S. BOCHNER. On Riemann's functional equation with multiple gamma factors. Ann. of Math. 67 (1958), 29-41.

[2] S. BOCHNER and K. CHANDRASEKHARAN. On Riemann's functional equation. Ann. of Math. 63 (1956), 336-360.

[3] Z.I. BOREVITCH et I.R. CHAFAREVITCH. Théorie des nombres. Gauthier-Villars (1967).

[4] H. HAMBURGER. Über die Riemannsche Funktionalgleichung der ζ-Funktion. Math. Zeit. 10 (1921), 240-254.

[5] N. LEVINSON. Gap and density Theorems. Amer. Math. Soc. Coll. Publ. 26 (1940).

[6] J.-P. SERRE and H. STARK. Modular forms of weight ½. Ce volume.

[7] G. SHIMURA. On modular forms of half integral weight. Ann. of Math. 97 (1973), 440-481.

[8] C.L. SIEGEL. Bemerkungen zu einem Satz von Hamburger über die Funktionalgleichung der Riemannschen Zetafunktion. Gesammelte Abhandlungen. Springer-Verlag (1966), Band I, 154-156.

[9] A. WEIL. Remarks on Hecke's Lemma and its use. Colloque de Kyoto (1976).

[10] E.T. WHITTAKER and G.N. WATSON. A Course of Modern Analysis. Cambridge University Press (1935).

International Summer School on Modular Functions

BONN 1976

MODULAR FORMS WHOSE FOURIER COEFFICIENTS INVOLVE ZETA-FUNCTIONS OF QUADRATIC FIELDS

D. Zagier

TABLE OF CONTENTS

Introduction

For over a hundred years it has been known that there exist identities expressing the coefficients of certain modular forms as finite sums involving class numbers of imaginary quadratic fields; these identities, the so-called "class number relations," arose classically in the theory of complex multiplication but have reappeared since in several other contexts, e.g. in the Eichler-Selberg formula for the traces of Hecke operators and in the calculation of intersection numbers of curves on Hilbert modular surfaces [8]. Recently Cohen [3], using Shimura's theory of modular forms of half-integral weight, constructed modular forms whose Fourier coefficients are given by finite sums similar to those occurring in the class number relations, but with the class numbers replaced by values of Dirichlet L-series (or equivalently, of zeta functions of quadratic number fields) at integral arguments. In this paper we construct modular forms whose Fourier coefficients are given by infinite sums of zeta functions of quadratic fields, now at an arbitrary complex argument. The result includes both the classical class number relations and the modular forms constructed by Cohen, and further provides an expression for the latter as linear combinations of Hecke eigenfunctions $f(z)$, the coefficients being certain values of the associated Rankin zeta functions $\sum_{n=1}^{\infty} \frac{a(n)^2}{n^s}$ (where $f|T(n) = a(n)f$). From this we obtain formulas for the values of the Rankin zeta function at integral values within the critical strip, a typical identity being

$$(1) \qquad \sum_{n=1}^{\infty} \frac{\tau(n)^2}{n^{20}} = \frac{2}{245} \; \frac{4^{20}}{20!} \; \pi^{29} \; \frac{\zeta(9)}{\zeta(18)} \; (\Delta,\Delta),$$

where $\zeta(s)$ is the Riemann zeta function, $\Delta(z) = \sum_{n=1}^{\infty} \tau(n)\, e^{2\pi i n z}$ the discriminant function, and (Δ,Δ) the Petersson product of Δ with itself. As another corollary of the main identity we obtain a new proof of a recent result of Shimura [21] on the holomorphy of the Rankin zeta function. Finally, by combining the method developed in this paper with the results of [24] we obtain applications to the Doi-Naganuma lifting from modular forms of Nebentypus to Hilbert modular forms in two variables and also to the mapping in the other direction which was constructed in [8] in terms of the intersection numbers of modular curves on Hilbert modular surfaces. In particular, we give partial results in the direction of the conjecture made in [8] that these two maps are adjoint to one another with respect to the Petersson scalar product.

In § 1 we describe the main result of the paper, namely the construction of a modular form whose Fourier coefficients are infinite linear combinations of zeta functions of quadratic fields (with Legendre functions as coefficients) and whose Petersson product with an arbitrary Hecke eigenform is the corresponding Rankin zeta function. We also show how this can be used to obtain identities for special values of the Rankin zeta function like the one cited above and discuss the relationship between these identities and other known or conjectured results on the values at integral arguments of Dirichlet series associated to cusp forms. In § 2 we reduce the proof of the main result to the evaluation of an integral involving kernel functions for Hecke operators. This integral is calculated in § 3, while § 4 contains the properties of zeta-functions and Legendre functions which are needed to deduce identities like (1) above. In § 5 we describe an alternate method for proving such identities by expressing the product of a theta series and an Eisenstein series of half-integral weight as an infinite linear combination of Poincaré series. The applications to Hilbert modular forms are contained in § 6.

<u>Note</u>: The identities expressing $\sum a(n)^2 n^{-s}$ for special integral values of s in terms of (f,f) and values of the Riemann zeta function have been discovered independently by Jacob Sturm (Thesis, Princeton 1977).

§ 1 Identities for the Rankin zeta function

We use the following notation:

$H = \{z = x + iy \mid y > 0\}$ the upper half-plane, $dV = \frac{dx\,dy}{y^2}$ the invariant metric on H,

$j_k(\gamma,z) = (cz+d)^{-k}$ $\quad (\gamma = \begin{pmatrix} a & b \\ c & d \end{pmatrix} \in SL_2(\mathbb{R}),\ k \in \mathbb{Z},\ z \in H)$,

$(f|_k\gamma)(z) = j_k(\gamma,z)\, f(\gamma z)$ $\quad$ (f any function on H).

Throughout §§ 1 - 4 we restrict ourselves to modular forms for the full modular group $\Gamma = SL_2(\mathbb{Z})/\{\pm 1\}$; the results could be generalized to arbitrary congruence subgroups, but this would involve considerable technical complication and no essentially new ideas. We denote by k an even integer >2, by S_k the space of cusp forms of weight k on Γ, equipped with the Petersson scalar product

$$(f,g) = \int_{\Gamma\backslash H} f(z)\,\overline{g(z)}\,y^k\,dV \qquad (f, g \in S_k),$$

and by $\{f_i\}_{1 \le i \le \dim S_k}$ the basis of S_k consisting of normalized Hecke eigenforms, with

$$f_i(z) = \sum_{n=1}^{\infty} a_i(n)\,q^n, \quad a_i(1) = 1, \quad f_i|T(n) = a_i(n)\,f_i$$

(where as usual $q = e^{2\pi i z}$). For each normalized Hecke eigenform $f(z) = \sum a(n)\,q^n$ we set

$$(2) \qquad D_f(s) = \prod_p (1 - \alpha_p^2 p^{-s})^{-1}(1 - \alpha_p\bar{\alpha}_p p^{-s})^{-1}(1 - \bar{\alpha}_p^2 p^{-s})^{-1} \qquad (\mathrm{Re}(s) > k),$$

where the product is over all primes and α_p, $\bar{\alpha}_p$ are defined by

$$\alpha_p + \bar{\alpha}_p = a(p), \quad \alpha_p\,\bar{\alpha}_p = p^{k-1}$$

(by Deligne's theorem, previously the Ramanujan-Petersson conjecture, the numbers α_p and $\bar{\alpha}_p$ are complex conjugates). The function $D_f(s)$ is related to the Rankin zeta function by

(3) $$D_f(s) = \frac{\zeta(2s-2k+2)}{\zeta(s-k+1)} \sum_{n=1}^{\infty} \frac{a(n)^2}{n^s}$$

and hence, by the results of Rankin [17], has a meromorphic continuation to the entire complex plane, satisfies the functional equation

(4) $$D_f^*(s) = 2^{-s}\pi^{-3s/2}\,\Gamma(s)\,\Gamma\!\left(\frac{s-k+2}{2}\right)D_f(s) = D_f^*(2k-1-s),$$

and is related to the norm of f in the Petersson metric by

(5) $$(f,f) = \frac{(k-1)!}{2^{2k-1}\,\pi^{k+1}}\,D_f(k).$$

For the statement of the main identity we will also need a certain zeta function, defined as follows. Let Δ be any discriminant, i.e. $\Delta \in \mathbb{Z}$ and $\Delta \equiv 0$ or $1 \pmod 4$. We consider binary quadratic forms

$$\phi(u,v) = au^2 + buv + cv^2 \qquad (a,\,b,\,c \in \mathbb{Z})$$

with discriminant $|\phi| = b^2 - 4ac = \Delta$. The group Γ operates on the set of such forms by $\gamma\circ\phi(u,v) = \phi(au+cv,\,bu+dv)$ $(\gamma = \begin{pmatrix} a & b \\ c & d\end{pmatrix} \in \Gamma)$, the number of equivalence classes being finite if $\Delta \neq 0$. We define

(6) $$\zeta(s,\Delta) = \sum_{\substack{\phi \bmod \Gamma \\ |\phi| = \Delta}} \; \sum_{\substack{(m,n)\in \mathbb{Z}^2/\mathrm{Aut}(\phi) \\ \phi(m,n)>0}} \frac{1}{\phi(m,n)^s} \qquad (\mathrm{Re}(s) > 1),$$

where the first sum is over all Γ-equivalence classes of forms ϕ of discriminant Δ and the second over inequivalent pairs of integers with respect to the group of units $\mathrm{Aut}(\phi) = \{\gamma\in\Gamma \mid \gamma\phi = \phi\}$ of the form. If Δ is the discriminant of a (real or imaginary) quadratic field K, then $\zeta(s,\Delta)$ coincides with the Dedekind zeta function $\zeta_K(s)$ (the first sum corresponds to the ideal classes of K, the second to the ideals in a given class, with $\phi(m,n) = N(\mathfrak{n})$), while $\zeta(s,\Delta)$ for $\Delta = 1$ and $\Delta = 0$ is equal to $\zeta(s)^2$ and to $\zeta(s)\,\zeta(2s-1)$, respectively. If $\Delta = Df^2$, with D equal either to 1 or to the discriminant of a quadratic field and f a natural number, then $\zeta(s,\Delta)$ differs from $\zeta(s,D)$ only by a finite Dirichlet series. Thus in all cases $\zeta(s,\Delta)$ is divisible by the Riemann zeta function, i.e.

(7) $\zeta(s,\Delta) = \zeta(s)\, L(s,\Delta)$

where $L(s,\Delta)$ is an entire function of s (unless Δ is a perfect square, in which case $L(s,\Delta)$ has a simple pole at $s = 1$ with residue $\frac{1}{2}$ if $\Delta = 0$ and residue 1 otherwise).

Finally, for real numbers Δ and t satisfying $\Delta < t^2$ and $s \in \mathbb{C}$ with $\frac{1}{2} < \mathrm{Re}(s) < k$ we define

$$(8)\quad \begin{aligned} I_k(\Delta,t;s) &= \int_0^\infty \int_{-\infty}^\infty \frac{y^{k+s-2}}{(x^2+y^2+ity-\frac{1}{4}\Delta)^k}\,dx\,dy \\ &= \frac{\Gamma(k-\frac{1}{2})\Gamma(\frac{1}{2})}{\Gamma(k)} \int_0^\infty \frac{y^{k+s-2}}{(y^2+ity-\frac{1}{4}\Delta)^{k-\frac{1}{2}}}\,dy\,, \end{aligned}$$

where the second integral converges absolutely for $1-k < \mathrm{Re}(s) < k$ (unless $\Delta = 0$, in which case we need $\frac{1}{2} < \mathrm{Re}(s) < k$) and can be expressed in terms of the associated Legendre function $P^{k-1}_{s-1}(\frac{t}{\sqrt{\Delta}})$ (see § 4). We can now formulate the main result.

<u>Theorem 1</u>: <u>Let</u> $k > 2$ <u>be an even integer</u>. <u>For</u> $m = 1,2,\ldots$ <u>and</u> $s \in \mathbb{C}$ <u>set</u>

$$(9)\quad c_m(s) = m^{k-1} \sum_{t=-\infty}^{\infty} \left[I_k(t^2-4m,t;s) + I_k(t^2-4m,-t;s)\right] L(s,t^2-4m) + \begin{cases} (-1)^{k/2} \dfrac{\Gamma(k+s-1)\,\zeta(2s)}{2^{2s+k-3}\,\pi^{s-1}\,\Gamma(k)}\, u^{k-s-1} & \text{if } m = u^2,\ u > 0, \\ 0 & \text{if } m \text{ is not a perfect square,} \end{cases}$$

<u>where</u> $L(s,\Delta)$ <u>and</u> $I_k(\Delta,t;s)$ <u>are defined by equations</u> (6), (7) <u>and</u> (8). <u>Then</u>

i) <u>The series</u> (9) <u>converges absolutely and uniformly for</u> $2-k < \mathrm{Re}(s) < k-1$;

ii) <u>The function</u>

(10) $$\Phi_s(z) = \sum_{m=1}^{\infty} c_m(s)\, e^{2\pi i m z} \qquad (z \in H,\ 2-k < \mathrm{Re}(s) < k-1)$$

<u>is a cusp form of weight</u> k <u>for the full modular group</u>;

iii) <u>Let</u> $f \in S_k$ <u>be a normalized Hecke eigenform. Then the Petersson product</u>

<u>of</u> Φ_s <u>and</u> f <u>is given by</u>

$$(\Phi_s, f) = C_k \frac{\Gamma(s+k-1)}{(4\pi)^{s+k-1}} D_f(s+k-1), \tag{11}$$

<u>where</u> $D_f(s)$ <u>is defined by</u> (2) <u>and</u>

$$C_k = \frac{(-1)^{k/2}\pi}{2^{k-3}(k-1)}. \tag{12}$$

We must say a few words concerning assertion i). If $t^2 - 4m$ is a perfect square, then $L(s, t^2 - 4m)$ has a pole at $s = 1$, as mentioned above. However, for $t^2 - 4m \geq 0$ the coefficient $[I_k(t^2 - 4m, t; s) + I_k(t^2 - 4m, -t; s)]$ has a simple zero at $s = 1$ (or any other odd integral value between 0 and k), as we will show in § 4, so the expression $[I_k(t^2 - 4m, t; s) + I_k(t^2 - 4m, -t; s)]$ $L(s, t^2 - 4m)$ makes sense even at $s = 1$, and the sum of these numbers as t runs from $-\infty$ to ∞ is absolutely convergent. Similarly, if m is a square then the second member of (9) has a simple pole at $s = \frac{1}{2}$, but in this case the terms $t = \pm 2\sqrt{m}$ in the first sum involve the function

$$I_k(0;t;s) + I_k(0,-t;s) = 2\pi(-1)^{k/2} \cos\frac{\pi s}{2} \frac{\Gamma(s-\frac{1}{2})\,\Gamma(k-s)}{\Gamma(k)\,\Gamma(\frac{1}{2})} |t|^{s-k} \tag{13}$$

which also has a simple pole at $s = \frac{1}{2}$, and the two poles cancel; then i) states that the sum of the other terms of the series (9) is finite. Thus the expression defining $c_m(s)$ is holomorphic in the region $2-k < \mathrm{Re}(s) < k-1$. From equation (11) we deduce that $D_f(s+k-1)$ is also holomorphic in this region. On the other hand, the Euler product defining $D_f(s)$ is absolutely convergent for $\mathrm{Re}(s) > k$, so $D_f(s)$ is certainly holomorphic in this half-plane and, by the functional equation, also in the half-plane $\mathrm{Re}(s) < k-1$. Theorem 1 therefore implies the following result, which was proved by Shimura [21] in 1975 by a different method.

<u>Corollary 1</u> (Shimura): <u>The function</u> $D_f(s)$ <u>defined by</u> (2) <u>has a holomorphic continuation to the whole complex plane</u>.

Secondly, we observe that statement iii) of Theorem 1 characterizes the cusp form Φ_s, since the space S_k is complete with respect to the Petersson metric. Indeed, since the eigenfunctions f_i form an orthogonal basis of S_k, equation (11) is equivalent to

$$(14)\qquad \Phi_s(z) = C_k \frac{\Gamma(s+k-1)}{(4\pi)^{s+k-1}} \sum_{i=1}^{\dim S_k} \frac{1}{(f_i,f_i)} D_{f_i}(s+k-1)\, f_i(z)$$

or to

$$(15)\qquad c_m(s) = C_k \frac{\Gamma(s+k-1)}{(4\pi)^{s+k-1}} \sum_{i=1}^{\dim S_k} \frac{a_i(m)}{(f_i,f_i)} D_{f_i}(s+k-1)\ .$$

In particular, if we take $s = 1$ and use formula (5), we find

$$(16)\qquad c_m(1) = \frac{\pi}{2}\ C_k \sum_{i=1}^{\dim S_k} a_i(m).$$

On the other hand, the Fourier coefficients $a_i(m)$ of the functions f_i are at the same time their eigenvalues for the mth Hecke operator $T(m)$, so

$$\sum_{i=1}^{\dim S_k} a_i(m) = \operatorname{Tr}(T(m), S_k).$$

Thus Theorem 1 includes as a special case a formula for the trace of $T(m)$. To see that this agrees with the well-known formula of Selberg and Eichler, we must investigate the various terms of (9) for $s = 1$. If $t^2 - 4m$ is negative, then (as we will show in § 4)

$$(17)\qquad \begin{aligned} &C_k^{-1}\, m^{k-1} \left[I_k(t^2-4m,t;1) + I_k(t^2-4m,t;1)\right] \\ &\qquad = -\frac{1}{4}\sqrt{4m-t^2}\ p_{k,1}(t,m), \end{aligned}$$

where

$$(18)\qquad \begin{aligned} p_{k,1}(t,m) &= \text{coefficient of } x^{k-2} \text{ in } \frac{1}{1-tx+mx^2} \\ &= \frac{\rho^{k-1}-\bar\rho^{k-1}}{\rho-\bar\rho} \qquad (\rho+\bar\rho = t,\ \rho\bar\rho = m) \end{aligned}$$

and

$$(19)\qquad L(1,t^2-4m) = \pi(4m-t^2)^{-\frac{1}{2}}\, H(4m-t^2),$$

where (with the same notation as in (6))

$$\text{(20)} \qquad H(n) = \sum_{\substack{\phi \bmod \Gamma \\ |\phi| = -n}} \frac{1}{\mathrm{Aut}(\phi)} \qquad (n > 0).$$

(This equals $2\sum \frac{h(-n/f^2)}{w(-n/f^2)}$, where the sum ranges over $f > 0$ such that $f^2|n$ and $-n/f^2$ is congruent to 0 or 1 modulo 4 and $h(\Delta)$, $w(\Delta)$ denote the class number and number of units, respectively, of the order in $\mathbb{Q}(\sqrt{\Delta})$ of discriminant Δ.) If $t^2 - 4m > 0$, then the coefficient $I_k(t^2-4m,t;s) + I_k(t^2-4m,-t;s)$ vanishes at $s = 1$, as mentioned above, so the contribution of (9) is 0 unless $L(s,t^2-4m)$ has a pole at $s = 1$, i.e. unless t^2-4m is a perfect square. In this case, we will show that

$$\text{(21)} \qquad \begin{aligned} &\lim_{s\to 1} C_k^{-1} m^{k-1} \left[I_k(t^2-4m,t;s) + I_k(t^2-4m,-t;s)\right] L(s,t^2-4m) \\ &= -\frac{\pi}{4}\left(\frac{|t| - u}{2}\right)^{k-1} \qquad (t^2-4m = u^2,\ u > 0). \end{aligned}$$

Notice that there are only finitely many t with t^2-4m a perfect square, and that they are in 1:1 correspondence with the positive divisors of m:

$$t^2 - 4m = u^2 \iff m = dd',\ d,\ d' = \frac{|t| \pm u}{2}.$$

Therefore the series (9) for $s = 1$ becomes a finite sum and we obtain

<u>Corollary 2</u> (Eichler, Selberg): <u>For</u> $k > 2$ <u>an even integer</u>, $m > 1$,

$$\mathrm{Tr}(T(m), S_k) = -\frac{1}{2}\sum_{\substack{t\in\mathbb{Z} \\ t^2<4m}} p_{k,1}(t,m)\, H(4m-t^2) - \frac{1}{2}\sum_{\substack{dd'=m \\ d,d'>0}} \min(d,d')^{k-1}$$
$$+ \begin{cases} \frac{k-1}{12}\, u^{k-2} & \text{if } m = u^2,\ u > 0, \\ 0 & \text{if } m \text{ is not a perfect square,} \end{cases}$$

<u>where</u> $p_{k,1}(t,m)$ <u>and</u> $H(4m-t^2)$ <u>are defined by equations</u> (18) <u>and</u> (20).

It is perhaps worth remarking that we could have obtained the trace formula by specializing Theorem 1 to $s = 0$ instead of $s = 1$. At $s = 0$, the coefficient $[I_k(t^2-4m,t;s) + I_k(t^2-4m,-t;s)]$ does not vanish for any t, but

$L(t^2-4m,0)$ is zero whenever t^2-4m is positive and not a square, so again we get a finite sum.

If we specialize Theorem 1 to $s = r$ (or $s = 1 - r$), where r is an odd integer between 1 and $k - 1$, then again the terms with $t^2-4m > 0$ vanish (including those for which t^2-4m is a perfect square, if $r > 1$), and the series defining $c_m(s)$ reduces to a finite sum. In this case we recover the modular forms constructed by Cohen [3]. We recall his result.

For $r \geqslant 1$, r odd, Cohen defines an arithmetical function $H(r, N)$ which generalizes the class number function $H(N) = H(1, N)$ introduced above. The function $H(r, N)$ is defined as $\zeta(1 - 2r)$ if $N = 0$ and as a simple rational multiple of $\pi^{-r} \sum_{n=1}^{\infty} (\frac{-N}{n})\, n^{-r}$ if $N > 0$, $N \equiv 0$ or $3 \pmod 4$. It is related to the function $L(s,\Delta)$ defined above by

$$H(r, N) = \frac{(-1)^{(r-1)/2}\,(r-1)!}{2^{r-1}\,\pi^r}\, N^{r-\frac{1}{2}}\, L(r,-N) \qquad (r \geqslant 1 \text{ odd}, \quad N \geqslant 0) \tag{22}$$

or, even more simply, by

$$H(r, N) = L(1 - r, -N) \qquad (r \geqslant 1 \text{ odd}, \quad N \in \mathbb{Z}). \tag{23}$$

Then:

<u>Theorem</u> (Cohen [3], Theorem 6.2): <u>Let</u> $3 \leqslant r \leqslant k - 1$, r <u>odd</u>, k <u>even</u>, <u>and set</u>

$$C_{k,r}(z) = \sum_{m=0}^{\infty} \left(\sum_{\substack{t \in \mathbb{Z} \\ t^2 \leqslant 4m}} p_{k,r}(t,m)\, H(r,4m-t^2) \right) e^{2\pi i m z} \qquad (z \in H), \tag{24}$$

<u>where</u> $p_{k,r}(t,m)$ <u>is the polynomial defined by</u>

$$p_{k,r}(t,m) = \text{coefficient of } x^{k-r-1} \text{ in } \frac{1}{(1 - tx + mx^2)^r} \tag{25}$$

<u>(Gegenbauer polynomial)</u>. <u>Then</u> $C_{k,r}$ <u>is a modular form of weight</u> k <u>for the full modular group</u>. <u>If</u> $r < k - 1$, <u>it is a cusp form.</u>

We shall show in § 4 that, for $r = 1,3,5,\ldots,k-1$,

$$C_k^{-1}\, m^{k-1}\, \left[I_k(t^2-4m,t;r) + I_k(t^2-4m,-t;r) \right]$$

(26)

$$= \begin{cases} (-\tfrac{1}{4})^{(r+1)/2}\, (4m-t^2)^{r-\frac{1}{2}}\, \dfrac{\Gamma(k-r)\,\Gamma(r)}{\Gamma(k-1)}\, p_{k,r}(t,m) & \text{if } t^2 < 4m, \\ 0 & \text{if } t^2 \geqslant 4m. \end{cases}$$

Together with (22), this shows that the series $\Phi_r(z)$ defined by (10) is a multiple of the function (24) if $3 \leqslant r \leqslant k-3$. For $r = k-1$ we are on the edge of the strip in which the series (9) is absolutely convergent. We will show in § 4 that

(27)

$$\lim_{s \to k-1} c_m(s) = \frac{(-1)^{\frac{k}{2}+1}\, \pi^k}{2^{k-1}\,(k-1)!} \sum_{t^2 \leqslant 4m} H(k-1, 4m-t^2) - \frac{2\pi}{k-1}\, \frac{\Gamma(k-\frac{1}{2})\,\Gamma(\frac{1}{2})}{\Gamma(k)}\, \frac{\zeta(2k-2)}{\zeta(k)}\, \sigma_{k-1}(m)$$

(where $\sigma_{k-1}(m) = \sum_{d \mid m} d^{k-1}$ as usual), so that in this case the cusp form $\tilde{\Phi}_{k-1}(z) = \lim_{s \to k-1} \Phi_s(z)$ is a linear combination of Cohen's function $C_{k,k-1}$ and the Eisenstein series of weight k. Thus Cohen's theorem is a consequence of statement ii) of Theorem 1, while statement iii) implies the following result:

<u>Theorem 2:</u> <u>Let</u> r, k <u>be integers with</u> $3 \leqslant r \leqslant k-1$, r <u>odd,</u> k <u>even.</u> <u>The Petersson product of the modular form</u> $C_{k,r}$ <u>defined by</u> (24) <u>with an arbitrary Hecke eigenform</u> $f \in S_k$ <u>is given by</u>

$$(28) \qquad (f, C_{k,r}) = -\frac{(r+k-2)!\,(k-2)!}{(k-r-1)!}\, \frac{1}{4^{r+k-2}\, \pi^{2r+k-1}}\, D_f(r+k-1),$$

<u>where</u> $D_f(s)$ <u>is the function defined by</u> (2).

Since the Fourier coefficients of $C_{k,r}$ are rational numbers, $C_{k,r}$ is a linear combination of eigenforms with algebraic coefficients, and we deduce:

<u>Corollary:</u> <u>Let</u> f <u>be a Hecke eigenform in</u> S_k. <u>The values of</u> $D_f(s)/\pi^{2s-k+1}$ <u>for</u> $s = k, k+2, k+4, \ldots, 2k-2$ <u>are algebraic multiples of</u> (f, f).

(The case $s = k$ is a consequence of equation (5) rather than (28).) By virtue of the functional equation (4), the numbers $D_f(s)/\pi^s (f,f)$ $(s = 1,3,5,\ldots,k-1)$

are also algebraic.

<u>Example</u>: For $k = 12$, the only normalized eigenform in S_k is the discriminant function

$$\Delta(z) = q \prod_{n=1}^{\infty} (1 - q^n)^{24} = \sum_{n=1}^{\infty} \tau(n)\, q^n.$$

The number r in Theorem 2 must be 3, 5, 7, 9 or 11. By computing the first few Fourier coefficients of Cohen's functions $C_{k,r}$ we find

$$C_{12,3} = -\frac{180}{7}\Delta, \quad C_{12,5} = -210\Delta, \quad C_{12,7} = -1120\Delta,$$

$$C_{12,9} = -20736\Delta, \quad C_{12,11} = -\frac{77683}{12\times 23} E_{12} - \frac{7\times 10!}{23\times 691}\Delta$$

where $E_{12} = 1 + \frac{65520}{691} \sum_{n=1}^{\infty} \sigma_{11}(n)\, q^n$ is the normalized Eisenstein series. Thus from (28) we get five identities like (1), namely

$$(29) \qquad \sum_{n=1}^{\infty} \frac{\tau(n)^2}{n^s} = a_s \frac{4^s}{s!} \pi^{2s-11} \frac{\zeta(s-11)}{\zeta(2s-22)} (\Delta,\Delta) \qquad (s = 14, 16, 18, 20, 22)$$

with

$$a_{14} = 1, \quad a_{16} = \frac{1}{6}, \quad a_{18} = \frac{1}{30}, \quad a_{20} = \frac{2}{245}, \quad a_{22} = \frac{77}{31786} = \frac{7\times 11}{2\times 23\times 691}.$$

The numerical values of the series on the left-hand side of (29), calculated by taking 250 terms of the series, are

1.06544, 1.0109865184, 1.00239992152, 1.00056976587, 1.00013948615.

Substituting any of these values (except the first, where the series converges too slowly to give 12-digit accuracy) into (29) we obtain the numerical value

$$(\Delta,\Delta) = 1.035\ 362\ 056\ 79 \times 10^{-6}$$

for (the square of) the norm of Δ in the Petersson metric. The previously published valued $1.035\ 290\ 481\ 79 \times 10^{-6}$ (Lehmer [12]), obtained by integrating $|\Delta(z)|^2 y^{10}$ numerically, is false in the 5th decimal place.

Finally, we make a few general remarks about values of Dirichlet series attached to modular forms. The series $D_f(s)$ can be thought of as the "symmetric square" of the Mellin transform

$$(30) \qquad L_f(s) = \sum_{n=1}^{\infty} a(n)\, n^{-s} = \prod_p (1-\alpha_p p^{-s})^{-1}(1-\bar{\alpha}_p p^{-s})^{-1} \quad (\mathrm{Re}(s) >> 0)$$

of f, which is an entire function of s with the functional equation

$$(31) \qquad L_f^*(s) = (2\pi)^{-s}\,\Gamma(s)\,L_f(s) = (-1)^{k/2}\,L_f^*(k-s).$$

By the theorem of Eichler-Shimura-Manin on periods of cusp forms (cf. Chapter V of [11]), the ratios $L_f^*(1) : L_f^*(3) : \ldots : L_f^*(k-1)$ and $L_f^*(2) : L_f^*(4) : \ldots : L_f^*(k-2)$ are algebraic (and in fact belong to the number field generated by the Fourier coefficients of f). For $f = \Delta$, for example, there are real numbers ω_+ and ω_- with

$$(32) \qquad \begin{aligned} &L_\Delta^*(1) = L_\Delta^*(11) = \frac{192}{691}\,\omega_+,\; L_\Delta^*(3) = L_\Delta^*(9) = \frac{16}{135}\,\omega_+,\; L_\Delta^*(5) = L_\Delta^*(7) = \frac{8}{105}\,\omega_+,\\ &L_\Delta^*(2) = L_\Delta^*(10) = \frac{384}{5}\,\omega_-,\; L_\Delta^*(4) = L_\Delta^*(8) = 40\,\omega_-,\; L_\Delta^*(6) = 32\,\omega_-; \end{aligned}$$

where by calculating the values of $L_\Delta(10)$ and $L_\Delta(11)$ (which are the most rapidly convergent of the series) numerically we find

$$\omega_+ = 2.144\,606\,670\,68 \times 10^{-2}, \quad \omega_- = 4.827\,748\,001 \times 10^{-5}.$$

On the other hand, Rankin ([18], Theorem 4) showed that for any normalized eigenform $f \in S_k$ and any even integer q with $\frac{k}{2} + 2 \leq q \leq k-4$ one has

$$(33) \qquad L_f^*(q)L_f^*(k-1) = (-1)^{q/2}\,2^{k-3}\,\frac{B_q}{q}\,\frac{B_{k-q}}{k-q}\,(f, E_q\,E_{k-q}),$$

where E_i is the normalized Eisenstein series and the B_i are Bernoulli numbers, so the product of the two independent periods of L_f^* is an algebraic multiple of (f,f). For $f = \Delta$, for example, (33) says

$$L^*(11)\,L^*(8) = \frac{7680}{691}\,(\Delta,\Delta)$$

or, using (32), that

$$\omega_+\,\omega_- = (\Delta,\Delta).$$

We can therefore restate the Corollary to Theorem 2 as saying that the values of $D_f(s)$ for $s = 1,3,5,\ldots,k-1,\ k,\ k+2,\ldots,2k-2$ are of the form

$$(\text{algebraic number})\cdot\omega_+\omega_-\pi^n,$$

while the result of Eichler-Shimura-Manin says that the values of $L_f(s)$ for $s = 1,2,\ldots,k-1$ are of the form $(\text{alg.})\cdot\omega_+\pi^n$ or $(\text{alg.})\cdot\omega_-\pi^n$. Both statements fit into a general philosophy of Deligne that, if $L(s) = \sum c_n n^{-s}$ is any "motivated" Dirichlet series (i.e. one arising from a natural mathematical object such as a number field, a Galois representation, an algebraic variety, or a modular form) and satisfies a functional equation of the form

$$L^*(s) = \gamma(s)\,L(s) = w\,L^*(C-s)$$

with some Γ-factor $\gamma(s)$, then the value of $L(s)$ at any integral value of s for which neither s nor $C-s$ is a pole of $\gamma(s)$ should be given by a "closed formula" $L(s) = A\cdot\omega$, where A is algebraic and ω is a "period" about which something nice can be said (for instance, the twisted functions $L_\chi(s) = \sum c_n\chi(n)\,n^{-s}$ should have values $A_\chi\cdot\omega$ with the <u>same</u> period ω, and the algebraic numbers A_χ should have nice p-adic properties as χ varies). Now the series $L_f(s)$ and $D_f(s)$ are just the first two cases of the Dirichlet series

$$L_{m,f}(s) = \prod_p \prod_{i=0}^{m} (1 - \alpha_p^{\,i}\,\bar{\alpha}_p^{\,i}\,p^{-s})^{-1} \qquad (\mathrm{Re}(s) \gg 0)$$

attached to the symmetric powers of the representation associated to f, and these functions are conjectured [19] to be holomorphic and to satisfy the functional equations

$$L^*_{m,f}(s) = \gamma_m(s)\,L_{m,f}(s) = \pm L^*_{m,f}((k-1)m+1-s),$$

$$\gamma_m(s) = \begin{cases} (2\pi)^{-rs} \displaystyle\prod_{j=0}^{r-1} \Gamma(s-j(k-1)) & \text{if } m = 2r-1, \\ \pi^{-s/2}\,\Gamma\!\left(\dfrac{s}{2} - \left[\dfrac{r(k-1)}{2}\right]\right)\gamma_{2r-1}(s) & \text{if } m = 2r. \end{cases}$$

In a letter to the author (February 1976), Serre suggested that, in accordance with the above philosophy, the values of $L_{m,f}(s)$ may be given by a formula of the type $L_{m,f}(s) = (\text{alg.}).\omega_+^a\,\omega_-^b\,\pi^n$, probably with a+b=m, possibly with $|a-b| \leq 1$, for those integral values of s for which $\gamma_m(s)$ and $\gamma_m((k-1)m+1-s)$ are finite. For f=Δ and m=3 or 4 this would mean that there are identities

$$L_{3,\Delta}(s) = A\omega_\pm^2\omega_\mp\pi^n \quad (s=18,19,20,21,22), \quad L_{4,\Delta}(s) = A\omega_+^2\omega_-^2\pi^n \quad (s=24,26,28,30,32)$$

with $A \in \mathbb{Q}$, $n \in \mathbb{N}$. (We have given only those values of s for which the Dirichlet series converge absolutely.) However, the numerical computation of the values in question (done by G.Köckritz and R.Schillo on the IBM 370/168 at Bonn University, using 32-digit accuracy and over 1000 terms of the Euler products) did not lead to any simple values of A and n satisfying these formulas. At the Corvallis conference (July 1977), Deligne gave a revised and sharper conjecture for the values of $L_{m,f}(s)$: if f is an eigenform with rational Fourier coefficients (i.e. k=12,16,18,20,22 or 26), then one should have

$$L_{2r-1,f}(s) = (\text{rat.}).(2\pi)^{rs-\frac{r(r-1)}{2}(k-1)}\,C_\pm^{\frac{r(r+1)}{2}}\,C_\mp^{\frac{r(r-1)}{2}} \qquad \left(r-1<\frac{s}{k-1}\leq r, (-1)^s=\pm 1\right),$$

$$L_{2r,f}(s) = \begin{cases} (\text{rat.}).(2\pi)^{rs-\frac{r(r-1)}{2}(k-1)}\,(C_+C_-)^{\frac{r(r+1)}{2}} & \left(r-1<\frac{s}{k-1}\leq r,\ s \text{ odd}\right), \\ (\text{rat.}).(2\pi)^{(r+1)s-\frac{r(r+1)}{2}(k-1)}\,(C_+C_-)^{\frac{r(r+1)}{2}} & \left(r<\frac{s}{k-1}\leq r+1,\ s \text{ even}\right), \end{cases}$$

where C_+ and C_- are real numbers depending on f but not on r or s. For k=12, f= Δ, and m=1 or 2, for instance, we have

s	$(2\pi)^{-s}\Gamma(s)L_{1,\Delta}(s)$		s	$(2\pi)^{-2s+11}\Gamma(s)L_{2,\Delta}(s)$	
6	$1/2\times3\times5$	C_+	12	$1/2$	C_+C_-
7	$1/2^2\times7$	C_-	14	$1/2\times7$	C_+C_-
8	$1/2^3\times3$	C_+	16	$1/2^5\times3$	C_+C_-
9	$1/2\times3^2$	C_-	18	$1/2^2\times3^3\times5$	C_+C_-
10	$2/5^2$	C_+	20	$1/2\times5^2\times7^2$	C_+C_-
11	$2\times3^2\times5/691$	C_-	22	$7/2^2\times23\times691$	C_+C_-

where $C_+ = 2^6\times3\times5\ \omega_-$, $C_- = 2^5/3\times5\ \omega_+$. The computer calculation gives

$C_+ \approx 0.046\ 346\ 380\ 811\ 850\ 816\ 182\ 4$, $C_- \approx 0.045\ 751\ 608\ 975\ 539\ 581\ 74$,

$C_+C_- = 2^{11}(\Delta,\Delta) \approx 0.002\ 120\ 421\ 492\ 335\ 249\ 248\ 968\ 328\ 831\ 438$

and suggests overwhelmingly the following identities (in accordance with Deligne's general conjecture) for $m = 3$ and 4 :

s	$(2\pi)^{-2s+11}\Gamma(s)L_{3,\Delta}(s)$	
18	$2^2/5$	$c_+^3 c_-$
19	$3/7$	$c_+ c_-^3$
20	$1/5$	$c_+^3 c_-$
21	$5/7^2$	$c_+ c_-^3$
22	$2\times 3/5\times 23$	$c_+^3 c_-$

s	$(2\pi)^{-3s+33}\Gamma(11)^{-1}\Gamma(s)\Gamma(s-11)L_{4,\Delta}(s)$	
24	$2^5\times 3^2$	$c_+^3 c_-^3$
26	$2^5\times 3\times 5$	$c_+^3 c_-^3$
28	$2^2\times 23\times 691/7^2$	$c_+^3 c_-^3$
30	$2^3\times 653$	$c_+^3 c_-^3$
32	$2\times 3\times 34891/7$	$c_+^3 c_-^3$

§ 2. An integral representation for the coefficients $c_m(s)$

In proving Theorem 1, we will reverse the order of the statements i) - iii). For $s \in \mathbb{C}$ with $\mathrm{Re}(s) > 1$ the numbers $D_f(s+k-1)$ are finite (since the series in (3) is absolutely convergent in the half-plane $\mathrm{Re}(s) > k$) and so there exists a unique cusp form $\tilde{\Phi}_s \in S_k$ satisfying

$$(34) \qquad (\tilde{\Phi}_s, f) = C_k \frac{\Gamma(s+k-1)}{(4\pi)^{s+k-1}} D_f(s+k-1)$$

for all eigenforms $f \in S_k$, namely the function given by the right-hand side of equation (14). We define $\tilde{c}_m(s)$ $(m = 1,2,\dots.)$ as the mth Fourier coefficient of $\tilde{\Phi}_s$ (= the expression on the right-hand side of (15)) and must show that $\tilde{c}_m(s) = c_m(s)$. To do this, we will write $\tilde{c}_m(s)$ as an integral involving a certain kernel function ω_m which was first introduced by Petersson.

We recall the definition of the kernel function. As in § 1, we fix an even integer $k > 2$ which will be omitted from the notations. For $m = 1,2,\dots$ set

$$(35) \qquad \omega_m(z,z') = \sum_{\substack{a,b,c,d\in\mathbb{Z} \\ ad-bc=m}} \frac{1}{(czz'+dz'+az+b)^k} \qquad (z,\ z' \in H).$$

The series converges absolutely and therefore defines a function holomorphic in both variables, and one can see easily that it transforms like a modular form of weight k with respect to the action of Γ on each variable separately. One also checks easily that ω_m is a cusp form.

Proposition 1 (Petersson [16]): The function $C_k^{-1} m^{k-1} \omega_m(z, -\overline{z'})$ (C_k as in equation (3)) is the kernel function for the mth Hecke operator with respect to the Petersson metric, i.e.

$$(36)\qquad C_k^{-1} m^{k-1} \int_{\Gamma\backslash H} f(z)\, \overline{\omega_m(z, -\overline{z'})}\, y^k\, dV = (f|T(m))(z') \qquad (\forall f \in S_k,\ z' \in H).$$

Equivalently, $\omega_m(z,z')$ has the following representation as a linear combination of Hecke eigenforms:

$$(37)\qquad m^{k-1} \omega_m(z,z') = C_k \sum_{i=1}^{\dim S_k} \frac{a_i(m)}{(f_i,f_i)} f_i(z) f_i(z').$$

Proof: The equivalence of (36) and (37) is immediate from the fact that the eigenforms f_i form an orthogonal basis of S_k. Also, it is easily seen that $m^{k-1} \omega_m(z,z')$ is obtained from $\omega_1(z,z')$ by applying the Hecke operator $T(m)$ with respect to (say) the first variable, so it suffices to prove (36) for $m = 1$.

We can write (35) for $m = 1$ in the form

$$\omega_1(z,z') = \sum_{ad-bc=1} \frac{1}{\left(z' + \frac{az+b}{cz+d}\right)^k} (cz+d)^{-k}.$$

For fixed $c, d \in \mathbb{Z}$ with $(c,d) = 1$, the pairs of integers a, b with $ad-bc = 1$ are all of the form $a_0 + nc$, $b_0 + nd$ ($n \in \mathbb{Z}$), where a_0, b_0 is any fixed solution. Thus

$$\omega_1(z,z') = \sum_{\substack{c,d\in\mathbb{Z}\\(c,d)=1}} \frac{1}{(cz+d)^k} \sum_{n=-\infty}^{\infty} \left(z' + \frac{a_0 z + b_0}{cz+d} + n\right)^{-k}.$$

Using the identity

$$(38)\qquad \sum_{n=-\infty}^{\infty} (\tau + n)^{-k} = \frac{(2\pi i)^k}{(k-1)!} \sum_{r=1}^{\infty} r^{k-1} e^{2\pi i r\tau} \qquad (\tau \in H),$$

we find

$$(39)\qquad \omega_1(z,z') = 2\, \frac{(2\pi i)^k}{(k-1)!} \sum_{r=1}^{\infty} r^{k-1} G_r(z)\, e^{2\pi i r z'},$$

where $G_r(z)$ is the Poincaré series

$$(40) \qquad G_r(z) = \frac{1}{2} \sum_{\substack{c,d\in\mathbb{Z}\\(c,d)=1}} \frac{1}{(cz+d)^k} e^{2\pi i r \frac{a_0 z+b_0}{cz+d}} \qquad (r = 1,2,\ldots, \quad z \in H)$$

(with a_0, b_0 again representing any integers with $a_0 d - b_0 c = 1$; in a more invariant notation $G_r(z) = \sum_{\gamma\in\Gamma_\infty\backslash\Gamma} j_k(\gamma,z)\, e^{2\pi i r\gamma z}$, where the summation is over representatives for the right cosets of $\Gamma_\infty = \{\pm \begin{pmatrix}1 & n\\ 0 & 1\end{pmatrix} \mid n \in \mathbb{Z}\}$ in Γ). But, as is well known (see, for example [6], p. 37), G_r is a cusp form of weight k and satisfies

$$(41) \qquad (f,G_r) = \frac{(k-2)!}{(4\pi r)^{k-1}}\, a(r) \qquad \text{for } f(z) = \sum_{n=1}^{\infty} a(n)\, q^n \in S_k$$

(this is proved in the same way as Rankin's identity below). Equation (36) for $m = 1$ follows immediately from equation (39) and (41). (For a different proof of Proposition 1, not using Poincaré series, see [25].)

The other main ingredient for the proof of Theorem 1 is Rankin's integral representation of the function (3), namely

$$(42) \qquad \zeta(2s)\, \frac{\Gamma(s+k-1)}{(4\pi)^{s+k-1}} \sum_{n=1}^{\infty} \frac{|a(n)|^2}{n^{s+k-1}} = \int_{\Gamma\backslash H} |f(z)|^2\, E(z,s)\, y^k\, dV$$

(valid for any cusp form $f(z) = \sum a(n)\, q^n \in S_k$ and $s \in \mathbb{C}$ with $\mathrm{Re}(s) > 1$), where $E(z,s)$ is the Epstein zeta-function

$$(43) \qquad E(z,s) = \frac{1}{2}\, y^s \sum_{m,n\in\mathbb{Z}}{}' \frac{1}{|mz+n|^{2s}} \qquad (z = x+iy \in H,\ s \in \mathbb{C},\ \mathrm{Re}(s) > 1)$$

(here $\sum'$ denotes a sum over non-zero pairs of integers). This is a special case of a more general identity, namely that

$$(44) \qquad \int_{\Gamma\backslash H} h(z)\, E(z,s)\, dV = \zeta(2s) \int_0^{\infty}\int_0^1 h(x+iy)\, y^{s-2}\, dx\, dy$$

for any Γ-invariant function h on the upper half-plane for which the integrals in question converge absolutely. To see this, we write each pair of integers m, n in (43) as rc, rd with $r \geq 1$ and $(c,d) = 1$ and note that there is a 2 : 1 correspondence between the pairs c, d and the right cosets of Γ_∞ in Γ, so

$$E(z,s) = \frac{1}{2} \sum_{r=1}^{\infty} \sum_{\substack{c,d \in \mathbb{Z} \\ (c,d)=1}} \frac{y^s}{r^{2s}|cz+d|^{2s}} = \zeta(2s) \sum_{\gamma \in \Gamma_\infty \backslash \Gamma} \mathrm{Im}(\gamma z)^s .$$

Also, if F is a fundamental domain for the action of Γ on H, then $\bigcup_{\gamma \in \Gamma_\infty \backslash \Gamma} \gamma F$ is a fundamental domain for the action of Γ_∞. Hence

$$\int_{\Gamma \backslash H} h(z)\, E(z,s)\, dV = \zeta(2s) \int_F \sum_{\gamma \in \Gamma_\infty \backslash \Gamma} \mathrm{Im}(\gamma z)^s\, h(\gamma z)\, dV$$

$$= \zeta(2s) \sum_{\gamma \in \Gamma_\infty \backslash \Gamma} \int_{\gamma F} \mathrm{Im}(z)^s\, h(z)\, dV$$

$$= \zeta(2s) \int_{\Gamma_\infty \backslash H} h(z)\, \mathrm{Im}(z)^s\, dV,$$

and (44) follows by choosing the fundamental domain $\{z \in H \mid 0 \leq x < 1\}$ for the action of Γ_∞. Equation (44) says that $\int_{\Gamma \backslash H} h(z)\, E(z,s)\, dV$ is $\zeta(2s)$ times the Mellin transform $\int_0^\infty h_o(y)\, y^{s-2}\, dy$ of the "constant term" $h_o(y)$ in the Fourier expansion

$$h(z) = \sum_{n=-\infty}^{\infty} h_n(y)\, e^{2\pi i n x}$$

of the function h (which is Γ-invariant and hence periodic). Equation (42) now follows by taking for h the Γ-invariant function

$$h(z) = y^k |f(z)|^2 = y^k \sum_{m=1}^{\infty} \sum_{n=1}^{\infty} a(n)\, \overline{a(m)}\, e^{2\pi i(n-m)x} e^{-2\pi(n+m)y}$$

with $h_o(y) = y^k \sum |a(n)|^2\, e^{-4\pi n y}$.

If f is a Hecke eigenform, then the series in (42) is related to $D_f(s)$ by equation (3) (note that a(n) is real in this case, so $a(n)^2 = |a(n)|^2$). Therefore (42) permits us to deduce the meromorphy of $D_f(s)$ and the two formulas (4) and (5) from the corresponding properties of E(z,s), namely that E(z,s) extends meromorphically to the whole s-plane with a simple pole of residue $\frac{\pi}{2}$

(independent of z!) at $s = 1$ as its only singularity and satisfies the functional equation

$$E^*(z,s) = \pi^{-s}\,\Gamma(s)\,E(z,s) = E^*(z,1-s). \tag{45}$$

Putting together equations (37), (3) and (42), we obtain the integral representation

$$\zeta(s)\,\tilde{c}_m(s) = m^{k-1}\int_{\Gamma\backslash H}\omega_m(z,-\bar z)\,E(z,s)\,y^k\,dV \qquad (m = 1,2,\ldots,\quad s\in\mathbb{C}) \tag{46}$$

for the function $\tilde{c}_m(s)$ defined by the right-hand side of (15). In the next paragraph we will compute the integral on the right-hand side of (46), thereby completing the proof of Theorem 1.

§ 3. Calculation of $\int_{\Gamma\backslash H}\omega_m(z,-\bar z)\,E(z,s)\,y^k\,dV$

The computation of the integral in equation (46) will be carried out by a method similar to that used in [25] for the simpler integral

$$\int_{\Gamma\backslash H}\omega_m(z,-\bar z)\,y^k\,dV$$

(which, by virtue of Proposition 1 above, equals $C_k m^{-k+1}$ times the trace of the Hecke operator $T(m)$ on S_k). The extra factor $E(z,s)$ in the integrand will actually simplify both the formal calculation and the treatment of convergence, which was handled incorrectly in [25] (see Correction following this paper).

The definition of $\omega_m(z,z')$, equation (35), involves a sum over all matrices of determinant m. We split up this sum according to the value of the trace of the matrix and observe that there is a $1:1$ correspondence between matrices of trace t and determinant m and binary quadratic forms of discriminant t^2-4m, given by

$$\begin{pmatrix} a & b\\ c & d\end{pmatrix} \mapsto \phi(u,v) = cu^2 + (d-a)\,uv - bv^2,$$
$$\phi(u,v) = au^2 + buv + cv^2 \longmapsto \begin{pmatrix} \frac12(t-b) & -c\\ a & \frac12(t+b)\end{pmatrix}.$$

Therefore

$$(47)\qquad \begin{aligned} y^k\,\omega_m(z,-\bar z) &= \sum_{t=-\infty}^{\infty}\ \sum_{\substack{a,b,c,d\in\mathbb{Z}\\ ad-bc=m\\ a+d=t}} \frac{y^k}{(c|z|^2+d\bar z-az-b)^k}\\ &= \sum_{t=-\infty}^{\infty}\ \sum_{|\phi|=t^2-4m} R_\phi(z,t), \end{aligned}$$

where the inner sum is over all quadratic forms ϕ of discriminant t^2-4m and where we have written

$$(48)\qquad R_\phi(z,t) = \frac{y^k}{(a|z|^2+bx+c-ity)^k} \qquad (z=x+iy\in H,\ \ t\in\mathbb{R})$$

for a form ϕ, $\phi(u,v) = au^2+buv+cv^2$. The sum (47) converges absolutely for all $z\in H$, and we have

Proposition 2: *For $s\in\mathbb{C}$ with $s\neq 1$ and $2-k<\mathrm{Re}(s)<k-1$, we have*

$$\sum_{t=-\infty}^{\infty}\int_{\Gamma\backslash H} \left|E(z,s)\right| \left|\sum_{|\phi|=t^2-4m} R_\phi(z,t)\right| dV < \infty .$$

By virtue of this proposition, which we will prove at the end of the section, we may substitute (47) into (46) and interchange the order of summation and integration to obtain

$$\zeta(s)\,\tilde c_m(s) = m^{k-1}\sum_{t=-\infty}^{\infty}\int_{\Gamma\backslash H}\sum_{|\phi|=t^2-4m} R_\phi(z,t)\,E(z,s)\,dV \quad (2-k<\mathrm{Re}(s)<k-1).$$

Theorem 1 is then a consequence of the following result, which is of interest in its own right.

Theorem 3: *Let k be an even integer >2, Δ a discriminant (i.e. $\Delta\in\mathbb{Z}$, $\Delta\equiv 0$ or $1 \pmod 4$), t a real number with $t^2>\Delta$. For each binary quadratic form ϕ of discriminant Δ let $R_\phi(t,z)$ $(z\in H)$ be the function defined by (48). Then for $s\in\mathbb{C}$ with $s\neq 1$, $1-k<\mathrm{Re}(s)<k$,*

$$\int_{\Gamma\backslash H}\left(\sum_{|\phi|=\Delta} R_\phi(z,t)\right) E(z,s)\, dV$$

(49)

$$= \zeta(s,\Delta)\,\{I_k(\Delta,t;s) + I_k(\Delta,-t;s)\} \;+$$

$$\begin{cases} (-1)^{k/2}\, \dfrac{\Gamma(s+k-1)\zeta(s)\zeta(2s)}{(2\pi)^{s-1}\Gamma(k)}\, |t|^{-s-k+1} & \text{if } \Delta = 0, \\ 0 & \text{if } \Delta \neq 0, \end{cases}$$

__where__ $\zeta(s,\Delta)$ __and__ $I_k(\Delta,t;s)$ __are given by__ (6) __and__ (8), __respectively.__

__Proof:__ We observe first that

$$R_{\gamma\phi}(z,t) = R_\phi({}^t\gamma z,t) \qquad (\gamma\in\Gamma,\ {}^t\gamma = \text{transpose of } \gamma),$$

so that the (absolutely convergent) series $\sum_{|\phi|=\Delta} R_\phi(z,t)$ defines a function in the upper half-plane which is invariant under Γ. Moreover, this function is $O(y^{1-k})$ as $y = \text{Im}(z) \to \infty$, as we will show in the proof of Proposition 2 below, while $E(z,s) = O(y^{\max(\sigma,1-\sigma)})$ for $y \to \infty$ ($\sigma = \text{Re}(s)$). Hence the integral on the left hand side of (49) makes sense and is holomorphic (for $s \neq 1$) in the range specified. On the other hand, $\zeta(s,\Delta)$ also has a holomorphic continuation for all $s \neq 1$ and the integral defining $I_k(\Delta,t;s)$ converges for $1-k < \sigma < k$ (unless $\Delta = 0$, in which case the integral has a pole at $s = \frac{1}{2}$ compensating the pole coming from $\zeta(2s)$ in the expression on the right-hand side of (49)). It therefore suffices to prove (49) under the assumption $1 < \sigma < k$ and then extend the result to $1-k < \sigma < k$ by analytic continuation.

Suppose, then, that $\text{Re}(s) > 1$. Written out in full, the expression on the left-hand side of (49) is

$$(50)\qquad \frac{1}{2}\int_{\Gamma\backslash H}\ \sum_{|\phi|=\Delta}\ \sum_{m,n\in\mathbb{Z}}{}^{'}\ R_\phi(z,t)\ \frac{y^s}{|mz+n|^{2s}}\, dV.$$

The action of Γ on $z \in H$ permutes the terms of this sum, transforming the form ϕ and the pair $\pm(m,n) \in (\mathbb{Z}^2 - \{(0,0)\})/\{\pm 1\}$ in such a way that $\phi(n,-m)$ remains invariant. In particular, the sum of the terms with $\phi(n,-m) > 0$ in the integrand of (50) is Γ-invariant. Also, the group Γ acts freely on the set of pairs $(\phi, (m,n))$ with $\phi(n,-m) > 0$. Therefore, ignoring convergence for the moment, we have

$$\frac{1}{2} \int_{\Gamma\backslash H} \sum_{|\phi|=\Delta} \sum_{\substack{m,n \\ \phi(n,-m)>0}}{}' R_\phi(z,t) \frac{y^s}{|mz+n|^{2s}} \, dV$$

(51)

$$= \sum_{\substack{\phi \\ \text{mod } \Gamma}} \sum_{\substack{\pm(m,n) \\ \phi(n,-m)>0}}{}' \int_H R_\phi(z,t) \frac{y^s}{|mz+n|^{2s}} \, dV.$$

Making the substitution $z \mapsto \dfrac{nz - \frac{1}{2}bn + cm}{-mz + an - \frac{1}{2}bm}$ (which maps H to H if $an^2 - bnm + cm^2 > 0$), we find

$$\int_H \frac{y^k}{(a|z|^2+bx+c-ity)^k} \frac{y^s}{|mz+n|^{2s}} \, dV = \frac{1}{(an^2-bnm+cm^2)^s} \int_H \frac{y^{k+s} \, dV}{(|z|^2-\frac{1}{4}\Delta-ity)^k}$$

so that the right-hand side of (51) is equal to $\zeta(s,\Delta)\, I_k(\Delta,t;s)$. Since the sum defining $\zeta(s,\Delta)$ and the integral defining $I_k(\Delta,t;s)$ converge absolutely for $1 < \mathrm{Re}(s) < k$, it follows a posteriori that the expression on the left-hand side of (51) was absolutely convergent in this range. The terms with $\phi(n,-m) < 0$ can be treated in a similar manner (or simply by observing that $R_{-\phi}(z,t) = R_\phi(z,-t)$) and contribute $\zeta(s,\Delta)\, I_k(\Delta,-t;s)$.

Finally, we must treat the terms in (50) with $\phi(n,-m) = 0$. They occur only if Δ is a perfect square. These terms are not absolutely convergent in (50) (if we replace each $R_\phi(z,t)$ by its absolute value, then the sum in the integrand converges for each z but the integral diverges). We argue as in the proof of equation (44). First, by removing the greatest common divisor of m and n, we can write (50) as $\zeta(2s)$ times the corresponding sum with the extra condition $(m,n) = 1$. Since any relatively prime pair of integers (m,n) is Γ-equivalent to the pair $(0,1)$ by an element of Γ which is well-defined up to left multiplication by an element of Γ_∞, the terms of (50) with $\phi(n,-m) = 0$ give

$$\zeta(2s) \int_{\Gamma_\infty \backslash H} \sum_{\substack{|\phi|=\Delta \\ \phi(1,0)=0}} R_\phi(z,t)\, y^s\, dV$$

$$= \zeta(2s) \int_{\Gamma_\infty \backslash H} \sum_{\substack{a,b,c \in \mathbb{Z} \\ b^2-4ac=\Delta \\ a=0}} \frac{y^k}{(a|z|^2+bx+c-ity)^k}\, y^s\, dV$$

$$(52) \qquad = \zeta(2s) \int_0^\infty \int_0^1 \sum_{b^2=\Delta} \sum_{c=-\infty}^{\infty} \frac{1}{(bx-ity+c)^k}\, y^{s+k-2}\, dx\, dy.$$

The sum over c can be evaluated using (38) and equals

$$\frac{(2\pi i)^k}{(k-1)!} \sum_{r=1}^{\infty} r^{k-1}\, e^{2\pi i r(\pm bx + i|t|y)}$$

if $\pm t > 0$ (note that $t^2 > \Delta$ and Δ is a square, so $t \neq 0$). If $\Delta \neq 0$, then this expression involves only terms $e^{2\pi i n x}$ with $n \neq 0$, so the integral (52) is identically zero. For $\Delta = 0$, the expression (52) becomes

$$\zeta(2s)\, \frac{(2\pi i)^k}{(k-1)!} \int_0^\infty \int_0^1 \sum_{r=1}^{\infty} r^{k-1} e^{-2\pi r|t|y} y^{s+k-2} dx\, dy = \frac{(2\pi i)^k}{(k-1)!} \zeta(2s)\zeta(s)\, \frac{\Gamma(s+k-1)}{(2\pi|t|)^{s+k-1}}$$

This completes the proof of Theorem 3.

<u>Proof of Proposition 2:</u> We choose for $\Gamma \backslash H$ the standard fundamental domain $\{z \,\big|\, |z| \geqslant 1,\ |x| \leqslant \frac{1}{2}\}$. Since $E(z,s) = O(y^\sigma + y^{1-\sigma})$ as $y = \mathrm{Im}(z) \to \infty$, where $\sigma = \mathrm{Re}(s)$ it will suffice to show that

$$(53) \qquad \sum_{t=-\infty}^{\infty} \int_1^\infty \int_{-\frac{1}{2}}^{\frac{1}{2}} \left| \sum_{|\phi|=t^2-4m} R_\phi(z,t) \right| y^{\sigma-2}\, dx\, dy < \infty,$$

for $\sigma < k-1$. Also, the above proof shows that the integral occurring in (53) is finite for each fixed value of t (even in the larger range $\sigma < k$), so we can ignore the finitely many values of t for which t^2-4m is a perfect square. If t^2-4m is not a square, then

$$\sum_{|\phi|=t^2-4m} R_\phi(z,t) = 2\,\mathrm{Re}\left(\sum_{\substack{b^2-4ac=t^2-4m\\ a>0}} \frac{y^k}{(a|z|^2+bx+c-ity)^k}\right)$$

(54)

$$= 2y^k\,\mathrm{Re}\left(\sum_{a=1}^{\infty}\frac{1}{a^k}\sum_{\substack{b \pmod{2a}\\ b^2\equiv t^2-4m \pmod{4a}}}\sum_{n=-\infty}^{\infty}\left[(x+\tfrac{b}{2a}+n)^2+y^2-\tfrac{ity}{a}-\tfrac{t^2-4m}{4a^2}\right]^{-k}\right).$$

But it is easily shown that

$$\sum_{n=-\infty}^{\infty}\left[(x+n)^2+L^2\right]^{-k} = O(L^{1-2k})$$

uniformly for $x \in \mathbb{R}$ and $L \in \mathbb{C}$ with $\mathrm{Re}(L)$ bounded away from 0. We apply this with $L^2 = y^2 - ity - \frac{t^2-4m}{4a^2} = (y - \frac{it}{2a})^2 + \frac{m}{a^2}$. Since m is fixed and $a \geqslant 1$, $t \to \infty$, $y \to \infty$ in the sum (54), we can write

$$\sum_{n=-\infty}^{\infty}\left[(x+\frac{b}{2a}+n)^2+y^2-\frac{ity}{a}-\frac{t^2-4m}{4a^2}\right]^{-k} = O\left((y-\frac{it}{2a})^{-2k+1}\right) = O\left((y^2+t^2/a^2)^{-k+\frac{1}{2}}\right).$$

Also, the number of solutions $b \pmod{2a}$ of the congruence $b^2 \equiv t^2 - 4m \pmod{4a}$ is $O(a^\varepsilon)$ as $a \to \infty$ for any $\varepsilon > 0$. Therefore (54) gives the estimate

$$\sum_{|\phi|=t^2-4m} R_\phi(z,t) = O\left(y^k \sum_{a=1}^{\infty} a^{\varepsilon-k}(y^2+t^2/a^2)^{-k+\frac{1}{2}}\right),$$

where the constant implied by $O(\ )$ depends on m and k but not on y or t. This expression is $O(y^{1-k})$ in the range $y \geqslant t$ and $O(y^{-\varepsilon}t^{-k+1+\varepsilon})$ for $t > y$, as one checks by splitting up the sum according as $a \leqslant t/y$ or $a > t/y$. Hence

$$\int_1^\infty \int_{-\frac{1}{2}}^{\frac{1}{2}} \left|\sum_{|\phi|=t^2-4m} R_\phi(z,t)\right| y^{\sigma-2}\,dx\,dy = O\left(t^{1-k+\varepsilon}\int_1^t y^{\sigma-2+\varepsilon}dy + \int_t^\infty y^{-k+\sigma-1}dy\right)$$

$$= O(t^{\sigma-k}),$$

so the sum (53) converges for $k - \sigma > 1$.

§ 4. Properties of the functions $\zeta(s,\Delta)$ and $I_k(\Delta,t;s)$

In order to deduce from Theorem 1 the various corollaries discussed in § 1, in particular the trace formula and the formula for the Petersson product of an eigenform with the modular forms constructed by Cohen, we will need various properties of the functions $\zeta(s,\Delta)$ and $I_k(\Delta,t;s)$ defined by Equations (6) and (8). We begin with the zeta-function.

Proposition 3: *Let $\zeta(s,\Delta)$ be the zeta-function defined by (6), where $\Delta \in \mathbb{Z}$, $s \in \mathbb{C}$, $\mathrm{Re}(s) > 1$. Then*

i) $\zeta(s,\Delta) = \zeta(2s) \sum_{a=1}^{\infty} \frac{n(a)}{a^s}$, *where* $n(a)$ *is the number of solutions* $b \pmod{2a}$ *of the congruence* $b^2 \equiv \Delta \pmod{4a}$.

ii) $\zeta(s,\Delta)$ *has a meromorphic continuation to the whole complex plane and, if* $\Delta \neq 0$, *satisfies the functional equation*

$$\gamma(s,\Delta)\,\zeta(s,\Delta) = \gamma(1-s,\Delta)\,\zeta(1-s,\Delta), \quad \textit{where}$$

$$\gamma(s,\Delta) = \begin{cases} (2\pi)^{-s}\,|\Delta|^{s/2}\,\Gamma(s) & \textit{if } \Delta < 0 \\ \pi^{-s}\,\Delta^{s/2}\,\Gamma(\frac{s}{2})^2 & \textit{if } \Delta > 0 \end{cases}$$

iii) $\zeta(s,\Delta)$ *can be expressed in terms of standard Dirichlet series as follows:*

$$\zeta(s,\Delta) = \begin{cases} 0 & \textit{if } \Delta \equiv 2 \text{ or } 3 \pmod 4 \\ \zeta(s)\,\zeta(2s-1) & \textit{if } \Delta = 0 \\ \zeta(s) L_D(s) \sum_{d|f} \mu(d) \left(\frac{D}{d}\right) d^{-s} \sigma_{1-2s}\left(\frac{f}{d}\right) & \textit{if } \Delta \equiv 0 \text{ or } 1 \pmod 4, \Delta \neq 0, \end{cases}$$

where if $\Delta \equiv 0$ *or* $1 \pmod 4$, $\Delta \neq 0$ *we have written* $\Delta = Df^2$ *with* $f \in \mathbb{N}$, D *the discriminant of* $\mathbb{Q}(\sqrt{\Delta})$, $\left(\frac{D}{\cdot}\right)$ *the Kronecker symbol,* $L_D(s) = \sum_{n=1}^{\infty} \left(\frac{D}{n}\right) n^{-s}$ *the associated L-series, and* $\sigma_\nu(m) = \sum_{\substack{d|m \\ d>0}} d^\nu$ $(m \in \mathbb{N})$, $\nu \in \mathbb{C}$). *In particular, the function* $L(s,\Delta)$ *defined by* (7) *is entire except for a simple pole (of residue* $\frac{1}{2}$ *if* $\Delta = 0$ *and* 1 *if* $\Delta \neq 0$) *if* Δ *is a square.*

iv) <u>For</u> $\Delta < 0$, <u>the values of</u> $L(s,\Delta)$ <u>at</u> $s = 1$ <u>and</u> $s = 0$ <u>are given by</u>

$$L(1,\Delta) = \frac{\pi}{\sqrt{|\Delta|}}, \qquad L(0,\Delta) = \frac{\pi}{\sqrt{|\Delta|}} H(|\Delta|),$$

<u>where</u> $H(n)$ <u>is the class number defined by equation</u> (20). <u>More generally, if</u> r <u>is a positive odd integer then</u> $L(r,\Delta)$ <u>and</u> $L(1-r,\Delta)$ <u>are given by equations</u> (22) <u>and</u> (23), <u>where</u> $H(r,N)$ <u>is the function defined in</u> [3].

<u>Proof</u>: i) This identity is equivalent to the main theorem of the theory of binary quadratic forms (cf. [10], Satz 203), according to which $n(a)$ is the number of $SL_2(\mathbb{Z})$ - inequivalent primitive representations of a by binary quadratic forms of discriminant Δ. We can prove it directly by arguing as for the proof of (44) or (52): Let Φ denote the set of binary quadratic forms of discriminant Δ and $X = (\mathbb{Z}^2 - \{(0,0)\})/\{\pm 1\}$. For $\phi \in \Phi$ and $\pm(m,n) \in X$ set $\phi \cdot x = \phi(n,-m)$. Then Γ acts on $\Phi \times X$ preserving the pairing $\phi \cdot x$ and we can write (6) as

$$\zeta(s,\Delta) = \sum_{\phi \in \Phi/\Gamma} \sum_{x \in X/\Gamma_\phi} (\phi \cdot x)^{-s} = \sum_{(\phi,x) \in (\Phi \times X)/\Gamma} (\phi \cdot x)^{-s}$$

$$= \sum_{x \in X/\Gamma} \sum_{\phi \in \Phi/\Gamma_x} (\phi \cdot x)^{-s}$$

where Γ_ϕ, Γ_x denote the isotropy groups of ϕ and x in Γ. The orbits of X under Γ are in 1:1 correspondence with the natural numbers, since $\pm(m,n)$ is Γ-equivalent to $\pm(0,r)$ (r = g.c.d of m and n), and the isotropy group of $\pm(0,r)$ is Γ_∞. Hence

$$\zeta(s,\Delta) = \sum_{r=1}^{\infty} \sum_{\phi \in \Phi/\Gamma_\infty} \frac{1}{\phi(r,0)^s}$$

$$= \zeta(2s) \sum_{a=1}^{\infty} \sum_{\substack{b \,(\mathrm{mod}\ 2a) \\ b^2 \equiv \Delta \,(\mathrm{mod}\ 4a)}} a^{-s}.$$

ii) This follows from iii) and the functional equations of $\zeta(s)$ and $L_D(s)$. However, we can also deduce it from Theorem 3 together with the easily-proved functional equations of $E(z,s)$ (equation (45)) and $I_k(\Delta,t;s)$ (Proposition 4, iii)), so that Theorem 3 gives as a corollary new proofs for the functional equations of the zeta functions of both real and imaginary quadratic fields.

iii) This can be deduced without difficulty from i). The details are given in [8], Prop. 2, pp. 69 - 71 (our $n(a)$ is denoted there by $r_D^*(f,a)$, where $\Delta = Df^2$).

iv) From iii) and the Dirichlet class-number formula we get

$$\begin{aligned} L(1,\Delta) &= L_D(1) \sum_{d|f} \mu(d) \left(\tfrac{D}{d}\right) d^{-1} \sigma_{-1}(f/d) \\ &= \frac{2\pi}{\sqrt{|D|}} \frac{h(D)}{w(D)} \sum_{cd|f} \mu(d) \left(\tfrac{D}{d}\right) \frac{c}{f} \\ &= \frac{2\pi}{\sqrt{|\Delta|}} \frac{h(D)}{w(D)} \sum_{e|f} e \prod_{p|e} \left(1 - \left(\tfrac{D}{p}\right) p^{-1}\right) \qquad (e=cd) \\ &= \frac{2\pi}{\sqrt{|\Delta|}} \sum_{e|f} \frac{h(De^2)}{w(De^2)} = \frac{\pi}{\sqrt{|\Delta|}} H(|\Delta|). \end{aligned}$$

The general case follows similarly from ii) and iii) and the formula given by Cohen in [3], c), p. 273.

<u>Proposition 4</u>: <u>Let</u> Δ, t <u>be real numbers with</u> $\Delta < t^2$. <u>Then</u>

i) <u>The first integral in</u> (8) <u>converges absolutely for</u> $s \in \mathbb{C}$ <u>with</u>

$$k > \operatorname{Re}(s) > \begin{cases} 1-k & \text{if } \Delta < 0 \\ 0 & \text{if } \Delta > 0 \\ 1/2 & \text{if } \Delta = 0 \end{cases}$$

<u>and is then equal to the second integral.</u>

ii) <u>For</u> $\Delta \neq 0$, <u>the second integral in</u> (8) <u>converges for</u> $s \in \mathbb{C}$ <u>with</u> $1-k < \operatorname{Re}(s) < k$. <u>The function</u> $I_k(\Delta,t;s)$ <u>which it defines has a meromorphic continuation to all</u> s <u>whose only singularities are simple poles at</u>

$s = k, k+1, k+2, \ldots$ <u>and</u> $s = -k+1, -k, -k-1, \ldots,$
<u>and which satisfies the functional equation</u>

$$I_k(\Delta,t;s) = \left(\tfrac{1}{4}|\Delta|\right)^{s-\frac{1}{2}} I_k(\Delta,t;1-s) \qquad \text{if } \Delta < 0,$$

$$\left[I_k(\Delta,t;s) + I_k(\Delta,-t;s)\right] = \cot\frac{\pi s}{2} \left(\tfrac{1}{4}\Delta\right)^{s-\frac{1}{2}} \left[I_k(\Delta,t;1-s) + I_k(\Delta,-t;1-s)\right] \quad \text{if } \Delta > 0.$$

iii) <u>For</u> $\Delta = 0,\ \pm t > 0$ <u>one has</u>

$$I_k(0,t;s) = e^{\pm\frac{i\pi}{2}(s-k)} \frac{\Gamma(\frac{1}{2})\,\Gamma(s-\frac{1}{2})\,\Gamma(k-s)}{\Gamma(k)} |t|^{-k+s}$$

iv) <u>For</u> $\Delta < 0$ <u>and</u> $0 < r < k$, <u>one has</u>

$$\left(\frac{t^2-\Delta}{4}\right)^{k-1}\left[I_k(\Delta,t;1-r) + I_k(\Delta,-t;1-r)\right] = \left(-\frac{1}{4}\right)^{\frac{k-r-1}{2}} \pi \frac{\Gamma(k-r)\Gamma(r)}{\Gamma(k)}\, p_{k,r}\!\left(t,\frac{t^2-\Delta}{4}\right)$$

<u>where</u> $p_{k,r}$ <u>is the polynomial defined by</u> (25).

v) <u>For</u> $\Delta > 0$

$$\frac{\sqrt{\Delta}}{2} I_k(\Delta,t;0) = i \operatorname{sign}(t)\, I_k(\Delta,t;1) = \frac{(-1)^{k/2}\pi}{k-1} \frac{i}{(|t| + \sqrt{\Delta})^{k-1}} .$$

<u>Proof</u>: i) The integrand $y^{k+s-2}(|z|^2 + ity - \frac{1}{4}\Delta)^{-k}$ $(z = x + iy \in H)$ has no poles in the upper half-plane H but grows on the boundary of H like

$$\begin{cases} |z|^{-k+\sigma-2} & \text{as } z \to i\infty \\ y^{k+\sigma-2} & \text{as } y \to 0 \\ |z-a|^{\sigma-2} & \text{as } z \to a \text{ if } \Delta = 4a^2 > 0 \\ y^{k+\sigma-2}(x^2+y)^{-k} & \text{as } z \to 0 \text{ if } \Delta = 0 \end{cases}$$

where $\sigma = \mathrm{Re}(s)$. The assertions about the convergence follow. The equality of the two integrals in (8), granted the convergence, is a consequence of the identity

$$(55)\qquad \int_{-\infty}^{\infty} (x^2+a)^{-\nu}\,dx = \frac{\Gamma(\nu-\frac{1}{2})\,\Gamma(\frac{1}{2})}{\Gamma(\nu)} a^{\frac{1}{2}-\nu} \qquad (a \in \mathbb{C} \smallsetminus (-\infty,0],\ \mathrm{Re}(\nu) > \tfrac{1}{2}).$$

ii) Set

$$(56)\qquad I_{k,s}(z) = \int_0^{\infty} \frac{x^{k+s-2}\,dx}{(x^2+2xz+1)^{k-1/2}} \qquad (1-k < \mathrm{Re}(s) < k,\ z \in \mathbb{C} - (-\infty,-1]) ;$$

this is related to the standard Legendre function $\mathfrak{P}_\nu^\mu(z)$ ("associated Legendre function of the first kind") by

$$I_{k,s}(z) = \frac{2^{1-k}\,\Gamma(\frac{1}{2})}{\Gamma(k-\frac{1}{2})}\,\Gamma(k-1+s)\,\Gamma(k-s)\,(z^2-1)^{-\frac{k-1}{2}}\,P_{-s}^{1-k}(s) \qquad (z \in \mathbb{C}-(-\infty,+1])$$

(cf. [5], 3.7 (33), p. 160). For $\Delta < 0$ the substitution $y = \frac{1}{2}\sqrt{|\Delta|}\,x$ in (8) gives

$$(57) \qquad I_k(\Delta,t;s) = (\tfrac{1}{4}|\Delta|)^{\frac{s-k}{2}}\,\frac{\Gamma(k-\frac{1}{2})\,\Gamma(\frac{1}{2})}{\Gamma(k)}\,I_{k,s}\left(\frac{it}{\sqrt{|\Delta|}}\right).$$

For $\Delta > 0$ we can also express $I_k(\Delta,t;s)$ in terms of $I_{k,s}(z)$. Indeed, since $t^2 > \Delta > 0$ we have $t \neq 0$. If t is positive, then the poles of the integrand in the second integral of (8) lie on the negative real axis, and by shifting the path of integration to the positive imaginary axis and substituting $y = \frac{1}{2}\sqrt{\Delta}\,x$ we obtain

$$(58a) \quad I_k(\Delta,t;s) = (\tfrac{1}{4}\Delta)^{\frac{s-k}{2}}\,e^{\frac{i\pi}{2}(s-k)}\,\frac{\Gamma(k-\frac{1}{2})\Gamma(\frac{1}{2})}{\Gamma(k)}\,I_{k,s}\left(\frac{t}{\sqrt{\Delta}}\right) \qquad (\Delta > 0,\ t > 0).$$

Similarly, if $t < 0$

$$(58b) \quad I_k(\Delta,t;s) = (\tfrac{1}{4}\Delta)^{\frac{s-k}{2}}\,e^{-\frac{i\pi}{2}(s-k)}\,\frac{\Gamma(k-\frac{1}{2})\,\Gamma(\frac{1}{2})}{\Gamma(k)}\,I_{k,s}\left(\frac{|t|}{\sqrt{\Delta}}\right) \qquad (\Delta > 0,\ t < 0).$$

The assertions about $I_k(\Delta,t;s)$ ($\Delta \neq 0$) now follow at once from the corresponding properties of $I_{k,s}(z)$: the function $I_{k,s}(z)$ satisfies the functional equation $I_{k,s}(z) = I_{k,1-s}(z)$ (as one sees by making the substitution $x \mapsto x^{-1}$ in (56)) and has a meromorphic continuation to the whole s-plane whose only singularities are simple poles of residue $-d_n(z)$ and $+d_n(z)$ at $s = k+n$ and $s = 1-k-n$, $n \geq 0$, where $d_n(z)$ is the polynomial of degree n defined by the asymptotic expansion

$$(1 + 2xz + x^2)^{-k+\frac{1}{2}} \sim \sum_{n=0}^{\infty} d_n(z)\,x^n \qquad (x \to 0).$$

iii) For $\Delta = 0$, the same argument as for $\Delta > 0$ gives

$$I_k(0,t;s) = e^{(\mathrm{sign}\ t)\cdot i\pi(s-k)/2}\,\frac{\Gamma(k-\frac{1}{2})\,\Gamma(\frac{1}{2})}{\Gamma(k)}\int_0^\infty \frac{x^{s-3/2}\,dx}{(x+|t|)^{k-\frac{1}{2}}}$$

which is equivalent to the formula given.

iv) We have to prove that

$$\left(\frac{t^2-\Delta}{4}\right)^{k-1}\int_{-\infty}^{\infty}\frac{y^{k-r-1}}{(y^2+iyt-\frac{1}{4}\Delta)^{k-3/2}}\,dy = (-\tfrac{1}{4})^{k-r-1}\,\frac{\Gamma(\frac{1}{2})\Gamma(r)\Gamma(k-r)}{\Gamma(k-\frac{1}{2})}\,p_{k,r}\left(t,\frac{t^2-\Delta}{4}\right)$$

This follows by comparing the coefficients of u^{k-r-1} in the two sides of the identity

$$\int_{-\infty}^{\infty}\left(y^2+iyt-\frac{\Delta}{4}+2iy\,\frac{t^2-\Delta}{4}u\right)^{-r-\frac{1}{2}}dy = \frac{\Gamma(r)\Gamma(\frac{1}{2})}{\Gamma(r+\frac{1}{2})}\left(\frac{t^2-\Delta}{4}\right)^{-r}\left(1+tu+\frac{t^2-\Delta}{4}u^2\right)^{-r},$$

which in turn can be proved by taking

$$a = \frac{t^2-\Delta}{4}\left(1+tu+\frac{t^2-\Delta}{4}u^2\right),\qquad \nu = r+\frac{1}{2}$$

in (55) and making the substitution $x = y+i(\frac{t}{2}+\frac{t^2-\Delta}{4}u)$.

v) These formulas (which are equivalent to one another by virtue of (58) and the functional equation $I_{k,s}(z) = I_{k,1-s}(z)$) follow from the identities

$$\int_0^{\infty}\frac{dy}{(y^2+ity-\frac{1}{4}\Delta)^{3/2}} = \frac{4}{t^2-\Delta}\,\frac{y-\frac{1}{2}it}{\sqrt{y^2+ity-\frac{1}{4}\Delta}}\Bigg|_0^{\infty} = -\frac{4}{\sqrt{\Delta}}\,\frac{1}{\sqrt{\Delta}+|t|}$$

$$\int_0^{\infty}\frac{y\,dy}{(y^2+ity-\frac{1}{4}\Delta)^{3/2}} = \frac{4}{t^2-\Delta}\,\frac{-\frac{1}{2}ity+\frac{1}{4}\Delta}{\sqrt{y^2+ity-\frac{1}{4}\Delta}}\Bigg|_0^{\infty} = \frac{-2i\,\mathrm{sign}\,(t)}{\sqrt{\Delta}+|t|}$$

by differentiating $k-2$ times with respect to t. This completes the proof of Proposition 4.

We can now prove the various assertions made in § 1 about special values of the series $c_m(s)$ defined in (9). Consider first $s = 1$. The contribution of the (finitely many) terms in (9) with $t^2 < 4m$ can be calculated from equations (17) and (19), which are special cases of Prop. 4, iv), and Prop. 3, iii), respectively. The contribution of the (finitely many) terms with $t^2 - 4m$ a non-zero square is given by (21), which is a consequence of Prop. 3, iii) and Prop. 4, ii) and v). The contribution of the two terms with $t^2 - 4m = 0$ when m is a square can be calculated from equation (13) (which follows from Prop. 4, iii)) and the equation $L(s, 0) = \zeta(2s-1)$ (Prop. 3, iii)). Finally, the (infinitely many) terms with $t^2 - 4m$ a positive non-square in (9) give 0 for $s = 1$ because $L(1, t^2 - 4m)$ is finite and $I_k(t^2-4m,t;1) + I_k(t^2-4m,-t;1)$ vanishes (by virtue of the functional equation, Prop.4,ii)). Putting all of this into the formula for $c_m(1)$ we obtain from (16) the Eichler-Selberg trace formula.

For $s = r \in \{3,5,7,\ldots, k-3\}$ the calculation is even easier, since the terms in (9) with $t^2 - 4m = u^2 > 0$ now give no contribution (the factor $I_k(t^2-4m,t;s) + I_k(t^2-4m,-t;s)$ is again 0 because of the functional equation, but $L(s, t^2-4m)$ is now finite). From equations (9), (22) (= Prop. 3, iv)) and (26) (which is a consequence of ii) and iv) of Prop. 4) we obtain

$$c_m(r) = C_k \cdot (-\tfrac{1}{4})^{\frac{r+1}{2}} \frac{\Gamma(k-r)\Gamma(r)}{\Gamma(k-1)} (-1)^{\frac{r-1}{2}} \frac{2^{r-1}\pi^r}{\Gamma(r)} \sum_{t^2 < 4m} p_{k,r}(t,m)\, H(r,4m-t^2)$$

$$(59) \qquad + \begin{cases} (-1)^{\frac{k}{2}+1} \dfrac{(k+r-2)!\,\pi^{r+1} u^{k-r-1}}{2^{k-2}(k-1)!\,(2r-1)!} \zeta(1-2r) & \text{if } m = u^2 \\ 0 & \text{if } m \neq \text{square} \end{cases}$$

$$= -\tfrac{1}{4} C_k \frac{\Gamma(k-r)}{\Gamma(k-1)} \pi^r \sum_{t^2 \leq 4m} p_{k,r}(t,m)\, H(r,4m-t^2)$$

or (with the notations of (10) and (24))

$$\Phi_r(z) = -\tfrac{1}{4} C_k \frac{\Gamma(k-r)}{\Gamma(k-1)} \pi^r C_{k,r}(z) \qquad (r = 3,5,\ldots,k-3)$$

This together with Theorem 1 shows that $C_{k,r}$ is a cusp form of weight k whose Petersson product with an arbitrary Hecke eigenform f is given by (28).

For $s = r = k - 1$ the same calculation shows that the value of the series (9) is given by equation (59), but the function (10) is no longer a modular form since we have left the region of convergence. On the other hand, it follows from ii) and iii) of Theorem 1 that the function

$$\tilde{\Phi}_{k-1}(z) = \lim_{s\to k-1} \Phi_s(z)$$

is a cusp form of weight k satisfying

$$(60) \qquad (\tilde{\Phi}_{k-1}, f) = C_k \frac{\Gamma(2k-2)}{(4\pi)^{2k-2}} D_f(2k-2)$$

for each Hecke eigenform $f \in S_k$. We want to show that the mth Fourier coefficient $\tilde{c}_m(k-1)$ of $\tilde{\Phi}_{k-1}$ is given by equation (27). Each term of the series (9) is continuous at $s = k-1$, and each term with $|t| > 2\sqrt{m}$ has the limit 0 as $s \to k-1$. Therefore

$$(61) \quad \tilde{c}_m(k-1) = c_m(k-1) + \lim_{\varepsilon\to 0}\left(m^{k-1} \sum_{|t|>2\sqrt{m}} \left[I_k(t^2-4m,t;k-1-\varepsilon)+I_k(t^2-4m,-t;k-1-\varepsilon)\right] \times L(k-1-\varepsilon,t^2-4m)\right).$$

By (59) the first term on the right is equal to the first term on the right-hand side of (27). From (58) we find

$$I_k(t^2-4m,t;k-1-\varepsilon) + I_k(t^2-4m,-t;k-1-\varepsilon) =$$

$$= 2\cos\frac{\pi(1+\varepsilon)}{2}\cdot\left(\frac{t^2-4m}{4}\right)^{-\frac{1+\varepsilon}{2}} \frac{\Gamma(k-\frac{1}{2})\Gamma(\frac{1}{2})}{\Gamma(k)} I_{k,k-1-\varepsilon}\left(\frac{|t|}{\sqrt{t^2-4m}}\right)$$

$$= -2\pi \frac{\Gamma(k-\frac{1}{2})\Gamma(\frac{1}{2})}{\Gamma(k)} I_{k,k-1}(1) \quad \varepsilon t^{-1-\varepsilon}(1+O(\varepsilon) + O(t^{-1})),$$

with

$$I_{k,k-1}(1) = \int_0^\infty \frac{x^{2k-3}dx}{(x^2+2x+1)^k} = \int_0^1 u^{2k-3}\,du = \frac{1}{2k-2} \qquad \left(u = \frac{x}{x+1}\right).$$

Also $L(t^2-4m,k-1-\varepsilon) = L(t^2-4m, k-1) + O(\varepsilon)$, with both terms uniformly bounded in t. Therefore the second term in (61) equals

$$(62) \qquad -\frac{\pi}{k-1}\frac{\Gamma(k-\frac{1}{2})\Gamma(\frac{1}{2})}{\Gamma(k)} m^{k-1} \lim_{\varepsilon\to 0}\left(\varepsilon \sum_{|t|>2\sqrt{m}} \frac{1}{t^{1+\varepsilon}} L(k-1,\ t^2-4m)\right).$$

On the other hand, Prop. 3)i) gives

$$L(k-1,t^2-4m) = \frac{\zeta(2k-2)}{\zeta(k-1)} \sum_{a=1}^{\infty} \frac{1}{a^{k-1}} \#\{b \pmod{2a} \mid b^2 \equiv t^2-4m \pmod{4a}\}$$

$$= \frac{\zeta(2k-2)}{\zeta(k-1)} \sum_{a=1}^{\infty} \frac{1}{a^{k-1}} \#\{d \pmod{a} \mid d(t-d) \equiv m \pmod{a}\},$$

where in the last line we have set $d = \frac{t-b}{2}$. The condition $d(t-d) \equiv m \pmod{a}$ depends only on the residue class of $t \pmod{a}$, and for a fixed residue class $t_o \pmod{a}$ one has

$$\lim_{\varepsilon\to 0} \left(\varepsilon \sum_{\substack{|t|>2\sqrt{m} \\ t\equiv t_o \pmod{a}}} \frac{1}{t^{1+\varepsilon}} \right) = \frac{2}{a} \lim_{\varepsilon\to 0} \varepsilon\zeta(1+\varepsilon) = \frac{2}{a},$$

so (62) equals

$$(63) \qquad -\frac{2\pi}{k-1} \frac{\Gamma(k-\frac{1}{2})\Gamma(\frac{1}{2})}{\Gamma(k)} \frac{\zeta(2k-2)}{\zeta(k-1)} m^{k-1} \sum_{a=1}^{\infty} \frac{\rho(a)}{a^k},$$

where $\rho(a) = \#\{d,d' \pmod{a} \mid dd' \equiv m \pmod{a}\}$. The function $a \mapsto \rho(a)$ is multiplicative and for a prime power $a = p^\nu$ is given by $\rho(p^\nu) = (p-1)p^{\nu-1}$ if $p \nmid m$ and by

$$\rho(p^\nu) = \begin{cases} (\nu+1)\, p^\nu - \nu p^{\nu-1} & \text{if } 0 \leqslant \nu \leqslant \mu \\ (\mu+1)\,(p^\nu - p^{\nu-1}) & \text{if } \nu > \mu \end{cases}$$

in general, where p^μ is the largest power of p dividing m. Hence

$$\sum_{a=1}^{\infty} \frac{\rho(a)}{a^k} = \prod_{p\nmid m} (1+ \frac{p-1}{p^k} + \frac{(p-1)p}{p^{2k}} + \dots) \prod_{\substack{p^\mu \| m \\ \mu\geqslant 1}} (1 + \frac{2p-1}{p^k} + \frac{3p^2-2p}{p^{2k}} + \dots + \frac{(\mu+1)p^\mu - \mu p^{\mu-1}}{p^{\mu k}}$$

$$+ \frac{(\mu+1)p^{\mu+1} - (\mu+1)p^\mu}{p^{(\mu+1)k}} + \dots)$$

$$= \prod_p \frac{1-p^{-k}}{1-p^{1-k}} \cdot \prod_{\substack{p^\mu \| m \\ \mu\geqslant 1}} (1 + p^{1-k} + p^{2(1-k)} + \dots + p^{\mu(1-k)})$$

$$= \frac{\zeta(k-1)}{\zeta(k)} \sigma_{1-k}(m),$$

and substituting this into (63) we obtain the second term in equation (27). Therefore

$$\tilde{\Phi}_{k-1}(z) = \sum_{m=1}^{\infty} \tilde{c}_m(k-1)e^{2\pi imz} = \frac{(-1)^{k/2+1}\pi}{2^{k-1}(k-1)!} C_{k,k-1}(z)$$

$$(64) \qquad - \frac{(-1)^{k/2}(2\pi)^{1-k}}{k-1} \Gamma(k-\tfrac{1}{2})\Gamma(\tfrac{1}{2})\ \zeta(2k-2)\ E_k(z)$$

$$= \frac{(-1)^{k/2+1}\pi^k}{2^{k-1}(k-1)!} \left[C_{k,k-1}(z) - \zeta(3-2k)\ E_k(z) \right],$$

where $C_{k,k-1}(z)$ is defined by (24) and

$$E_k(z) = 1 + \frac{(-1)^{k/2}(2\pi)^k}{\Gamma(k)\zeta(k)} \sum_{m=1}^{\infty} \sigma_{k-1}(m)\ e^{2\pi imz}$$

is the normalized Eisenstein series of weight k (the formula $H(k-1,0) = \zeta(3-2k)$ implies that the constant term on the right-hand side of (64) is zero). Equations (60) and (64) and the fact that E_k is orthogonal to all cusp forms imply that equation (28) holds even in the case $r = k - 1$, when $C_{k,r}$ is not a cusp form.

Finally, we should say something about the case $k = 2$. Up to now we have excluded this case because it presents the most awkward convergence questions and because there are no cusp forms of weight 2 on $SL_2(\mathbb{Z})$ anyway. However, the case $k = 2$ is also important, both for the generalization of Theorems 1 and 2 to congruence subgroups and for the applications to Hilbert modular forms given in § 6. For $k > 2$ the interesting range of values for s was $1 < Re(s) < k - 1$, and the two extreme values $s = 1$ and $s = k - 1$ created extra terms (and extra difficulties) as given by formulas (21) and (27). For $k = 2$, the only interesting value is $s = 1$, and one has all of the convergence difficulties which occured previously for $s = 1$ and for $s = k - 1$, and some new difficulties due to the fact that the series expression (35) for the kernel function $\omega_m(z,z')$ is no longer absolutely convergent, so that at first sight the whole method of proof appears to break down. To get around this, one must define ω_m as $\lim_{\varepsilon\to 0} \omega_{m,\varepsilon}$, where

$$\omega_{m,\varepsilon}(z,z') = \sum_{ad-bc=m} (czz'+dz'+az+b)^{-k} \, |czz'+dz'+az+b|^{-\varepsilon}$$

("Hecke's trick"). As in Appendix 2 of [24], one can show that ω_m is a cusp form of weight 2 with the properties given by Proposition 1 (of course for $SL_2(\mathbb{Z})$ this simply means $\omega_m = 0$). Then one carries out the whole calculation of §§ 2 - 4 with $\omega_{m,\varepsilon}$ instead of ω_m, taking in Theorem 1 a value of s with $1 < \mathrm{Re}(s) < 1 + \varepsilon$, and at the end lets ε tend to 0. I omit the calculation, which is awful. The result is as simple as one could hope: for $k = 2$ and $s = 1$ the mth Fourier coefficient $\tilde{c}_m(1)$ of the cusp form $\tilde{\Phi}_1$ defined by (34) is given by the sum of the expression previously obtained for $k > 2$, $s = 1$ (i.e. for the trace formula) and of the extra contribution previously obtained for $k > 2$, $s = k - 1$ (second term of (27)), i.e.

$$\tilde{c}_m(1) = \frac{\pi^2}{2} \left\{ \sum_{t^2 \leqslant 4m} H(4m-t^2) + \sum_{\substack{dd'=m \\ d,d'>0}} \min(d,d') - 2\sigma_1(m) \right\}.$$

Since $\tilde{\Phi}_1(z)$ is a cusp form of weight 2 on $SL_2(\mathbb{Z})$ all coefficients must be zero and we obtain the class number relation

$$\sum_{t^2 \leqslant 4m} H(4m-t^2) = \sum_{\substack{dd'=m \\ d>0}} \max\,(d,d')$$

due to Hurwitz [9]. If, however $\Gamma' \subset \Gamma$ is a congruence subgroup for which there are cusp forms of weight 2, then we obtain an expression for the trace of the Hecke operator $T(m)$ on $S_2(\Gamma')$.

§ 5. The series $\sum n^{k-1-s} G_{n^2}(z)$ and the convolution of L-series associated to modular forms

Let s be a complex number with $\mathrm{Re}(s) > 1$ and $\tilde{\Phi}_s$ the unique cusp form in S_k satisfying (34) for all normalized Hecke eigenforms $f \in S_k$. Our starting point for the calculation of the Fourier coefficients of $\tilde{\Phi}_s$ in §§ 2 - 3 was the identity (3) expressing $D_f(s)$ in terms of the Rankin zeta-function $\sum_{n=1}^{\infty} a(n)^2 n^{-s}$ (where $f = \sum a(n) q^n$). But $D_f(s)$ also satisfies the identity

$$D_f(s) = \zeta(2s-2k+2) \sum_{n=1}^{\infty} a(n^2) n^{-s},$$

as is well-known and easily verified using the multiplicative properties of the $a(n)$. Thus the equation defining $\tilde{\Phi}_s$ is equivalent to

$$(65) \qquad (\tilde{\Phi}_s, f) = C_k \frac{\Gamma(s+k-1)}{(4\pi)^{s+k-1}} \zeta(2s) \sum_{n=1}^{\infty} a(n^2) n^{-s-k+1}$$

and since this equation is linear in the coefficients $a(n)$, it must hold for all cusp forms $f = \sum a(n) q^n \in S_k$, not just for eigenforms. Equation (65) determines $\tilde{\Phi}_s$ uniquely, and by comparing it with equation (41) we obtain the identity

$$(66) \qquad \tilde{\Phi}_s(z) = C_k \frac{\Gamma(s+k-1)}{(4\pi)^s \, \Gamma(k-1)} \zeta(2s) \sum_{n=1}^{\infty} n^{k-1-s} G_{n^2}(z)$$

expressing $\tilde{\Phi}_s$ as an infinite linear combination of Poincaré series.

It is now natural to ask whether one can obtain a proof of Theorem 1 (which states that $\tilde{\Phi}_s = \Phi_s$ for $\mathrm{Re}(s) < k - 1$, where Φ_s is defined by (9) and (10)) by combining (66) with known facts about Poincaré series. Two methods suggest themselves:

1. One can substitute into (66) the formula for the mth Fourier coefficient g_{rm} of $G_r(z)$, namely

$$g_{rm} = \delta_{rm} + 2\pi(-1)^{k/2}(m/r)^{\frac{k-1}{2}} \sum_{c=1}^{\infty} H_c(r,m)\, J_{k-1}\left(\frac{4\pi}{c}\sqrt{rm}\right)$$

(where δ_{rm} is the Kronecker delta, J_{k-1} a Bessel function, and

$$H_c(r,m) = \frac{1}{c} \sum_{\substack{a,d \pmod c \\ ad\equiv 1 \pmod c}} e^{2\pi i(ra+md)/c}$$

a Kloosterman sum), and try to show directly that the sum equals $c_m(s)$. I do not know whether this can be done, but it is amusing to note that the term δ_{rm} in the formula for g_{rm} produces in (66) exactly the extra contribution to $c_m(s)$ occurring in (9) when m is a square.

2. One can substitute into (66) the defining equation (40) of the Poincaré series and interchange the order of summation to obtain

$$(67) \qquad \tilde{\Phi}_s(z) = C_k \frac{\Gamma(s+k-1)}{(4\pi)^s\, \Gamma(k-1)} \zeta(2s) \sum_{\gamma\in\Gamma_\infty\backslash\Gamma} j_k(\gamma,z) \sum_{n=1}^{\infty} n^{k-1-s} e^{2\pi i n^2 \gamma z}$$

(this is certainly legitimate for $\mathrm{Re}(s) > 2$, since the double series is absolutely convergent in that region). Again, I have not been able to deduce from this that $\tilde{\Phi}_s = \Phi_s$ in general. But if $k - 1 - s$ is a non-negative even integer, then the series $\sum n^{k-1-s} e^{2\pi i n^2 z}$ occurring in (67) is (up to a factor) a derivative of the theta-series

$$\Theta(z) = \sum_{n=1}^{\infty} e^{2\pi i n^2 z}.$$

and therefore transforms nicely under the action of the modular group, and in this case it <u>is</u> possible to deduce from (67) the expression for $c_m(s)$ as a finite sum of values of zeta-functions, thus obtaining a different (and conceptually simpler) proof of Theorem 2 and of the identities for special values of $D_f(s)$ discussed in § 1. To present the idea as clearly as possible, we begin with the special case $s = k - 1$.

For $r = k - 1$, the modular form (24) figuring in Theorem 2 is given by

$$(68)\qquad C_{k,k-1}(z) = \sum_{m=0}^{\infty}\left(\sum_{\substack{t\in\mathbb{Z}\\ t^2\leqslant 4m}} H(k-1,\, t^2-4m)\right) e^{2\pi imz} = (\Theta\mathcal{H}_{k-1})|U_4,$$

where $\mathcal{H}_r$ $(r \geqslant 1)$ is defined by

$$\mathcal{H}_r(z) = \sum_{N=0}^{\infty} H(r,N)\, q^N \qquad (q = e^{2\pi iz})$$

and U_4 is the operator which sends $\sum a(n)\, q^n$ to $\sum a(4n)\, q^n$. In [3], Cohen proved that $\mathcal{H}_r$ is a modular form of weight $r + \frac{1}{2}$, namely

$$(69)\qquad \mathcal{H}_r(z) = \zeta(1-2r)\left[E^{(4)}_{r+\frac{1}{2}}(z) + (1-i)(4z)^{-r-\frac{1}{2}}\, E^{(4)}_{r+\frac{1}{2}}\left(\frac{-1}{4z}\right)\right],$$

where

$$E^{(4)}_{r+\frac{1}{2}}(z) = \sum_{\left(\begin{smallmatrix} a & b\\ c & d\end{smallmatrix}\right)\in\Gamma_\infty\backslash\Gamma_o(4)} \frac{\left(\frac{c}{d}\right)\left(\frac{-4}{d}\right)^{-1/2}}{(cz+d)^{r+1/2}}$$

is the Eisenstein series of weight $r + \frac{1}{2}$ on $\Gamma_o(4)$ (for conventions concerning modular forms of half-integral weight, see [20]). It follows that $\mathcal{H}_{k-1}(z)\,\Theta(z)$ is a modular form of weight k on $\Gamma_o(4)$ having the property that its mth Fourier coefficient is 0 for all $m \equiv 2 \pmod 4$, and since one easily shows (directly or using results in [13]), that U_4 maps all forms on $\Gamma_o(4)$ with this property to forms on the full modular group, one obtains $C_{k,k-1} \in M_k(SL_2(\mathbb{Z}))$. We want to show how (67) implies that the cusp form $C_{k,k-1} - \zeta(3-2k)\, E_k$ is a multiple of $\widetilde{\Phi}_{k-1}$.

Equation (67) for $s = k - 1$ can be written

$$\widetilde{\Phi}_{k-1}(z) = C_k \frac{\Gamma(2k-2)}{(4\pi)^{k-1}\,\Gamma(k-1)}\, \zeta(2k-2) \sum_{\gamma\in\Gamma_\infty\backslash\Gamma} j_k(\gamma,z)\left(\tfrac{1}{2}\Theta(\gamma z) - \tfrac{1}{2}\right)$$

$$= \frac{(-1)^{k/2}\pi^k}{2^{k-1}\,\Gamma(k)}\, \zeta(3-2k)\left[E_k(z) - \sum_{\gamma\in\Gamma_\infty\backslash\Gamma} j_k(\gamma,z)\,\Theta(\gamma,z)\right]$$

$$= \frac{(-1)^{k/2}\pi^k}{2^{k-1}\,\Gamma(k)}\, \zeta(3-2k)\left[E_k(z) - \mathrm{Tr}_1^4\left(\sum_{\gamma\in\Gamma_\infty/\Gamma_o(4)} j_k(\gamma,z)\Theta(\gamma z)\right)\right].$$

where $\mathrm{Tr}_1^4 : M_k(\Gamma_o(4)) \to M_k(SL_2(\mathbb{Z}))$ is the map defined by

$$\mathrm{Tr}_1^4(f) = \sum_{\gamma\in\Gamma_o(4)\backslash\Gamma} f|_k\gamma = \sum_{n \pmod 4} f|_k\begin{pmatrix}1&0\\n&1\end{pmatrix} + \sum_{n \pmod 2} f|_k\begin{pmatrix}0&-1\\1&2n\end{pmatrix}$$

(cf. [13]). Also, for $\gamma = \begin{pmatrix}a&b\\c&d\end{pmatrix} \in \Gamma_o(4)$,

$$\Theta(\gamma z) = \left(\frac{c}{d}\right)\left(\frac{-4}{d}\right)^{-1/2} (cz+d)^{1/2}\ \Theta(z),$$

so

$$\sum_{\gamma\in\Gamma_\infty\backslash\Gamma_o(4)} j_k(\gamma,z)\Theta(\gamma z) = \Theta(z)\ E^{(4)}_{k-\frac{1}{2}}(z).$$

Therefore

$$\mathrm{Tr}_1^4\ (\Theta E^{(4)}_{k-\frac{1}{2}}) = E_k + (-1)^{\frac{k}{2}+1} \frac{2^{k-1}\ \Gamma(k)}{\pi^k\zeta(3-2k)}\ \tilde{\Phi}_{k-1}.$$

Equation (64) can now be obtained from this by using equations (68) and (69) and the explicit description of the way the series Θ and $E^{(4)}_{k-1/2}$ transform under the operation of $\begin{pmatrix}0&-1\\4&0\end{pmatrix}$ and of the matrices involved in the definitions of Tr_1^4 and U_4. We omit the details.

We observe that the argument used here for $\sum G_{n^2}(z)$ would apply to any series $\sum b(n)\ G_n(z)$, where the $b(n)$ are the Fourier coefficients of a modular form (here $\Theta(z)$). Since this principle is not very well known (although it was already used by Rankin in 1952), we give a general formulation of it, applicable also to forms of non-integral weight.

Let $\Gamma' \subset \Gamma$ be a congruence subgroup and $k > 0$ a real number. We consider multiplier systems $v : \Gamma' \to \{t \in \mathbb{C}\ |\ |t| = 1\}$ such that the automorphy factor

$$J(\gamma,z) = v(\gamma)^{-1}(cz+d)^{-k} \quad (\gamma = \begin{pmatrix}a&b\\c&d\end{pmatrix}\in\Gamma',\ \ z\in H)$$

satisfies the cocycle condition

$$J(\gamma_1\gamma_2,z) = J(\gamma_1,\gamma_2 z)\ J(\gamma_2,\ z) \qquad (\gamma_1,\gamma_2 \in \Gamma',\ z\in H)$$

and such that $v(\gamma) = 1$ for $\gamma \in \Gamma'_\infty = \Gamma' \cap \Gamma_\infty$ (then $J(\gamma,z)$ depends only on the coset of γ in $\Gamma'_\infty\backslash\Gamma'$, i.e. on the second row of γ). We write $M_k(\Gamma',v)$

$(S_k(\Gamma',v))$ for the spaces of modular forms (cusp forms) on Γ' which transform by

$$f(z) = J(\gamma,z)\, f(\gamma z) \qquad (\gamma \in \Gamma',\ z \in H).$$

(If $k \in \mathbb{Z}$, then v is a character on Γ' and this agrees with the usual notation; if $k \in \mathbb{Z} + \frac{1}{2}$, then our notation conflicts with that of [21] but has the advantage that the product of forms in $M_{k_1}(\Gamma',v_1)$ and $M_{k_2}(\Gamma',v_2)$ lies in $M_{k_1+k_2}(\Gamma',v_1v_2)$.) If $k > 2$, we have the Eisenstein series

$$E_k'(z) = \sum_{\gamma\in\Gamma_\infty'\backslash\Gamma'} J(\gamma,z) \in M_k(\Gamma',v)$$

and for each natural number n the Poincaré series

$$G_n'(z) = \sum_{\gamma\in\Gamma_\infty'\backslash\Gamma'} J(\gamma,z)\, e^{2\pi i n\gamma z/w} \in S_k(\Gamma',v),$$

where $w = [\Gamma_\infty : \Gamma_\infty']$ is the width of Γ'; the same proof as for (41) shows that

$$(70) \qquad (f,G_n') = \int_{\Gamma'\backslash H} f(z)\overline{G_n'(z)}\, y^k\, dV = \frac{\Gamma(k-1)w^k}{(4\pi n)^{k-1}}\, a(n)$$

for any form $f = \sum_{n=0}^{\infty} a(n)e^{2\pi i n z/w}$ in $M_k(\Gamma',v)$. With these notations we have:

<u>Proposition 5</u>: <u>Let</u> $J_i(\gamma,z) = v_i(\gamma)^{-1}(cz+d)^{-k_i}$ $(i = 1,2)$ <u>be two automorphy factors on</u> Γ', <u>where</u> k_1,k_2 <u>are real numbers with</u> $k_2 \geq k_1 + 2 > 2$. <u>Let</u> $f(z) = \sum_{n=1}^{\infty} a(n)\, e^{2\pi i n z/w}$ <u>and</u> $g(z) = \sum_{n=0}^{\infty} b(n)\, e^{2\pi i n z/w}$ <u>be modular forms in</u> $S_{k_1+k_2}(\Gamma',v_1v_2)$ <u>and</u> $M_{k_1}(\Gamma',v_1)$, <u>respectively, and</u> $E_{k_2}'(z)$ <u>the Eisenstein series in</u> $M_{k_2}(\Gamma',v_2)$ <u>as defined above. Then the Petersson product of</u> $g(z)\, E_{k_2}'(z)$ <u>and</u> $f(z)$ <u>is given by</u>

$$(71) \qquad (f,gE_{k_2}') = \frac{\Gamma(k_1+k_2-1)}{(4\pi)^{k_1+k_2-1}}\, w^{k_1+k_2} \sum_{n=1}^{\infty} \frac{a(n)b(n)}{n^{k_1+k_2-1}}\ .$$

<u>Proof</u>:
Set $k = k_1+k_2$, $v = v_1v_2$, $J(\gamma,z) = v(\gamma)(cz+d)^{-k} = J_1(\gamma,z)\, J_2(\gamma,z)$.

If $k_2 > k_1 + 2$, then

$$\begin{aligned} g(z)\,E'_{k_2}(z) &= \sum_{\gamma\in\Gamma'_\infty\backslash\Gamma'} J_2(\gamma,z)\,g(z) \\ &= \sum_{\gamma\in\Gamma'_\infty\backslash\Gamma'} J_2(\gamma,z)\,J_1(\gamma,z)\,g(\gamma z) \\ &= \sum_{\gamma\in\Gamma'_\infty\backslash\Gamma'} \sum_{n=0}^{\infty} b(n)\,J(\gamma,z)\,e^{2\pi i n\gamma z/w} \\ &= b(0)\,E'_k(z) + \sum_{n=1}^{\infty} b(n)\,G'_n(z), \end{aligned}$$

because the double series is absolutely convergent; if $k_2 = k_1 + 2$, we obtain the same equation by multiplying $J_2(\gamma,z)$ by $y^\varepsilon\,|cz+d|^{-2\varepsilon}$ and letting $\varepsilon \to 0$ (Hecke's trick). Equation (70) now implies the statement of the proposition. (The series in (71) converges for $k_2 > 2$ because $a(n) = O(n^{k_2/2})$ and $b(n) = O(n^{k_1-1})$.)

The method we have just described was used by Rankin (for forms on the full modular group) in [18]; his identity (33) is obtained by taking $k_1 = q$, $k_2 = k - q$, $g(z) = E_q(z) = 1 - \frac{2q}{B_q}\sum_{n=1}^{\infty}\sigma_{q-1}(n)\,e^{2\pi i n z}$ and $f = \sum a(n)\,e^{2\pi i n z} \in S_k$ an eigenform and using the identity

$$(72)\qquad \sum_{n=1}^{\infty}\frac{\sigma_r(n)\,a(n)}{n^s} = \frac{L_f(s)\,L_f(s-r)}{\zeta(2s-r-k+1)} \qquad (\mathrm{Re}(s) > r + \tfrac{k+1}{2}).$$

We remark that equation (71) is in fact true under weaker restrictions than those given. For example, Rankin's identity (33) is still valid for $q=\frac{k}{2}$ (the reader can check this for $k=12$, $f=\Delta$, using (32)). It is also worth remarking that the identity (33), together with the non-vanishing of $L_f(s)$ in the region of absolute convergence and the fact that the Hecke algebra acts on M_k with multiplicity 1, imply that for $q \geq \frac{k}{2} + 1$ the modular form $E_q(z)\,E_{k-q}(z)$ generates M_k as a module over the Hecke algebra. I do not know any elementary proof of this fact; a direct proof in the case $q = \frac{k}{2}$ would imply the non-vanishing of $L_f(k/2)$.

We now prove a generalization of Prop. 5 which can be used to give another proof of Theorem 2 (i.e. of the proportionality of $\tilde{\Phi}_r$ and $C_{k,r}$) and hence of the identities for $D_f(r+k-1)$, where r is an odd number satisfying

$1 < r < k-1$. To prove that $C_{k,r}$ is a modular form when $r < k-1$, Cohen defined bilinear operators $F_\nu = F_\nu^{k_1,k_2}$ ($\nu \in \mathbb{N}$, $k_1,k_2 \in \mathbb{R}$) on smooth functions by the formula

$$(73)\qquad F_\nu(f_1,f_2)(z) = \sum_{\mu=0}^{\nu} (-1)^{\nu-\mu} \binom{\nu}{\mu} \frac{\Gamma(k_1+\nu)\,\Gamma(k_2+\nu)}{\Gamma(k_1+\mu)\,\Gamma(k_2+\nu-\mu)} \frac{\partial^\mu f_1}{\partial z^\mu} \frac{\partial^{\nu-\mu} f_2}{\partial z^{\nu-\mu}}$$

and showed ([3], Theorem 7.1) that

$$(74)\qquad F_\nu(f_1|_{k_1}\gamma,\ f_2|_{k_2}\gamma) = F_\nu(f_1,f_2)|_{k_1+k_2+2\nu}\,\gamma$$

for all $\gamma \in GL_2^+(\mathbb{R})$. From this it follows that if f_1 and f_2 are modular forms on some group Γ', with weights k_1 and k_2 and multiplier systems v_1 and v_2, respectively, then $F_\nu(f_1,f_2)$ is a modular form on Γ' of weight $k_1+k_2+2\nu$ and with multiplier system v_1v_2 and is a cusp form if $\nu > 0$. (Of course $F_0(f_1,f_2) = f_1f_2$.) The fact that $C_{k,r}$ $(r < k-1)$ is a cusp form of weight k on $SL_2(\mathbb{Z})$ then follows by the same argument as in the case $r = k-1$ from the identity

$$(75)\qquad C_{k,r} = (2\pi i)^{-\nu} \frac{\Gamma(\nu+r)}{\Gamma(r)\,\Gamma(k-r)} F_\nu(\Theta, \mathcal{H}_r)|U_4 \qquad \left(\nu = \frac{k-r-1}{2}\right)$$

We now give a formula for $F_\nu(f_1,f_2)$ when either F_1 or F_2 is an Eisenstein series; this proposition in conjunction with (69) and (75) can be used to give another proof of the identity

$$\tilde{\Phi}_r(z) = -\frac{1}{4} C_k \frac{\Gamma(k-r)}{\Gamma(k-1)} \pi^r\, C_{k,r}(z) \qquad (r = 3,5,\ldots,k-3),$$

which is equivalent to Theorem 2.

<u>Proposition 6</u>: <u>Let</u> k_1, k_2 J_1, J_2, g <u>and</u> E_{k_2} <u>be as in Proposition</u> 5, ν <u>a non-negative integer, and</u> $f(z) = \sum_{n=1}^{\infty} a(n)\, e^{2\pi i n z/w}$ <u>a cusp form in</u> $S_k(\Gamma',v)$, <u>where</u> $k = k_1+k_2+2\nu$ <u>and</u> $v = v_1v_2$. <u>Define</u> $F_\nu(g, E'_{k_2})$ <u>as in</u> (73). <u>Then</u>

$$(f, F_\nu(g,E'_{k_2})) = (2\pi i)^\nu \frac{\Gamma(k-1)\,\Gamma(k_2+\gamma)}{(4\pi)^{k-1}\,\Gamma(k_2)} w^{k-\nu} \sum_{n=1}^{\infty} \frac{a(n)\,\overline{b(n)}}{n^{k_1+k_2+\nu-1}}$$

<u>Proof</u>: Let $g^{(\mu)}(z) = \frac{\partial^\mu g(z)}{\partial z^\mu}$. Then for $\gamma = \begin{pmatrix} a & b \\ c & d \end{pmatrix} \in \Gamma'$

$$(76)\quad g^{(\nu)}(\gamma z) = v_1(\gamma) \sum_{\mu=0}^{\nu} \binom{\nu}{\mu} \frac{\Gamma(k_1+\nu)}{\Gamma(k_1+\mu)} c^{\nu-\mu} (cz+d)^{k_1+\nu+\mu} g^{(\mu)}(z).$$

This can be proved by induction on ν (the case $\nu = 0$ is just the transformation law of g), using the identities

$$g^{(\nu+1)}(\gamma z) = (cz+d)^2 \frac{d}{dz} g^{(\nu)}(\gamma z)$$

and

$$\binom{\nu}{\mu} \frac{\Gamma(k_1+\nu)}{\Gamma(k_1+\mu)} (k_1+\nu+\mu) + \binom{\nu}{\mu-1} \frac{\Gamma(k_1+\nu)}{\Gamma(k_1+\mu-1)} = \binom{\nu+1}{\mu} \frac{\Gamma(k_1+\nu+1)}{\Gamma(k_1+\mu)};$$

we leave the verification to the reader. Let $G_n'(z)$ be the Poincaré series in $S_k(\Gamma', v)$. Using the Fourier expansion

$$g^{(\nu)}(z) = \left(\frac{2\pi i}{w}\right)^\nu \sum_{n=0}^{\infty} n^\nu b(n)\, e^{2\pi i n z/w}$$

and (76) we obtain

$$\begin{aligned}\left(\frac{2\pi i}{w}\right)^\nu \sum_{n=0}^{\infty} n^\nu b(n)\, G_n'(z) &= \sum_{\gamma\in\Gamma_\infty'\backslash\Gamma'} v(\gamma)^{-1} (cz+d)^{-k} g^{(\nu)}(\gamma z)\\
&= \sum_{\mu=0}^{\nu} \binom{\nu}{\mu} \frac{\Gamma(k_1+\nu)}{\Gamma(k_1+\mu)} g^{(\mu)}(z) \sum_{\gamma\in\Gamma_\infty'\backslash\Gamma'} v_2(\gamma)^{-1} c^{\nu-\mu} (cz+d)^{-k_2-\nu+\mu}\\
&= \sum_{\mu=0}^{\nu} \binom{\nu}{\mu} \frac{\Gamma(k_1+\nu)}{\Gamma(k_1+\mu)} g^{(\mu)}(z) \cdot (-1)^{\nu-\mu} \frac{\Gamma(k_2)}{\Gamma(k_2+\nu-\mu)} E_{k_2}'^{(\nu-\mu)}(z)\\
&= \frac{\Gamma(k_2)}{\Gamma(k_2+\nu)} F_\nu(g, E_{k_2}'),\end{aligned}$$

and this together with (70) implies the statement of the Proposition.

Applying Proposition 6 to the case $\Gamma' = \Gamma$, g an Eisenstein series, and f a Hecke eigenform, we obtain (using (72)) the following generalization of Rankin's identity (33):

<u>Corollary</u>: <u>Let</u> $k_1, k_2 \geq 4$ <u>be even integers with</u> $k_1 \neq k_2$ <u>and</u> E_{k_1}, E_{k_2}

the normalized Eisenstein series of weight k_1, k_2. Let ν be a non-negative integer and $f(z) = \sum a(n)\, e^{2\pi inz}$ a normalized eigenform in S_k, $k = k_1 + k_2 + 2\nu$. Then

$$(77) \qquad (2\pi i)^{-\nu}\,(f, F_\nu(E_{k_1}, E_{k_2})) = (-1)^{k_2/2}\,\frac{2k_1}{B_{k_1}}\,\frac{2k_2}{B_{k_2}}\,\frac{\Gamma(k-1)}{2^{k-1}\,\Gamma(k-\nu-1)} \times L_f^*(k-\nu-1)\,L_f^*(k_2+\nu),$$

where B_{k_1}, B_{k_2} are Bernoulli numbers and $L_f^*(s)$ is defined by equations (30) and (31).

Remarks: 1. If $k_2 > k_1 \geqslant 4$, we prove (77) by applying Proposition 6 directly; if $k_1 > k_2$, we interchange the roles of k_1 and k_2, using the functional equation (31). As in the case $\nu = 0$, we observe that (77) remains valid also when $k_1 = k_2$.

2. Since $(2\pi i)^{-\nu} F_\nu(E_{k_1}, E_{k_2})$ has rational coefficients, the left-hand side of (77) is equal to the product of (f,f) with an algebraic number lying in the field generated by the Fourier coefficients of f. For any $k \geqslant 16$ there are sufficiently many triples (k_1, k_2, ν) satisfying the conditions of the Corollary to deduce that $L_f^*(a)\,L_f^*(b)$ is an algebraic multiple of (f,f) whenever a and b are integers of opposite parity satisfying $\frac{k}{2} < a,b < k$ (or simply $0 < a,b < k$ if we use (77) for $k_1 = k_2$). For example, if $k = 16$ and $f = \Delta_{16}$ is the unique normalized eigenform in S_{16}, we find

$$\frac{L_f^*(13)}{L_f^*(11)} = \frac{L_f^*(13)\,L_f^*(10)}{2^{14}(f,f)} \times \frac{2^{14}(f,f)}{L_f^*(10)\,L_f^*(15)} \times \frac{L_f^*(15)\,L_f^*(12)}{2^{14}(f,f)}$$

$$\times \frac{2^{14}(f,f)}{L_f^*(12)\,L_f^*(9)} \times \frac{L_f^*(9)\,L_f^*(14)}{2^{14}(f,f)} \times \frac{2^{14}(f,f)}{L_f^*(14)\,L_f^*(11)}$$

$$= \frac{1}{2.3.5.13} \times \frac{3617}{2.3.7} \times \frac{3.5.7}{3617} \times \frac{2^3.3^2.13}{1} \times \frac{1}{5.7^2} \times \frac{5.7.11}{3}$$

$$= \frac{22}{7}\ .$$

Thus we obtain a different proof of the result of Eichler-Shimura-Manin on periods of cusp forms mentioned in § 1.

3. For the six values of k with $\dim S_k = 1$, equation (77) takes the form

$$L_f^*(k_2)\, L_f^*(k-1) = (-1)^{\frac{k_2}{2}-1}\, 2^{k-2} \left(\frac{B_{k_1}}{k_1} + \frac{B_{k_2}}{k_2} - \frac{k}{B_k}\, \frac{B_{k_1}}{k_1}\, \frac{B_{k_2}}{k_2} \right) (f,f)$$

if $\nu = 0$ (Rankin [18], Theorem 5) and

$$L_f^*(k_2+\nu)\, L_f^*(k-\nu-1) = (-1)^{\frac{k_2}{2}-1}\, 2^{k-2}\, \frac{\Gamma(k-\nu-1)}{\Gamma(k-1)}$$

$$\cdot \left((-1)^{\nu} \frac{\Gamma(k_1+\nu)}{\Gamma(k_1+1)}\, B_{k_1} + \frac{\Gamma(k_2+\nu)}{\Gamma(k_2+1)}\, B_{k_2} \right) (f,f)$$

if $\nu > 0$.

4. Proposition 6 is similar to a recent result of Shimura ([22], Theorem 2). Also, the method sketched in this section for proving Theorem 2 is related to the method used by J.Sturm (cf. note at the end of the introduction).

§ 6 The Doi-Naganuma lifting and curves on Hilbert modular surfaces

In 1969 Doi and Naganuma [4] constructed a "lifting" from modular forms on $SL_2(\mathbb{Z})$ to Hilbert modular forms on $SL_2(\mathcal{O})$, where $\mathcal{O}$ is the ring of integers of a real quadratic field $K = \mathbb{Q}(\sqrt{D})$. Four years later, Naganuma [14] defined a similar lifting from $S_k(\Gamma_o(D), (\frac{D}{\cdot}))$ to $S_k(SL_2(\mathcal{O}))$; together, these maps give the subspace of $S_k(SL_2(\mathcal{O}))$ generated by eigenforms which are invariant under the action of Gal(K/Q). In [24] the author constructed a "kernel function" for the Naganuma mapping, i.e. a function $\Omega(z, z'; \tau)$ of three variables which is a modular form of Nebentypus $(\frac{D}{\cdot})$ with respect to τ and a Hilbert modular form with respect to (z, z') and whose Petersson product with any modular form $f(\tau)$ of Nebentypus is the Naganuma lift $\hat{f}(z, z')$ of f. The mth Fourier coefficient of $\Omega(z, z'; \tau)$ (with respect to τ) is a Hilbert modular form $\omega_{m,D}(z,z')$ defined by a series similar to that defining the function $\omega_m = \omega_{m,1}$ of § 2. By replacing ω_m by $\omega_{m,D}$ in the calculations of §§ 2 - 3 of this paper, we will obtain a theorem generalizing Theorem 1 and, as corollaries,

i) new proofs that certain functions constructed in [3] and in [8], given by Fourier expansions whose coefficients involve finite sums of values of L-series at integer arguments, are modular forms;

ii) characterization of these forms in terms of their Petersson product with Hecke eigenforms;

iii) proof that $(\hat{f},\hat{f})/(f,f)^2$ is an algebraic number for any eigenform $f \in S_k(\Gamma_o(D), (\frac{D}{\cdot}))$;

iv) partial proof of a conjecture made in [8] expressing the adjoint map of the Naganuma lifting (w.r.t. the Petersson product) in terms of intersection numbers of curves on the Hilbert modular surface $H^2/SL_2(\mathcal{O})$.

Some of the results have been obtained independently by T. Asai and T. Oda in the period since the Bonn conference. In particular, both iii) and iv) overlap with work of Oda.

We recall the result of [24]. We suppose throughout that the discriminant

D of K is $\equiv 1 \pmod 4$ and denote by k some positive even integer. The forms $\omega_{m,D}$ are defined (if $k > 2$) by

(78) $$\omega_{m,D}(z,z') = \sum_{\substack{a,b\in\mathbb{Z}\\ \lambda\in\vartheta^{-1}\\ \lambda\lambda'-ab=m/D}} (azz' + \lambda z + \lambda' z' + b)^{-k} \quad (z,z'\in H,\ m=1,2,\ldots),$$

where λ' denotes the conjugate of λ and $\vartheta = (\sqrt{D})$ is the different of K; one checks without difficulty that $\omega_{m,D}$ is a cusp form of weight k for the Hilbert modular group $SL_2(\mathcal{O})$. The main result of [24] is that the function

(79) $$\Omega(z,z';\tau) = \sum_{m=1}^{\infty} m^{k-1}\,\omega_{m,D}(z,z')\,e^{2\pi i m\tau} \qquad (z,z',\tau\in H)$$

is a cusp form on $\Gamma_o(D)$ of weight k and Nebentypus $(\frac{D}{\cdot})$ with respect to the variable τ whose Petersson product with any other cusp form $f \in S_k(\Gamma_o(D);(\frac{D}{\cdot}))$ is given by

(80) $$\int_{\Gamma\backslash H} f(\tau)\,\Omega(z,z';-\bar\tau)(\mathrm{Im}\ \tau)^k dV = C_k \sum_{\substack{\nu\in\vartheta^{-1}\\ \nu\gg 0}} c((\nu)\vartheta)\,e^{2\pi i(\nu z+\nu' z')},$$

where C_k is given by (12) and $c(\mathfrak{A})$ ($\mathfrak{A}$ an integral ideal) is an explicitly given finite linear combination of the Fourier coefficients of f at the various cusps of $\Gamma_o(D)$. It is also shown that, if D is prime and f a normalized Hecke eigenform, then $\sum c((\nu)\vartheta)\,e^{2\pi i(\nu z+\nu' z')}$ equals the Naganuma lift $\hat f$ of f i.e. the coefficients $c(\mathfrak{A})$ are multiplicative and satisfy

(81) $$c(\mathfrak{p}) = \begin{cases} a(p) & \text{if } \mathfrak{p}\mathfrak{p}' = (p),\ (\frac{D}{p}) = 1,\\ a(p)^2 + 2p^{k-1} & \text{if } \mathfrak{p} = (p),\ (\frac{D}{p}) = -1,\\ a(p) + \overline{a(p)} & \text{if } \mathfrak{p}^2 = (p),\ (\frac{D}{p}) = 0,\end{cases}$$

(82) $$\sum_{\mathfrak{A}} c(\mathfrak{A})\,N(\mathfrak{A})^{-s} = \left(\sum_{n=1}^{\infty} a(n)n^{-s}\right)\left(\sum_{n=1}^{\infty}\overline{a(n)}n^{-s}\right).$$

Asai [1] has shown that equations (81) and (82) still hold (for f an eigenform) when D is not prime.

Finally, for our generalization of Theorem 1 we must define the analogue of (2) for forms of Nebentypus. Let $f = \sum a(n)q^n \in S_k(\Gamma_o(D), (\frac{D}{\cdot}))$ and set

(83) $$D_f(s) = \prod_p (1-\alpha_p^2 p^{-s})^{-1}(1-(\tfrac{D}{p})\alpha_p\bar{\alpha}_p p^{-s})^{-1}(1-\bar{\alpha}_p^2 p^{-s})^{-1},$$

where $\alpha_p, \bar{\alpha}_p$ are defined by

(84) $$\alpha_p + (\tfrac{D}{p})\bar{\alpha}_p = a(p), \qquad \alpha_p\bar{\alpha}_p = p^{k-1}$$

or equivalently by

(85) $$\sum_{n=1}^{\infty} a(n)n^{-s} = \prod_p \frac{1}{(1-\alpha_p p^{-s})(1-(\frac{D}{p})\bar{\alpha}_p p^{-s})}.$$

Then, with the same notations as in Theorem 1, we have:

<u>Theorem 4</u>: <u>Let</u> $D \equiv 1 \pmod 4$, $D > 1$, <u>be a square-free integer and</u> $k > 2$ <u>an even integer</u>. <u>For</u> $m = 1,2,\ldots$ <u>and</u> $s \in \mathbb{C}$, $2-k < \mathrm{Re}(s) < k-1$ <u>set</u>

(86) $$c_{m,D}(s) = m^{k-1}D^{\frac{1}{2}-s} \sum_{\substack{t \in \mathbb{Z} \\ t^2 \equiv 4m \pmod D}} \left[I_k(t^2-4m,t;s)+I_k(t^2-4m,-t;s)\right] L\left(s,\frac{t^2-4m}{D}\right)$$
$$+ \begin{cases} (-1)^{k/2} \dfrac{\Gamma(s+k-1)\,\zeta(2s)}{2^{2s+k-3}\pi^{s-1}\,\Gamma(k)}\, u^{k-s-1} & \text{if } m = u^2,\ u > 0, \\ 0 & \text{if } m \neq \text{square}. \end{cases}$$

<u>Then the function</u>

$$\Phi_{s,D}(z) = \sum_{m=1}^{\infty} c_{m,D}(s)\, e^{2\pi i m z}$$

<u>is a cusp form on</u> $\Gamma_o(D)$ <u>of weight</u> k <u>and Nebentypus</u> $(\frac{D}{\cdot})$ <u>and satisfies</u>

$$(\Phi_{s,D}, f) = C_k \frac{\Gamma(s+k-1)}{(4\pi)^{s+k-1}} D_f(s+k-1)$$

<u>for any normalized Hecke eigenform</u> $f \in S_k(\Gamma_o(D), (\frac{D}{\cdot}))$.

Proof: We would like to imitate the proof of Theorem 1 in § 2 - 3 with $\omega_{m,D}$ instead of ω_m. We cannot use $\omega_{m,D}(z, -\bar{z})$ for this purpose, because $(z,z') \to (z, -\overline{z'})$ is not compatible with the action of the Hilbert modular group. We can get around this by using $\omega_{m,D}(\varepsilon z, \varepsilon'\bar{z})$ if K has a unit ε with $\varepsilon > 0 > \varepsilon'$ and in general by using the function $\omega_{-m,D}(z, \bar{z})$, where $\omega_{-m,D}$ is obtained by replacing m by $-m$ in (78) and is defined for $(z,z') \in H \times H_-$ (H_- = lower half-plane). Writing c instead of b in (78) and setting $\lambda = \frac{1}{2}(b + \frac{t}{\sqrt{D}})$, we obtain

$$\omega_{-m,D}(z,\bar{z}) = \sum_{\substack{a,c\in\mathbb{Z}\\ \lambda\in\mathfrak{d}^{-1}\\ \lambda\lambda'-ac=-m/D}} (a|z|^2 + \lambda z + \lambda'\bar{z} + c)^{-k}$$

$$= \sum_{\substack{a,b,c,t\in\mathbb{Z}\\ \frac{1}{4}(b^2-t^2/D)-ac=-m/D}} (a|z|^2 + bx + c + \frac{ity}{\sqrt{D}})^{-k}$$

or, with the notation (48),

$$y^k\omega_{-m,D}(z,\bar{z}) = \sum_{\substack{t\in\mathbb{Z}\\ t^2\equiv 4m \pmod{D}}} \sum_{|\phi|=\frac{t^2-4m}{D}} R_\phi(z,t) .$$

Theorem 3 (§ 3) now implies

$$(87)\qquad c_{m,D}(s) = \zeta(s)^{-1}D^{-\frac{s+k-1}{2}} m^{k-1} \int_{\Gamma\backslash H} \omega_{-m,D}(z,\bar{z})\, E(z,s)\, y^k\, dV.$$

Hence

$$(88)\qquad \Phi_{s,D}(\tau) = \zeta(s)^{-1} D^{-\frac{s+k-1}{2}} \int_{\Gamma\backslash H} \Omega_-(z,\bar{z};\tau)\, E(z,s)\, y^k\, dV,$$

where

$$\Omega_-(z,z';\tau) = \sum_{m=1}^{\infty} m^{k-1}\, \omega_{-m,D}(z,z')\, e^{2\pi i m\tau} \qquad (z,\ \tau\in H,\ \ z'\in H_-).$$

The function Ω_- has properties like those of Ω, namely it is a cusp form of Nebentypus with respect to τ and satisfies an equation like (80) but with the summation running over all $\nu \in \mathfrak{d}^{-1}$ such that $\nu > 0 > \nu'$. (This follows directly from the results of [24] if K has a unit ε with $\varepsilon > 0 > \varepsilon'$, since then $\Omega_-(z,z';\tau)$ equals $\Omega(\varepsilon z, \varepsilon' z';\tau)$, and can be proved without this assumption by making the obvious modifications in the proofs given in [24].) Therefore (88) implies that $\Phi_{s,D} \in S_k(\Gamma_0(D), (\frac{D}{\cdot}))$ and that

$$(\Phi_{s,D}, f) = C_k \zeta(s)^{-1} D^{-\frac{s+k-1}{2}} \int_{\Gamma\backslash H} h_f(z)\, E(z,s)\, dV$$

for any $f \in S_k(\Gamma_0(D), (\frac{D}{\cdot}))$, where

$$h_f(z) = y^k \sum_{\substack{\nu\in\vartheta^{-1} \\ \nu > 0 > \nu'}} \overline{c((\nu)\vartheta)}\, e^{2\pi i(\nu z + \nu'\bar{z})} \qquad (z \in H).$$

The function $h_f(z)$ is $SL_2(\mathbb{Z})$-invariant because $\Omega_-(z,z';\tau)$ is a cusp form of weight k with respect to the action of $SL_2(O)$ on $(z,z') \in H \times H_-$. Therefore we can apply the general principle (44) to obtain

$$\int_{\Gamma\backslash H} h_f(z)\, E(z,s)\, dV = \zeta(2s) \int_0^\infty \int_0^1 \sum_\nu \overline{c((\nu)\vartheta)}\, e^{2\pi i(\nu+\nu')x - 2\pi(\nu-\nu')y}\, y^{k+s-2}\, dx\, dy.$$

The only terms that contribute to this integral are those with $\nu + \nu' = 0$, i.e. $\nu = \frac{n}{\sqrt{D}}$ with $n \in \mathbb{N}$, and we obtain the identity

$$(89) \qquad (\Phi_{s,D}, f) = C_k \frac{\Gamma(s+k-1)}{(4\pi)^{s+k-1}} \left(\frac{\zeta(2s)}{\zeta(s)} \sum_{n=1}^\infty \frac{\overline{c((n))}}{n^{s+k-1}} \right),$$

valid for all $f \in S_k(\Gamma_0(D), (\frac{D}{\cdot}))$. If f is an eigenform, then $\overline{c((n))} = c((n))$ and the series $\sum c((n))\, n^{-s}$ has an Euler product whose terms can be computed using (81); a short computation then shows that the expression in brackets in (89) equals $D_f(s+k-1)$.

We can now deduce several corollaries exactly as in the case $D = 1$. First of all, the functions $D_f(s)$ is entire (proved by Shimura, [21], Theorem 1) and satisfies a functional equation (proved by Asai [1], Theorem 3). Next, by taking $s = r \in \{3,5,\ldots, k-3\}$ and using (22) and (26), we find that

$$\Phi_{r,D}(z) = -\frac{\pi^r}{4} C_k \frac{\Gamma(k-r)}{\Gamma(k-1)}\, C_{k,r,D}(z),$$

where

$$C_{k,r,D}(z) = \sum_{m=0}^{\infty} \left(\sum_{\substack{t \in \mathbb{Z} \\ t^2 \leq 4m \\ t^2 \equiv 4m \pmod{D}}} p_{k,r}(t,m)\, H(r, \frac{4m-t^2}{D}) \right) e^{2\pi i m z}.$$

For $r = k-1$ we get an extra contribution which can be computed as in § 4, the only difference being that the multiplicative function $\rho(a)$ occurring in equation (63) must be replaced by the multiplicative function

$$\begin{aligned} \rho_D(a) &= \#\{b,t \pmod{2a} \mid b^2 \equiv \frac{t^2-4m}{D} \pmod{4a}\} \\ &= \# \quad \{\lambda \in \mathcal{O} / a\mathcal{O} \mid \lambda\lambda' \equiv m \pmod{a}\}, \end{aligned}$$

which is calculated in [24] (Lemma 3, p. 27). We obtain an equation similar to (27) but with $\sigma_{k-1}(m)$ replaced by the mth Fourier coefficient of the Eisenstein series $E_2^+(z)$ in the space $M_k^+(\Gamma_o(D),(\frac{D}{\cdot}))$ consisting of those modular forms in $M_k(\Gamma_o(D),(\frac{D}{\cdot}))$ whose m^{th} Fourier coefficient vanishes whenever m is not a quadratic residue of D ($M_k^+(\Gamma_o(D),(\frac{D}{\cdot}))$ is the subspace of $M_k(\Gamma_o(D),(\frac{D}{\cdot}))$ fixed under all Atkin-Lehner involutions). Therefore Theorem 4 implies the result of Cohen ([3], Theorem 6.2) that the functions $C_{k,r,D}$ are modular forms (cusp forms if $r < k-1$) and at the same time gives a formula for the Petersson product of these functions with Hecke eigenforms. In particular, since $D_f(r+k-1) \neq 0$ for $r > 1$, we deduce from the "multiplicity 1" theorem that each $C_{k,r,D}$ generates the whole of $M_k(\Gamma_o(D),(\frac{D}{\cdot}))$ (resp. of $S_k(\Gamma_o(D),(\frac{D}{\cdot}))$ if $r < k-1$) under the action of the Hecke algebra.

For $r = 1$, we again find an extra contribution (given by (21)) coming from the terms in (86) for which $\frac{t^2-4m}{D}$ is a perfect square. In contrast to the case $D = 1$, there are in general infinitely many such terms, in $1:1$ correspondence with the integers of norm m in K, and we find

(90) $$\Phi_{1,D}(z) = -\frac{\pi}{4} C_k \cdot C_{k,1,D}(z) \qquad (k > 2)$$

with

$$(91)\qquad C_{k,1,D}(z) = \sum_{m=0}^{\infty}\left(\sum_{\substack{t\in\mathbb{Z}\\ t^2\leq 4m\\ t^2\equiv 4m \pmod D}} p_{k,1}(t,m)H\left(\frac{4m-t^2}{D}\right) + \frac{1}{\sqrt{D}}\sum_{\substack{\lambda\in\mathcal{O}\\ \lambda>0\\ \lambda\lambda'=m}}\min(\lambda,\lambda')^{k-1}\right)e^{2\pi imz}.$$

Finally, if $k = 2$ then we find (as in the case $D = 1$), that $\Phi_{1,D}$ equals $-\frac{\pi}{4}C_{2,1,D}$ plus a multiple of the Eisenstein series $E_2^+(z)$. Thus $C_{2,1,D}\in M_2^+(\Gamma_o(D),(\frac{D}{\cdot}))$ and $C_{k,1,D}\in S_k^+(\Gamma_o(D),(\frac{D}{\cdot}))$ for $k > 2$. This result is considerably harder to prove directly than the modularity of $C_{k,r,D}$ for $r > 1$, because the function $\mathcal{H}_r(z) = \sum H(r,N)q^N$ used by Cohen is no longer a modular form of half-integral weight when $r=1$. The fact that $C_{2,1,D}\in M_2^+(\Gamma_o(D),(\frac{D}{\cdot}))$ was proved in [8], Chapter 2.

Equation (90) together with Theorem 4 characterize the function $C_{k,1,D}$ by the formula

$$(92)\qquad (C_{k,1,D},f) = -\frac{4}{\pi}\frac{\Gamma(k)}{(4\pi)^k}D_f(k)\qquad (f\in S_k(\Gamma_o(D),(\tfrac{D}{\cdot}))\text{ an eigenform}).$$

To interpret this, we need some other product representations of $D_f(s)$. Let $a(n)$ and $c(\mathfrak{n})$ be the Fourier coefficients of f and of the Naganuma lift $\hat{f}$, respectively, α_p, $\bar{\alpha}_p$ the numbers defined by (85), and $A_{\mathfrak{p}}$ ($\mathfrak{p}\subset\mathcal{O}$ a prime ideal) the numbers defined by

$$(93)\qquad A_{\mathfrak{p}} = \alpha_p^f \qquad \text{if } N(\mathfrak{p}) = p^f.$$

Then (81), (82) are equivalent to the Euler product expansion

$$(94)\qquad \sum_{\mathfrak{n}} c(\mathfrak{n})N(\mathfrak{n})^{-s} = \prod_{\mathfrak{p}}(1-A_{\mathfrak{p}}N(\mathfrak{p})^{-s})^{-1}(1-\overline{A_{\mathfrak{p}}}N(\mathfrak{p})^{-s})^{-1}$$

for the Mellin transform of $\hat{f}$, and by applying the identity (3) (or rather its analogues for forms of Nebentypus and for Hilbert modular forms) to f and $\hat{f}$ we obtain

$$(95)\qquad \sum_{\mathfrak{n}} c(\mathfrak{n})^2 N(\mathfrak{n})^{-s} = \frac{\zeta_K(s-k+1)}{\zeta_K(2s-2k+2)}D_{\hat{f}}(s),$$

and

$$(96)\quad \sum_{n=1}^{\infty} |a(n)|^2 n^{-s} = \frac{\zeta(s-k+1)}{\zeta(2s-2k+2)} \prod_{p|D} (1+p^{k-1-s})^{-1} D_f'(s)$$

where

$$D_{\hat{f}}(s) = \prod_{\wp} (1 - A_{\wp}^2 N(\wp)^{-s})^{-1} (1 - A_{\wp}\bar{A}_{\wp} N(\wp)^{-s})^{-1} (1 - \bar{A}_{\wp}^2 N(\wp)^{-s})^{-1},$$

and

$$D_f'(s) = \prod_{p} (1 - (\tfrac{D}{p}) a_p^2 p^{-s})^{-1} (1 - p^{k-1-s})^{-1} (1 - (\tfrac{D}{p}) \bar{a}_p^2 p^{-s})^{-1}$$

(= the "twist" of D_f by $(\frac{D}{\cdot})$). On the other hand, using (93) we deduce after some trivial manipulations

$$(97)\qquad D_{\hat{f}}(s) = D_f(s)\, D_f'(s).$$

Thus $D_f(s)$ is, up to a simple factor, the ratio of the Rankin zeta functions associated to $\hat{f}$ and to f. The above formulas (and more general ones corresponding to Hilbert eigenforms which are not liftings) have also been observed by Asai [2].

Using the analogue of formula (5) for f and $\hat{f}$ (i.e. comparing the residues on the two sides of equations (95) and (96) at $s = k$ by Rankin's method) we obtain

$$D_f'(k) = \frac{2^{2k-1}\pi^{k+1}}{\Gamma(k)} D^{-1} (f,f),$$

$$D_{\hat{f}}(k) = \frac{2^{4k-1}\pi^{2k+2}}{\Gamma(k)^2} D^{-k-1} (\hat{f},\hat{f}),$$

and hence, by (97),

$$D_f(k) = \frac{2^{2k}\pi^{k+1}}{\Gamma(k)} D^{-k} \frac{(\hat{f},\hat{f})}{(f,f)}.$$

Substituting this into (92), we obtain

<u>Theorem 5</u>: <u>Let</u> $D \equiv 1 \pmod 4$ <u>be a square-free integer</u> > 1 <u>and</u> k <u>an integer</u> $\geqslant 2$. <u>Then the function</u> $C_{k,1,D}(z)$ <u>defined by (91) is a modular form in</u> $M_k(\Gamma_o(D), (\frac{D}{\cdot}))$ <u>(a cusp form if</u> $k > 2$). <u>If</u> $f \in S_k(\Gamma_o(D), (\frac{D}{\cdot}))$ <u>is a Hecke</u>

eigenform and $\hat{f} \in S_k(SL_2(\mathcal{O}))$ its lift under the Naganuma mapping, then

$$(98) \qquad (C_{k,1,D}, f) = -\frac{4}{D^k} \frac{(\hat{f},\hat{f})}{(f,f)} .$$

Since $C_{k,1,D}$ has rational Fourier coefficients we deduce

Corollary 1: Let $D, f, \hat{f}$ be as in the Theorem. Then $(\hat{f},\hat{f})/(f,f)^2$ is an algebraic number belonging to the field generated by the Fourier coefficients of f.

Doi informs me that this result has also been obtained recently by T. Oda.

Secondly, since $(\hat{f},\hat{f})$ and (f,f) are non-zero, we deduce from the "multiplicity 1" principle:

Corollary 2: The modular form $C_{k,1,D}$ together with its images under all Hecke operators span the space $S_k(\Gamma_0(D),(\frac{D}{\cdot}))$ (respectively the space $M_2(\Gamma_0(D),(\frac{D}{\cdot}))$ if $k=2$).

This corollary was conjectured in the case D prime, $k = 2$ by Hirzebruch and the author ([8], Conjecture 1', p. 108) in connection with the intersection behaviour of modular curves on Hilbert modular surfaces. We devote the rest of this section to a discussion of the relation between the above results and the results of [8].

We suppose from now on that D is a prime. For each $m \geq 1$ define a curve $T_m \subset H \times H$ by

$$T_m = \{(z,z') \mid \exists\, a,b, \in \mathbb{Z},\ \lambda \in \mathfrak{d}^{-1} \text{ with } ab - \lambda\lambda' = \tfrac{m}{D}\} ,$$

i.e. T_m is the union of the divisors of all of the expressions figuring in the definition of $\omega_{-m,D}$. The curve T_m is invariant under $SL_2(\mathcal{O})$, its image on $X = SL_2(\mathcal{O})\backslash H \times H$ being an affine algebraic curve (also denoted by T_m) each of whose components is isomorphic to the quotient of H by some arithmetic group. It was shown in [8] that

$$T_m \circ T_1 = \sum_{\substack{t \in \mathbb{Z} \\ t^2 \leq 4mm \\ t^2 \equiv 4m \pmod D}} H\left(\frac{4m-t^2}{D}\right)$$

if m is not a norm in K. In general, we must compactify X to a smooth surface $\widetilde{X}$ (by adding finitely many "cusps" and resolving all singularities on the resulting surface). Then the closure of T_m represents a homology class in $H_2(\widetilde{X})$ which we decompose as the sum of a class T_m^c in $\operatorname{Im}(H_2(X) \to H_2(\widetilde{X}))$ and of a linear combination of the classes represented by the curves of the singularity resolutions, and one has

$$T_1^c \circ T_m^c = \sum_{\substack{t \in \mathbb{Z} \\ t^2 \leq 4m \\ t^2 \equiv 4m \pmod D}} H\left(\frac{4m-t^2}{D}\right) + \frac{1}{\sqrt{D}} \sum_{\substack{\lambda \in \mathcal{O} \\ \lambda > 0 \\ \lambda\lambda' = m}} \min(\lambda, \lambda').$$

Therefore we can write

$$C_{2,1,D}(z) = -\frac{1}{12} + \sum_{m=1}^{\infty} (T_1^c \circ T_m^c)\, e^{2\pi i m z}.$$

The formula for $T_n^c \circ T_m^c$ ($n, m \in \mathbb{N}$ arbitrary) was also given in [8] and can be compactly summarized by

$$(99) \qquad \sum_{m=0}^{\infty} (T_n^c \circ T_m^c)\, e^{2\pi i m z} = (C_{2,1,D} | T^+(n))(z),$$

where T_o^c is defined as a certain multiple of the volume form on $\widetilde{X}$ and $T^+(n)$ is the composition of the nth Hecke operator on $M_2(\Gamma_o(D), (\frac{D}{\cdot}))$ with the canonical projection $M_2(\Gamma_o(D), (\frac{D}{\cdot})) \to M_2^+(\Gamma_o(D), (\frac{D}{\cdot}))$. By using these intersection number formulas in combination with a direct analytical proof (by means of non-holomorphic modular forms of weight 3/2) of the fact that $C_{2,1,D} \in M_2(\Gamma_o(D), (\frac{D}{\cdot}))$, the following theorem was proved in [8]:

<u>Theorem</u> ([8], Chapter 3): <u>For each homology class</u> $K \in H_2(\widetilde{X}; \mathbb{C})$ <u>the series</u>

$$\Phi_K(z) = \sum_{m=0}^{\infty} (K \cdot T_m^c)\, e^{2\pi i m z} \qquad (z \in H)$$

<u>is a modular form in</u> $M_2^+(\Gamma_o(D), (\frac{D}{\cdot}))$. <u>The map</u> $\Phi : K \to \Phi_K$ <u>is injective on the</u>

<u>subspace of</u> $H_2(\tilde{X}; \mathbb{C})$ <u>generated by the classes</u> T_n^c.

On the other hand, the map Φ is zero on the orthogonal complement of a certain subspace U of $H_2(\tilde{X}; \mathbb{C})$ (defined in [7], p. 91) containing all the classes T_n^c, with

$$\dim U = \left[\frac{D+19}{24}\right] = \dim M_2^+(\Gamma_o(D), (\tfrac{D}{\cdot})).$$

On the basis of this and of numerical calculations for $D < 200$ it was conjectured ([8], Conjecture 1, p. 108) that the subspace of $H_2(\tilde{X}; \mathbb{C})$ spanned by the classes T_n^c coincides with U and that the map $\Phi : U \to M_2^+(\Gamma_o(D), (\frac{D}{\cdot}))$ is an isomorphism. But $\Phi(T_n^c) = C_{2,1,D}| T^+(n)$ by equation (99), so Corollary 2 implies that the restriction of Φ to the space generated by the T_n^c is surjective, thus proving the conjecture. We state this result as

<u>Corollary</u> 3: <u>The subspace of</u> $H_2(\tilde{X}; \mathbb{C})$ <u>generated by the homology classes</u> T_n^c <u>has dimension</u> $\left[\frac{D+19}{24}\right]$ <u>and is mapped isomorphically onto</u> $M_2^+(\Gamma_o(D), (\frac{D}{\cdot}))$ <u>by</u> Φ.

By associating to a Hilbert cusp form $F \in S_2(SL_2(\mathcal{O}))$ the differential form

$$\frac{i}{2}\left[F(\varepsilon z, \varepsilon'\overline{z'})\ dz \wedge d\overline{z'} + F(\varepsilon z', \varepsilon'\overline{z})\ dz' \wedge d\overline{z}\right]$$

(ε = fundamental unit) and then applying the Poincaré duality map to the cohomology class represented by this form, one obtains an injective map

$$j : S_2(SL_2(\mathcal{O})) \to H_2(\tilde{X}; \mathbb{C})$$

(see [7] or [8] for details). Under this map, the codimension 1 subspace $U^o \subset U$ consisting of classes x with $xT_o^c = 0$ corresponds to the subspace $S_2^{sym} \subset S_2(SL_2(\mathcal{O}))$ generated by Hecke eigenforms F with $F(z,z') = F(z',z)$. Thus Φ can be identified with a map from S_2^{sym} to $S_2^+(\Gamma_o(D), (\frac{D}{\cdot}))$. On the other hand, one has the Naganuma lifting $\iota: f \to \hat{f}$ going the other way. It was conjectured in [8] (Conjecture 2, p. 109) that the two maps $\Phi \circ j : S_2^{sym} \to$

$S_2^+(\Gamma_o(D), (\frac{D}{\cdot}))$ and $\iota : S_2^+(\Gamma_o(D), (\frac{D}{\cdot})) \to S_2^{sym}$ are, up to a constant, adjoint maps with respect to the Petersson scalar product. From the definition of Φ via intersection numbers with the classes T_m^c and of ι via the Petersson product with $\Omega = \sum m\omega_{m,D} e^{2\pi i m\tau}$ one sees that this is equivalent to the statement

$$(100) \qquad j(m\omega_{m,D}) = (\text{const}) \cdot T_m^{co} \qquad (m = 1,2,\ldots),$$

where

$$T_m^{co} = T_m^c - \frac{(T_m^c\, T_o^c)}{(T_o^c\, T_o^c)}\, T_o^c$$

is the component of T_m^c in U^o (equation (100) is conjecture 2' of [8], p. 110, except that there T_m^c was inadvertently written instead of T_m^{co}).

There are two partial results in the direction of (100) which can be deduced from Theorem 5. First of all, a formal calculation using (80) shows that for $k > 2$ the Petersson product of $m^{k-1}\omega_{m,D}$ and $n^{k-1}\omega_{n,D}$ equals $-\frac{c_k^2}{2} D^k$ times the mth Fourier coefficient of $C_{k,1,D}|T^+(n)$, where $T^+(n) = S_k^+(\Gamma_o(D), (\frac{D}{\cdot})) \longrightarrow S_k^+(\Gamma_o(D), (\frac{D}{\cdot}))$ is the modified Hecke operator introduced above, while for $k = 2$ the same is true if we remove from $C_{2,1,D}$ a multiple of the Eisenstein series $E_2^+(z)$ to get a cusp form. Using (99) and the equation $j(\omega_{m,D}) \circ j(\omega_{n,D}) = -2(\omega_{m,D}, \omega_{n,D})$ ([8], p. 109, equation (17)), we can state this result for $k = 2$ as

$$j(m\omega_{m,D}) \circ j(n\omega_{n,D}) = \pi^2 D^2\ T_m^{co} \circ T_n^{co},$$

which is compatible with (100) and gives the value of the constant occurring there as $\pm\pi D$. Secondly, by letting $s \to 1$ in (87) and using

$$\lim_{s\to 1} E(z,s)/\zeta(s) = \frac{\pi}{2}$$

we obtain the formula

$$c_{m,D}(1) = \frac{\pi}{2} D^{-k/2} m^{k-1} \int_{\Gamma\backslash H} \omega_{m,D}(\varepsilon z, \varepsilon'\bar{z})\ y^k\, dV,$$

and for $k = 2$ this can be interpreted (using (90) and (99) and the fact that T_1

is the curve $\Gamma\backslash H \subset SL_2(\mathcal{O})\backslash H^2$, embedded by the diagonal map) as the statement

$$\pi D\, T_1 \cdot T_m^{co} = T_1 \circ j(m\omega_{m,D}),$$

which is again compatible with (100) and now gives the value of the constant exactly as πD. It should be possible to prove this statement with T_1 replaced by T_n using similar methods; by virtue of Corollary 3, this would suffice to establish (100) in full generality.

We end this section by proving the analogue of (100) for forms of higher weight. The principle is very general and should be applicable to any cycles on automorphic varieties (i.e. quotients of bounded symmetric domains by arithmetic groups) and automorphic forms which have the same formal relation to one another as T_m has to $\omega_{m,D}$. The proof we give would probably be carried over to the case of weight 2 (proving (100)), by using the definition of $\omega_{m,D}$ as

$$\lim_{s\to 0} \sum_{a,b,\lambda} (azz' + \lambda z + \lambda' z' + b)^{-2}\, |azz' + \lambda z + \lambda' z' + b|^{-s}$$

([24], Appendix 1) and carrying out the limit in the integrals.

Equation (100) (with the constant equal to πD) is equivalent to the formula

$$(101) \qquad (m\omega_{m,D}, F) = -\frac{\pi D}{2} \cdot i \int_{T_m} F(\varepsilon z, \varepsilon' \overline{z'})\, dz \wedge d\overline{z'} \qquad (\forall F \in S_2(SL_2(\mathcal{O}))),$$

because the right-hand side is just $-\frac{1}{2}\pi D$ times the intersections number of $j(F)$ with T_m (we can write T_m instead of T_m^c or T_m^{co} because $j(F)$ is orthogonal to the curves of the singularity resolutions and to the volume form T_0^c). Let

$$\mathcal{A} = \{A \in M_2(\mathcal{O}) \mid A^* = A'\},$$

where A^* and A' are defined for $A = \begin{pmatrix} a & b \\ c & d\end{pmatrix} \in M_2(\mathcal{O})$ as $\begin{pmatrix} d & -b \\ -c & a\end{pmatrix}$ and $\begin{pmatrix} a' & b' \\ c' & d'\end{pmatrix}$, respectively. The group $G = SL_2(\mathcal{O})/\{\pm 1\}$ acts on $\mathcal{A}$ by $M \circ A = M^* A M'$. Each $A \in \mathcal{A}$ with $\det A > 0$ defines a curve in $H \times H$, namely its graph $\{(z, Az) \mid z \in H\}$, and T_m consists of the images in $SL_2(\mathcal{O})\backslash H^2$ of these graphs for all A

with $\det A = m$. The components of ω_m correspond to the $SL_2(\mathcal{O})$-equivalence classes of A with $\det A = m$. Let $A_i = \begin{pmatrix} a_i & b_i \\ c_i & d_i \end{pmatrix}$ $(i = 1,\ldots,r)$ denote representatives of these classes and G_i $(i = 1,\ldots,r)$ the isotropy groups

$$G_i = \{ M \in G \mid M^* A_i M' = \pm A_i \}.$$

Then

$$T_m = \bigcup_{i=1}^{r} G_i \backslash H,$$

where the ith component is embedded by $z \mapsto (z, A_i z)$, and $\omega_{m,D}$ is defined by

$$\omega_{m,D}(z,z') = D^{k/2} \sum_{i=1}^{r} \sum_{M \in G_i \backslash G} \phi_{M^* A_i M'}(\varepsilon z, \varepsilon' z')^{k/2} \qquad (k>2)$$

$$= \sum_{i=1}^{r} \omega_{m,D}^{(i)}(z,z')$$

where

$$\phi_{\left(\begin{smallmatrix} a & b \\ c & d \end{smallmatrix}\right)}(z,z') = (czz' - az + dz' - b)^{-2}$$

$$= (cz+d)^{-2} \left(z' - \frac{az+b}{cz+d}\right)^{-2}$$

(cf. [24], 11.4 - 5), and each function $\omega_{m,D}^{(i)}$ is in $S_k(SL_2(\mathcal{O}))$. For $k = 2$, one has a similar splitting of $\omega_{m,D}$ as $\sum \omega_{m,D}^{(i)}$. Then equation (100) states that

$$\sum_{i=1}^{r} (F, \omega_{m,D}^{(i)}) = -\pi D \sum_{i=1}^{r} \int_{G_i \backslash H} (c_i \bar{z} + d_i)^{-2} F(\varepsilon z, \varepsilon' \overline{A_i z}) \, dx\, dy$$

for all $F \in S_2(SL_2(\mathcal{O}))$. The analogue we prove for forms of higher weight is the following.

<u>Theorem 6</u>: <u>Let</u> $k > 2$, $m \geq 1$ <u>and</u> $\omega_{m,D}$ <u>the Hilbert modular form</u> (78) <u>Then</u>

$$(F, \omega_{m,D}) = \tfrac{1}{2} C_k D^{k/2} \sum_{i=1}^{r} \int_{G_i \backslash H} (c_i \bar{z} + d_i)^{-k} F(\varepsilon z, \varepsilon' \overline{A_i z}) \, y^k \, dV$$

<u>for all</u> $F \in S_k(SL_2(\mathcal{O}))$, i.e. <u>the Petersson product of</u> F <u>with</u> $m^{k-1}\omega_{m,D}$ <u>is proportional to the integral of</u> F <u>over the curve</u> T_m <u>in a suitable sense</u>.

<u>Proof</u>: We have

$$(F,\omega^{(i)}_{m,D}) = \iint_{G\backslash H^2} F(\varepsilon z, \varepsilon'\overline{z'})\ \omega^{(i)}_{m,D}(\varepsilon\bar{z}, \varepsilon'z')\quad y^k\, y'^k\ dV'\, dV$$

(102)

$$= D^{k/2} \iint_{G_i\backslash H^2} F(\varepsilon z, \varepsilon'\overline{z'})\quad \phi_{A_i}(\bar{z},z')^{k/2}\ \ y^k\, y'^k\ dV'dV\ .$$

Since G_i acts properly discontinuously on H, we can take for $G_i\backslash H^2$ a fundamental domain of the form $\mathcal{F} \times H$, where $\mathcal{F}$ is a fundamental domain for the action of G_i on H. Then the integral on the right-hand side of (102) equals

$$(103)\qquad \int_{G_i\backslash H} (c_i\bar{z}+d_i)^{-k}\left(\int_H F(\varepsilon z,\varepsilon'\overline{z'})(z'-\overline{A_i z})^{-k}\ y'^k\ dV'\right) y^k\ dV.$$

But one has the identity

$$\int_H f(z)(\bar{z}-a)^{-k}\ y^k\ dV = \tfrac{1}{2}\ C_k\ f(a) \qquad\qquad (a \in H)$$

for any holomorphic function f on H with $\int_H |f(z)|^2\ y^k\ dV < \infty$ (cf. [25], p. 46), so the inner integral in (103) equals

$$\tfrac{1}{2}\ C_k\ F(\varepsilon z,\ \varepsilon'\overline{A_i z}).$$

This proves the theorem.

<u>Remarks</u>. 1. Theorem 6 is contained in recent work of T. Oda [15]. However, the explicit working out of his very general results for the case of the curve T_m has, so far as I know, not yet been given in the literature.

2. In the theorem we describe a way of integrating cusp forms of weight k over certain curves of X, whereas one would expect such an integral to make sense only for $k = 2$. Presumably there is some appropriate homology theory $\mathcal{H}_k(\tilde{X})$ which has a natural pairing with $S_k(SL_2(\Theta))$ and such that the curves in question yield classes in $\mathcal{H}_k$. The bilinear form on $\mathcal{H}_k$ corresponding to the Petersson product in $S_k(SL_2(\Theta))$ should then have a geometrical interpretation, i.e. for two compact curves C_1 and C_2 which meet transversally the intersection number of

C_1 and C_2 in $\mathcal{H}_k$ should be a sum of local contributions from the intersection points of C_1 and C_2. The intersection number of T_n and T_m in $\mathcal{H}_k$ (assuming that n or m is not a norm and $(n,m) = 1$) must be given by

$$\sum_{\substack{t^2 < 4nm \\ t^2 \equiv 4nm \pmod D}} \frac{\rho^{k-1} - \bar{\rho}^{k-1}}{\rho - \bar{\rho}} \; H(\frac{4nm-t^2}{D}) ,$$

(cf. (91)), where $\rho + \bar{\rho} = \frac{t}{\sqrt{nm}}$, $\rho\bar{\rho} = 1$. Here $\sum H(\frac{4nm-t^2}{D})$ is the number of intersection points of T_n and T_m, and from the description of the local geometry near such an intersection point given in Chapter I of [8] we find that the number ρ at a point $x \in T_n \cap T_m$ has a local description as the cross-ratio of the four tangent directions in the tangent space $T_x X$ given by $\partial_{z_1}, \partial_{z_2}$ and the directions of T_n and T_m at x. (This was suggested to me by Atiyah.)

3. For $k \geq 12$ the space $S_k^{sym} \subset S_k(SL_2(\mathcal{O}))$ is no longer the image of the Naganuma lifting ι but the direct sum of this image with the image of the Doi-Naganuma lifting

$$\iota_0 \; : \; S_k(SL_2(\mathbb{Z})) \rightarrow S_k(SL_2(\mathcal{O})).$$

We can give a description of the adjoint map (w.r.t. the Petersson product) of ι_0 in terms of intersection numbers as follows: The curve T_m is $\bigcup_{d^2|m} F_{m/d^2}$, where F_m is defined in the same way as T_m but with the condition that the triple (a, b, λ) not be divisible by any natural number > 1. In a recent thesis ("Kurven in Hilbertschen Modulflächen und Humbertsche Flächen im Siegel-Raum", Bonn 1977), H.-G. Franke proved that, for prime discriminants D, the curve F_m is irreducible if $D^2 \nmid m$ and has exactly two components if $D^2 | m$. Call these two components F_m^+ and F_m^-; they are given by taking those triples (a, b, λ) in the definition of F_m for which $(\frac{a}{D}) + (\frac{b}{D})$ is positive or negative, respectively (note that $ab \equiv \lambda\lambda' \pmod D \Rightarrow (\frac{a}{D}) + (\frac{b}{D}) \neq 0$). Set

$$T^{\pm}_m = \bigcup_{f^2|m} F^{\pm}_{mD^2/f^2}$$

(so that T^+_m and T^-_m are unions of components of T_{mD^2}, with $T_{mD^2} = T_m + T^+_m + T^-_m$). We can break up $\omega_{mD^2,D}$ as $\omega_{m,D} + \omega^+_{m,D} + \omega^-_{m,D}$ in a parallel way, and the proof of Theorem 6 again gives an interpretation of $(F, \omega^{\pm}_{m,D})$ as an integral of F over $T^{\pm}_m$. The relation to the Doi-Naganuma mapping is given by

__Theorem 7:__ __Let__ $k > 2$. __The function__

$$\Omega^o(z,z';\tau) = \sum_{m=1}^{\infty} m^{k-1}\left[\omega^+_{m,D}(z,z') - \omega^-_{m,D}(z,z')\right] e^{2\pi i m\tau} \quad (z,z',\tau \in H)$$

__is a cusp form of weight__ k __on__ $SL_2(\mathbb{Z})$ __with respect to__ τ __and is, up to a factor, the kernel function of the Doi-Naganuma lifting__ $\iota_o : S_k(SL_2(\mathbb{Z})) \to S_k(SL_2(\mathcal{O}))$.

I omit the proof, which is analogous to that in [24].

References

[1] T. Asai, On the Fourier coefficients of automorphic forms at various cusps and some applications to Rankin's convolution, J. Math. Soc. Japan 28 (1976) 48 - 61

[2] T. Asai, On certain Dirichlet series associated with Hilbert modular forms and Rankin's method, Math. Ann. 226 (1977) 81 - 94.

[3] H. Cohen, Sums involving the values at negative integers of L functions of quadratic characters, Math. Ann. 217 (1975) 271 - 285.

[4] K. Doi, H. Naganuma, On the functional equation of certain Dirichlet series, Inv. Math. 9 (1969) 1 - 14

[5] A. Erdelyi, W. Magnus, F. Oberhettinger, F. Tricomi, Higher Transcendental Functions, Vol. I, Mc Graw-Hill, New York-Toronto-London 1953.

[6] R. Gunning, Lectures on Modular Forms, Ann. of Math. Studies No. 48, Princeton Univ. Press, Princeton 1962.

[7] F. Hirzebruch, Kurven auf den Hilbertschen Modulflächen und Klassenzahlrelationen, in Classification of algebraic varieties and compact complex manifolds, Springer Lecture Notes No. 412, pp. 75 - 93.

[8] F. Hirzebruch, D. Zagier, Intersection numbers of curves on Hilbert modular surfaces and modular forms of Nebentypus, Inv. Math. 36 (1976) 57 - 113

[9] A. Hurwitz, Über Relationen zwischen Klassenzahlen binärer quadratischer Formen von negativer Determinante, Math. Ann. 25 (1885). Math. Werke, Bd. II, pp. 8 - 50, Birkhäuser, Basel-Stuttgart 1963

[10] E. Landau, Vorlesungen über Zahlentheorie (Aus der elementaren Zahlentheorie), Hirzel, Leipzig 1927.

[11] S. Lang, Introduction to Modular Forms, Grundlehren der mathematischen Wissenschaften 222, Springer-Verlag, Berlin-Heidelberg-New York 1976

[12] D.H. Lehmer, Ramanujan's function $\tau(n)$, Duke Math. J. 10 (1943) 483 - 492.

[13] W. Li, New forms and functional equations, Math. Ann 212 (1975) 285 - 315.

[14] H. Naganuma, On the coincidence of two Dirichlet series associated with cusp forms of Hecke's "Neben"-type and Hilbert modular forms over a real quadratic

field, J. Math. Soc. Japan 25 (1973) 547 - 555.

[15] T. Oda, On modular forms associated with indefinite quadratic forms of signature (2, n-2), Math. Ann., to appear

[16] H. Petersson, Über eine Metrisierung der ganzen Modulformen, Jahresb. d. Deutschen Math. Verein. 49 (1939) 49 - 75.

[17] R. A. Rankin, Contributions to the theory of Ramanujan's function $\tau(n)$ and similar arithmetical functions, I, Proc. Camb. Phil. Soc. 35 (1939) 351 - 372.

[18] R.A. Rankin, The scalar product of modular forms, Proc. London Math. Soc. 2 (1952) 198 - 217

[19] J -P. Serre, Une interprétation des congruences relatives à la fonction τ de Ramanujan, Séminaire Delange-Pisot-Poitou, 9e année, 1967/8, n° 14

[20] G. Shimura, On modular forms of half integral weight, Ann of Math. 97 (1973) 440 - 481

[21] G. Shimura, On the holomorphy of certain Dirichlet series, Proc. Lond. Math. Soc. 31 (1975) 79 - 98

[22] G. Shimura, The special values of the zeta functions associated with cusp forms, Comm. on Pure and Applied Math. 29 (1976) 783 - 804

[23] D. Zagier, On the values at negative integers of the zeta-function of a real quadratic field, L'Ens. Math. 22 (1976) 55 - 95.

[24] D. Zagier, Modular forms associated to real quadratic fields, Inv. Math. 30 (1975) 1 - 46

[25] D. Zagier, Traces des opérateurs de Hecke, Séminaire Delange-Pisot-Poitou, 17e année, 1975/6, n° 23; = The Eichler-Selberg trace formula on $SL_2(\mathbb{Z})$, Appendix to Part I of [11], pp. 44 - 54.

International Summer School on Modular Functions
BONN 1976

CORRECTION TO "THE EICHLER-SELBERG TRACE FORMULA ON $SL_2(\mathbb{Z})$"

by D. Zagier

The paper in question is a translation of the author's paper "Trace des opérateurs de Hecke" (Séminaire Delange-Pisot-Poitou, 17e année, 1975/6, n° 23) and appeared as an appendix to Part I of Serge Lang's recent book Introduction to Modular Forms (Springer-Verlag, 1976, pp. 44-54; all page and equation numbers below refer to this appendix). Its purpose was to give a self-contained account, in the language of the classical theory of modular forms, of the formula of Selberg and Eichler for the trace of the Hecke operator $T(N)$ acting on $S_k(SL_2(\mathbb{Z}))$. Unfortunately, as several people have pointed out to me, the calculation of the contribution from the hyperbolic matrices with rational fixed points (Case 3 , p.53) is incorrect. The contribution from all such matrices with given determinant u^2 ($u \in \mathbb{Z}$, $u > 0$, $u^2+4N = t^2$ with $t \in \mathbb{Z}$) is given by

$$(1) \qquad \int_F R(z) \frac{dxdy}{y^2},$$

where $F = \{z=x+iy \in H \mid |z| \geq 1, |x| \leq \frac{1}{2}\}$ is a fundamental domain for the action of $SL_2(\mathbb{Z})$ on the upper half-plane H and

$$(2) \qquad R(z) = \sum_{\substack{a,b,c \in \mathbb{Z} \\ b^2-4ac = u^2}} \frac{y^k}{(a|z|^2+bx+c-ity)^k} \qquad (z=x+iy \in H).$$

Substituting (2) into (1) and interchanging the summation and integration gives

$$(3) \qquad \int_F R(z) \frac{dxdy}{y^2} = u\, I$$

with

$$(4) \qquad I = \int_H \frac{y^k}{(|z|^2 - ity - \frac{1}{4}u^2)^k} \frac{dxdy}{y^2}.$$

This integral is computed by integrating first over x and then over y, i.e. as

$$(5) \qquad \int_0^\infty \left(\int_{-\infty}^\infty (x^2+y^2-ity-\tfrac{1}{4}u^2)^{-k} dx \right) y^{k-2} dy ,$$

and it is claimed that this leads to the value

$$(6) \qquad (-1)^{(k-2)/2} \frac{2\pi}{k-1} u^{-1} (u+|t|)^{-k+1}$$

for I, which when multiplied by u gives the correct value for (1). However, the computation given contains a sign mistake (the integral (5) is correctly evaluated in the text as $\frac{\pi i^{k-2}}{2(k-1)!} \frac{d^{k-2}}{dt^{k-2}} (-\frac{4}{u} \frac{1}{u+|t|})$, which is the negative of (6)), and in fact the integral (1) is not equal to u times expression (5). The reason is that the expression obtained by substituting (2) into (1) is not absolutely convergent, so that the interchange leading to (3) is not justified; moreover, equation (4) makes no sense until we specify the order of integration over H, since the integral is not absolutely convergent.

The correct procedure is to replace the integral (1) by the limit as $C \to \infty$ of the integral over the compact region $F_C = \{z=x+iy \in F \mid y \geq C\}$, in which the convergence of the sum (2) is uniform. Then equations (3) and (4) remain valid if the integration over H in (4) is interpreted as $\lim_{\varepsilon \to 0} \int_{H_\varepsilon}$, where H_ε is obtained from H by removing all points which have imaginary part $> \frac{1}{\varepsilon}$ or which lie in the interior of a circle of radius $\frac{\varepsilon}{c}$ tangent to the real axis at any rational point $\frac{d}{c}$, $(c,d) = 1$. Since the only poles of the integrand in (4) are at $z = \pm\frac{1}{2}u$, we can shrink all of these circles to zero except those tangent to the real axis at $\frac{1}{2}u$ and $-\frac{1}{2}u$. Hence

(7) $$I = I_1 - \lim_{\varepsilon\to 0} I_{\varepsilon,\frac{1}{2}u} - \lim_{\varepsilon\to 0} I_{\varepsilon,-\frac{1}{2}u} ,$$

where I_1 is the integral (5) and

$$I_{\varepsilon,\pm\frac{1}{2}u} = \int_0^{2\varepsilon}\left(\int_{\pm\frac{1}{2}u-\sqrt{2\varepsilon y-y^2}}^{\pm\frac{1}{2}u+\sqrt{2\varepsilon y-y^2}} (x^2+y^2-ity-\tfrac{1}{4}u^2)^{-k}\, dx\right) y^{k-2}\, dy$$

is the integral over the circle tangent to the real axis at $\pm\frac{1}{2}u$, the integration being carried out first over x and then over y. Making the substitution $x=\pm\frac{1}{2}u+\varepsilon a$, $y=\varepsilon b$, we find

$$I_{\varepsilon,\pm\frac{1}{2}u} = \int_0^2 \int_{-\sqrt{2b-b^2}}^{\sqrt{2b-b^2}} \frac{b^{k-2}}{(\pm ua-itb+\varepsilon(a^2+b^2))^k}\, da\, db ,$$

so

$$\lim_{\varepsilon\to 0} I_{\varepsilon,\pm\frac{1}{2}u} = \int_0^2 \int_{-\sqrt{2b-b^2}}^{\sqrt{2b-b^2}} \frac{b^{k-2}}{(\pm ua-itb)^k}\, da\, db$$

$$= -\frac{1}{k-1} u^{-1} \int_0^2 \left\{(u\sqrt{2b-b^2}-itb)^{-k+1}+(u\sqrt{2b-b^2}+itb)^{-k+1}\right\} b^{k-2}\, db$$

$$= -\frac{2}{k-1} u^{-1} \int_{-\infty}^{\infty} (uv+it)^{-k+1} \frac{v\, dv}{v^2+1} ,$$

where in the last line we have made the substitution $b = \frac{2}{v^2+1}$. The latter integral can be evaluated easily by contour integration (for example, if $t>0$ then the only pole of the integrand in the upper half plane is at $v=i$) and equals the negative of expression (6). Since also I_1 equals the negative of (6), as stated above, the expressions (7) and (6) are equal.

A second and minor correction is that on page 49, line 3 and page 52, line 26, the 48 should be replaced by 24, and in the middle of page 52 the $\frac{\pi}{6}$ should be $\frac{\pi}{3}$.

International Summer School on Modular Functions
BONN 1976

A LIFTING OF MODULAR FORMS IN ONE VARIABLE TO HILBERT MODULAR FORMS IN TWO VARIABLES

by Henri COHEN

§. 1. - Introduction

Let K be a real quadratic field of discriminant D ; write O_K for the ring of integers of K, $\mathfrak{d}$ for the different, and x' for the conjugate of x in K.

When one computes the Fourier coefficients of the Hecke-Eisenstein series of weight k for $SL_2(O_K)$, one obtains the following (see [1]) :

PROPOSITION 1.1. - *Let $k \geq 2$ be an even integer. Set*

$$\mathcal{G}_k(z_1, z_2) = \frac{1}{4}\zeta_K(1-k) + \sum_{\substack{\nu \in \mathfrak{d}^{-1} \\ \nu \gg 0}} e^{2i\pi(\nu z_1 + \nu' z_2)} c(\nu\mathfrak{d})$$

where $$c(\nu\mathfrak{d}) = \sum_{\substack{d \mid (\nu\mathfrak{d}) \\ d \in \mathbb{N}^*}} d^{k-1}\left(\frac{D}{d}\right)\sigma_{k-1}(N_{K/\mathbb{Q}}((\nu\mathfrak{d})/d)) .$$

Then $\mathcal{G}_k$ is, up to a constant factor, the Hecke-Eisenstein series of weight k for K, and thus is a Hilbert modular form on $SL_2(O_K)$.

The appearance of σ_{k-1} makes one think about the one variable Eisenstein series :

$$g_k(z) = \frac{1}{2}\zeta(1-k) + \sum_{n \geq 1} \sigma_{k-1}(n) q^n .$$

Thus it is natural to ask if proposition 1.1 can be slightly modified so that it holds with $\sigma_{k-1}(n)$ replaced by the nth coefficient of an arbitrary modular form, for instance by $\tau(n)$.

The purpose of this paper is to show that this is indeed true, but with certain restrictions.

The result is as follows (A weaker result was given in [3] and [4]) :

THEOREM 1.2. - Let $f = \sum_{n\geq 1} a(n)q^n \in S_k(\Gamma_o(N), \chi)$ be a cusp form of weight k and character χ on $\Gamma_o(N)$. Set :

$$E_f^K(z_1, z_2) = \sum_{\substack{\nu\in\mathfrak{d}^{-1}\\ \nu\gg 0}} e^{2i\pi(\nu z_1+\nu' z_2)} c(\nu\mathfrak{d})$$

where

$$c(\nu\mathfrak{d}) = \sum_{\substack{d|(\nu\mathfrak{d})\\ d\in\mathbb{N}^*}} d^{k-1}\chi(d)\left(\frac{4D}{d}\right) a\left(N_{K/\mathbb{Q}}\left(\frac{\nu\mathfrak{d}}{d}\right)\right)$$

Then if $k\geq 3$ is an integer, E_f^K is a Hilbert modular form of weight k and character $\chi\circ N_{K/\mathbb{Q}}$ on a congruence subgroup Γ_{N_1} of $SL_2(O_K)$ of level dividing $N_1 w$, defined by

$$\Gamma_{N_1} = \{\begin{pmatrix}\alpha & \beta\\ \gamma & \delta\end{pmatrix}\in SL_2(O_K),\ \alpha,\delta\in\mathbb{Z}+N_1O_K,\ \gamma\in N_1(\mathbb{Z}+wO_K)\}$$

where $N_1 = 2N/(4, D, N)$, and $w = 2$ if $D\equiv 1 \pmod 8$, $w = 1$ otherwise.

§. 2. - Main tools for the proof

One of the first things to check before attempting to prove that E_f^K is a Hilbert modular form is that its restriction to the diagonal $E_f^K(z,z)$ is an ordinary modular form of weight $2k$ and character χ^2 for $\Gamma_{N_1}\cap SL_2(\mathbb{Z}) = \Gamma_o(N_1)$.

One finds easily that :

$$E_f^K(z,z) = \sum_{n\geq 1} q^n \sum_{d|n} d^{k-1}\chi(d)\left(\frac{4D}{d}\right)\sum_{s\in\mathbb{Z}} a\left(\frac{(n/d)^2D - s^2}{4}\right)$$

where here and elsewhere we set $a(x) = 0$ if $x\notin\mathbb{N}$. If for instance $D\equiv 1 \pmod 4$, then clearly :

$$f(4z)\,\theta(z) = \sum_{n\geq 1} q^n\left(\sum_{s\in\mathbb{Z}} a\left(\frac{n-s^2}{4}\right)\right)$$

and so the fact that $E_f^K(z,z)$ is a modular form of weight $2k$ and character χ^2 on $\Gamma_o(N_1)$ is a consequence of Shimura's theorem on forms of half integral weight ([11]) as improved by Niwa ([8]), which we state as follows :

THEOREM 2.1. (Shimura, Niwa) - Let $g = \sum_{n\geq 1} b(n)q^n \in S_{k+\frac{1}{2}}(\Gamma_o(4N), \chi)$ If $t\geq 1$ is squarefree, set

$$S(g;t)(z) = \sum_{n\geq 1} q^n \sum_{d|n} d^{k-1} \chi(d) \left(\frac{(-1)^k 4t}{d}\right) b((n/d)^2 t).$$

Then if $k \geq 3$ is an integer, $S(g;t) \in S_{2k}(\Gamma_0(2N), \chi^2)$.

So it is clear from above that Shimura's theorem is closely tied in with our mapping, and it will be (in a slightly generalized form) the first of our two main tools.

Our other main tool is a combinatorial characterization of Hilbert modular forms.

We need some notation.

Let $GL_2^+(\mathbb{R}) = \{ \begin{pmatrix} a & b \\ c & d \end{pmatrix} \in GL_2(\mathbb{R}) \mid ad - bc > 0 \}$ and let $r \geq 1$ be an integer.

If g is a function defined on the r-fold product of the upper half plane H^r, $k = (k_1, \dots, k_r) \in \mathbb{R}^r$, and $\gamma = (\gamma_1, \dots, \gamma_r)$ where $\gamma_i = \begin{pmatrix} a_i & b_i \\ c_i & d_i \end{pmatrix} \in GL_2^+(\mathbb{R})$ for $1 \leq i \leq r$, we set :

$$(g|_k \gamma)(z_1, \dots, z_r) = \prod_{i=1}^{r} (a_i d_i - b_i c_i)^{k_i/2} (c_i z_i + d_i)^{-k_i} g(\gamma_1 z_1, \dots, \gamma_r z_r).$$

Then we have the following theorem :

THEOREM 2.2. - For $m \in \mathbb{N}$ and G any analytic function on $H \times H$, set

$$(\Phi_m(G))(z) = \sum_{0 \leq \ell \leq m} (-1)^\ell e_{m,\ell} (\partial_{z_1}^\ell \partial_{z_2}^{m-\ell} G)(z, z)$$

where $e_{m,\ell} = (\ell!(m-\ell)!(k_1+\ell-1)!(k_2+m-\ell-1)!)^{-1}$.

(We write freely $x!$ in place of $\Gamma(x+1)$; furthermore if τ is any symbol, ∂_τ is an abbreviation for $\frac{\partial}{\partial \tau}$.) Then :

a) For any $\gamma \in GL_2^+(\mathbb{R})$

$$\Phi_m(G)|_{2m+k_1+k_2} \gamma = \Phi_m(G|_{(k_1,k_2)}(\gamma, \gamma)).$$

b) Given $\chi \in \mathbb{C}$ and $\gamma \in GL_2^+(\mathbb{R})$, the following two statements are equivalent :

i) $G|_{(k_1,k_2)}(\gamma, \gamma) = \chi G$

ii) For all $m \geq 0$

$$\Phi_m(G)|_{2m+k_1+k_2} \gamma = \chi \, \Phi_m(G)$$

c) When $k_1 = k_2 = k$ we have

$$\Phi_m(G) = (-1)^m \frac{(2k-2)!}{(k-1)!(k+m-1)!(2k+m-2)!} \varphi_m(G),$$

where

$$\varphi_m(G)(z) = \sum_{0 \leq \ell \leq m/2} (-1)^\ell \frac{(m+k-\ell-3/2)! \, 2^{m-2\ell}}{\ell!(m-2\ell)!(k-3/2)!} ((\partial_{z_1} + \partial_{z_2})^{2\ell} (\partial_{z_1} - \partial_{z_2})^{m-2\ell} G)(z, z).$$

It will be clear from the proof that this theorem can easily be generalized to any number of variables.

§. 3. - Proof of theorem 2. 2.

a) Set $E(z) = (z-\bar{z})^{-1} = 1/2iy$, where $y = \operatorname{Im} z$. The following facts are easily proved (see [2], §. 7) :

- If f is a C^∞ function on H, then for $\gamma \in GL_2^+(\mathbb{R})$:

$$(\partial_z f + kEf)|_{k+2}\gamma = \partial_z(f|_k\gamma) + kE(f|_k\gamma).$$

- For $m \geq 0$ set

$$F_m^{(z)}(f) = (\partial_z + (k+2m-2)E(z))(\partial_z + (k+2m-4)E(z)) \dots (\partial_z + kE(z))f$$

then :

$$\text{i)} \quad F_m^{(z)}(f) = \sum_{0\leq \ell\leq m} \binom{m}{\ell} \frac{(k+m-1)!}{(k+\ell-1)!} E(z)^{m-\ell} \partial_z^\ell f$$

and

$$\text{ii)} \quad F_m^{(z)}(f)|_{k+2m}\gamma = F_m^{(z)}(f|_k\gamma) .$$

Now let f be a C^∞ function on H^r. If $m = (m_1, \dots, m_r)$ and $k = (k_1, \dots, k_r)$, we can set

$$F_m^{(z)}(f) = F_{m_1}^{(z_1)}(F_{m_2}^{(z_2)} \dots (F_{m_r}^{(z_r)}(f))\dots) .$$

Then with the usual conventions for multiindices

$$(\binom{m}{\ell} = \prod_{i=1}^r \binom{m_i}{\ell_i}, \quad n! = \prod_{i=1}^r n_i!, \quad E(z)^{m-\ell} = \prod_{i=1}^r E(z_i)^{m_i-\ell_i},$$

$$\partial_z^\ell f = \partial_{z_1}^{\ell_1} \dots \partial_{z_r}^{\ell_r} f, \quad t^m = \prod_{i=1}^r t_i^{m_i}, \quad |m| = m_1 + \dots + m_r,$$

$m \geq 0 \Leftrightarrow \forall i,\ m_i \geq 0$) it is easy to see that i) and ii) are still valid.

Now for $z = (z_1, \dots, z_r)$, $t = (t_1, \dots, t_r)$ set :

$$F_f(z,t) = \sum_{m\geq 0} \frac{(2i\pi)^{|m|} t^{2m} F_m^{(z)}(f)}{m!\,(k+m-1)!}$$

$$G_f(z,t) = \sum_{m\geq 0} \frac{(2i\pi)^{|m|} t^{2m} \partial_z^m f}{m!\,(k+m-1)!} .$$

Then as in [2] one shows easily from i) that

$$F_f(z,t) = e^{2i\pi \sum_{i=1}^r t_i^2 E(z_i)} G_f(z,t) .$$

From this and ii) one deduces the following proposition :

PROPOSITION 3.1. - With the above notations,

$$G_f\left(\frac{az+b}{cz+d}, \frac{(ad-bc)^{\frac{1}{2}}t}{cz+d}\right) = (ad-bc)^{-k/2}(cz+d)^k \times S \times G_{(f|_k\gamma)}(z,t)$$

where

$$S = e^{2i\pi \sum_{i=1}^{r} t_i^2 c_i/(c_i z_i + d_i)} .$$

If we specialize to the case $r=2$ and take $t_1 = t = -it_2$ and $\gamma_1 = \gamma_2 = \gamma$, one has $S=1$ and theorem 2.2.a) follows.

b) The operator $\Phi_m(G)$ being linear in G, it is clear that i) ⇒ ii) follows from a). So we need to prove ii) ⇒ i). Replacing γ by $\gamma/(\det\gamma)^{\frac{1}{2}}$, we can clearly suppose $\gamma \in SL_2(\mathbb{R})$. Set :

$$h(z_1,z_2) = (G|_{(k_1,k_2)}(\gamma,\gamma))(z_1,z_2) - \chi G(z_1,z_2),$$

that is :

$$h(z_1,z_2) = (cz_1+d)^{-k_1}(cz_2+d)^{-k_2} G\left(\frac{az_1+b}{cz_1+d}, \frac{az_2+b}{cz_2+d}\right) - \chi G(z_1,z_2) .$$

Our hypothesis tells us that G is analytic on $H \times H$, and so h is also. To prove that $h \equiv 0$, it is thus enough to prove that $(\partial_{z_1}^{i_1} \partial_{z_2}^{i_2} h)(z,z) = 0$ $\forall z \in H$ and $\forall (i_1,i_2) \in \mathbb{N}\times\mathbb{N}$. Call $\mathcal{P}_\lambda$ the proposition :

$(\mathcal{P}_\lambda)$ - For all i_1, i_2 such that $i_1+i_2 \leq \lambda$ and all $z \in H$:

$$(\partial_{z_1}^{i_1} \partial_{z_2}^{i_2} h)(z,z) = 0 .$$

We shall prove $(\mathcal{P}_\lambda)$ by induction on λ, using ii) at each step.

- $\lambda = 0$.

$(\mathcal{P}_0)$ states that $h(z,z)=0$. But clearly

$$h(z,z) = (cz+d)^{-(k_1+k_2)} G(\gamma z, \gamma z) - \chi G(z,z)$$

and so

$$h(z,z) = (k_1-1)!\,(k_2-1)!\,(\Phi_0(G)|_{k_1+k_2}\gamma - \chi\Phi_0(G)) = 0$$ by ii), using the fact that

$$\Phi_0(G) = G(z,z)/((k_1-1)!\,(k_2-1)!) .$$

So now we assume $(\mathcal{P}_{\lambda-1})$ and we want to prove $(\mathcal{P}_\lambda)$, where $\lambda \geq 1$.

Differentiating with respect to z the known identities

$$\forall i \in [0, \lambda-1] : \quad \partial_{z_1}^{i} \partial_{z_2}^{\lambda-1-i} h(z,z) = 0 ,$$

we obtain :

$$\partial_{z_1}^{i+1} \partial_{z_2}^{\lambda-1-i} h(z,z) + \partial_{z_1}^{i} \partial_{z_2}^{\lambda-i} h(z,z) = 0 ,$$ and consequently :

$$\partial^i_{z_1} \partial^{\lambda-i}_{z_2} h(z,z) = (-1)^i \partial^\lambda_{z_2} h(z,z) .$$

To prove that all these partial derivatives vanish it is thus sufficient to find one non trivial independent linear relation between them, and this relation will be given to us by ii). First we need a lemma :

LEMMA 3.2. -

$$\partial^{i_1}_{z_1} \partial^{i_2}_{z_2} h(z_1, z_2) = \sum_{\substack{0 \le j_1 \le i_1 \\ 0 \le j_2 \le i_2}} (c z_1+d)^{-(k_1+i_1+j_1)} (c z_2+d)^{-(k_2+i_2+j_2)} \times$$

$$\times \frac{(i_1+k_1-1)!}{(j_1+k_1-1)!} \frac{(i_2+k_2-1)!}{(j_2+k_2-1)!} \binom{i_1}{j_1}\binom{i_2}{j_2}(-c)^{i_1+i_2-j_1-j_2} (\partial^{j_1}_{z_1} \partial^{j_2}_{z_2} G)(\gamma z_1, \gamma z_2) - \chi \partial^{i_1}_{z_1} \partial^{i_2}_{z_2} G(z_1, z_2) .$$

The proof of this lemma is by induction on i_1 and i_2 and simply checking that the coefficients are correct by differentiating with respect to i_1 (or i_2). The detailed calculation is left to the reader.

LEMMA 3.3. - <u>For any</u> $m \ge 0$

$$\sum_{0 \le \ell \le m} (-1)^\ell e_{m,\ell} (\partial^\ell_{z_1} \partial^{m-\ell}_{z_2} h)(z,z) = 0 .$$

<u>Proof</u>. - We replace $\partial^\ell_{z_1} \partial^{n-\ell}_{z_2} h$ by its expression given by lemma 3.2.

The coeffieint of

$$(cz+d)^{-(k_1+k_2+m+j_1+j_2)} (-c)^{m-j_1-j_2} (\partial^{j_1}_{z_1} \partial^{j_2}_{z_2} G)\left(\frac{az+b}{cz+d}, \frac{az+b}{cz+d}\right)$$

is equal to

$$\sum_{0 \le \ell \le m} (-1)^\ell e_{m,\ell} \frac{(\ell+k_1-1)! \, (m-\ell+k_2-1)! \, \ell! \, (m-\ell)!}{(j_1+k_1-1)! \, (j_2+k_2-1)! \, j_1! \, j_2! \, (\ell-j_1)! \, (m-\ell-j_2)!}$$

$$= \frac{1}{j_1! \, j_2! \, (j_1+k_1-1)! \, (j_2+k_2-1)!} \sum_{0 \le \ell \le m} \frac{(-1)^\ell}{(\ell-j_1)! \, (m-\ell-j_2)!}$$

and this last sum is equal to

$$\frac{(-1)^{j_1}}{(m-j_1-j_2)!} \sum_{0 \le \ell' \le m-j_1-j_2} (-1)^{\ell'} \binom{m-j_1-j_2}{\ell'}$$

and so vanishes if $j_1+j_2 < m$. Hence :

$$\sum_{0\leq \ell\leq m} (-1)^{\ell} e_{m,\ell} (\partial_{z_1}^{\ell} \partial_{z_2}^{m-\ell} h)(z,z) =$$

$$= \sum_{0\leq \ell\leq m} (-1)^{\ell} e_{m,\ell} (\partial_{z_1}^{\ell} \partial_{z_2}^{m-\ell} G(z,z)|_{k_1+k_2+2m}\gamma - \chi\, \partial_{z_1}^{\ell} \partial_{z_2}^{m-\ell} G(z,z))$$

$$= \Phi_m(G)|_{2m+k_1+k_2}\gamma - \chi\, \Phi_m(G) = 0$$

using ii), and so lemma 3.3 is proved.

Lemma 3.3 gives us the missing relation : since by the induction hypothesis we proved that

$$(\partial_{z_1}^{i} \partial_{z_2}^{\lambda-i} h)(z,z) = (-1)^{i} (\partial_{z_2}^{\lambda} h)(z,z)$$

we have from lemma 3.3 :

$$\Big(\sum_{0\leq \ell\leq \lambda} e_{\lambda,\ell}\Big)(\partial_{z_2}^{\lambda} h)(z,z) = 0$$

and since $e_{\lambda,\ell} > 0$ it follows that $(\partial_{z_2}^{\lambda} h)(z,z) = 0$ and so (P_λ) is true by induction. This ends the proof of theorem 2.2. b).

c) We first need a lemma :

LEMMA 3.4. -

i) $$(1-2st+pt^2)^{-(k-\frac{1}{2})} = \frac{1}{(k-3/2)!} \sum_{m\geq 0} t^m \sum_{0\leq \ell\leq m/2} (-1)^{\ell} \frac{(m+k-\ell-3/2)!}{\ell!\,(m-2\ell)!} (2s)^{m-2\ell} p^{\ell}$$

ii) $$(1-2(x-y)t+(x+y)^2t^2)^{-(k-\frac{1}{2})} = \sum_{m\geq 0} t^m \Big[\sum_{0\leq \ell\leq m} d_{m,\ell}\, x^{\ell} y^{m-\ell}\Big]$$

where $$d_{m,\ell} = (-1)^{m-\ell} \frac{(k-1)!\,(k+m-1)!\,(2k+m-2)!}{(2k-2)!\,\ell!\,(m-\ell)!\,(k+\ell-1)!\,(k+m-\ell-1)!}.$$

Proof. - i) The left hand side satisfies the differential equation :

$$(1-2st+pt^2)\frac{dY}{dt} + (k-\frac{1}{2})(2pt-2s)\,Y = 0\,.$$

Setting $Y = \sum_{m\geq 0} a_m t^m$ we obtain the recurrence relation

$$(m+1)a_{m+1} = 2s\,(m+k-\tfrac{1}{2})\,a_m - p(m+2k-2)\,a_{m-1}$$

with initial terms $a_o = 1$, $a_1 = (2k-1)s$.

It is now easy to see that the right hand side satisfies these conditions, so i) follows.

ii) In the recurrence relation above we substitute

$$\sum_{0\leq\ell\leq m} d_{m,\ell}\, x^{\ell} y^{m-\ell} \quad \text{for } a_m\,, \quad (x-y) \text{ for } s \text{ and } (x+y)^2 \text{ for } p\,.$$

Identifying the coefficients of $x^{\ell} y^{m+1-\ell}$ we obtain easily :

$$(m+1)d_{m+1,\ell} = (2m+2k-1)[d_{m,\ell-1} - d_{m,\ell}] - (m+2k-2)[d_{m-1,\ell-2} + 2d_{m-1,\ell-1} + d_{m-1,\ell}]$$

with initial terms

$$d_{0,0} = 1 \qquad d_{1,0} = -(2k-1) \qquad d_{1,1} = 2k-1\,.$$

Again it is an easy matter to prove that the expression given for $d_{m,\ell}$ satisfies all these conditions, which proves ii).

The proof of theorem 2.2. c) can now be easily completed :

From lemma 3.4. i) we have the formal identity

$$\sum_{n\geq 0} t^m \varphi_m = (1-2(\partial_{z_1} - \partial_{z_2})t + (\partial_{z_1} + \partial_{z_2})^2 t^2)^{-(k-\frac{1}{2})}$$

and so we deduce from lemma 3.4. ii) that

$$\varphi_m = \sum_{0\leq\ell\leq m} (-1)^{m-\ell} \frac{(k-1)!\,(k+m-1)!\,(2k+m-2)!}{(2k-2)!\,\ell!\,(m-\ell)!\,(k+\ell-1)!\,(k+m-\ell-1)!}\, \partial_{z_1}^{\ell} \partial_{z_2}^{m-\ell}$$

from which the result follows by definition of Φ_m.

§.4. - Proof of theorem 1.2. - Preliminaries

LEMMA 4.1. - Let f_1, f_2 be two C^∞ functions on $\mathbb{C}$, and k_1, k_2 be given real numbers. Set :

$$F_m(f_1,f_2) = \sum_{0\leq\ell\leq m} (-1)^{\ell} \binom{m}{\ell} \frac{(k_1+m-1)!\,(k_2+m-1)!}{(k_1+m-\ell-1)!\,(k_2+\ell-1)!}\, \partial_z^{m-\ell} f_1 \times \partial_z^{\ell} f_2\,.$$

Then : a) For all $\gamma \in GL_2^+(\mathbb{R})$ we have

$$F_n(f_1|_{k_1}\gamma, f_2|_{k_2}\gamma) = F_n(f_1,f_2)|_{k_1+k_2+2n}\gamma\,.$$

b) We have the identity

$$F_m(f_1,f_2) = \sum_{0\leq\ell\leq m} (-1)^{\ell}\binom{m}{\ell} \frac{(k_1+m-1)!\,(k_1+k_2+2m-\ell-2)!}{(k_1+m-\ell-1)!\,(k_1+k_2+m-2)!}\, \partial_z^{\ell}(f_2 \times \partial_z^{m-\ell} f_1)\,.$$

This lemma is proved in [2] theorem 7.1. Note that part a) follows easily from theorem 2.2. a) by taking

$$G(z_1,z_2) = f_1(z_1)\, f_2(z_2)\,.$$

Part b) can be proved by expanding $\partial_z^{\ell}(f_2 \times \partial_z^{m-\ell} f_1)$ with Leibnitz' formula.

LEMMA 4.2. - Let $f = \sum_{n\geq 1} a(n) q^n \in S_k(\Gamma_o(N), \chi)$ where $k \in \mathbb{N}$, $k \geq 3$. For every $m \in \mathbb{N}$, set

$$C_m^K(f) = \sum_{n\geq 1} q^n \left[\sum_{d|n} d^{m+k-1} \chi(d) \left(\frac{4D}{d}\right) \sum_{s\in\mathbb{Z}} P_m^K(n/d, s)\, a\left(\frac{(n/d)^2 D - s^2}{4}\right)\right]$$

where

$$P_m^K(n/d, s) = \frac{1}{(k-3/2)!} \sum_{0\leq \ell \leq m/2} (-1)^{\ell} \frac{(m+k-\ell-3/2)!}{\ell!\,(m-2\ell)!} \left(\frac{n}{d}\right)^{2\ell} D^{\ell} (2s)^{m-2\ell}.$$

Then $C_m^K(f) \in S_{2k+2m}(\Gamma_o(N_1), \chi^2)$ with $N_1 = 2N/(4, D, N)$.

Proof. - When m is odd, terms in s and $-s$ cancel and so $C_m^K(f) = 0$. So we may suppose m even.

Consider first the case where $D \equiv 0 \pmod 4$. Take

$f_1(z) = \theta(z)$, $f_2(z) = f(z)$, $k_1 = 1/2$, $k_2 = k$, where

$$\theta(z) = \sum_{s\in\mathbb{Z}} e^{2i\pi n^2 z}$$ is the usual theta function of weight $1/2$

on $\Gamma_o(4)$. Then using $n!\,(n-1/2)! = (2n)!\,2^{-2n}\sqrt{\pi}$, one sees from lemma 4.1 that

$$(2i\pi)^{-m/2} F_{\frac{m}{2}}(f_1, f_2) = \frac{m!\,2^{-m}}{(\frac{m}{2}+k-3/2)!} \sum_{n\geq 1} q^n \left[\sum_{s\in\mathbb{Z}} Q_m(n,s)\, a(n-s^2)\right]$$

where $Q_m(n,s) = \sum_{0\leq\ell\leq m/2} (-1)^{\ell} \frac{(m+k-\ell-3/2)!}{\ell!\,(m-2\ell)!} n^{\ell} (2s)^{m-2\ell}$

and $F_{m/2}(f_1, f_2) \in S_{m+k+\frac{1}{2}}(\Gamma_o([N,4]), \chi(\frac{-4}{\cdot})^k)$.

It is now easy to check that, with the notations of theorem 2.1 :

$$S((2i\pi)^{-m/2} F_{m/2}(f_1, f_2); \frac{D}{4}) = \frac{m!\,(k-3/2)!}{(m/2+k-3/2)!} 2^{-2m} C_m^K(f)$$

and thus lemma 4.2 follows from theorem 2.1 in the case $D \equiv 0 \pmod 4$.

If $D \equiv 1 \pmod 4$, we take $f_2(z) = f(4z)$ instead of $f(z)$ and one shows similarly that

$$S((2i\pi)^{-m/2} F_{m/2}(f_1, f_2); D) = \frac{m!\,(k-3/2)!}{(m/2+k-3/2)!} 2^{-m} C_m^K(f)$$

hence lemma 4.2 follows.

COROLLARY 4.3. - When $k\geq 3$ is an integer, E_f^K is a Hilbert modular form of weight k and character $\chi\circ N_{K/\mathbb{Q}}$ on the congruence subgroup Γ'_{N_1} of $GL_2^+(O_K)$ generated by $\Gamma_0(N_1)$ and the matrices $\begin{pmatrix}\varepsilon_1 & \beta\\ 0 & \varepsilon_2\end{pmatrix}$ where $\varepsilon_1,\varepsilon_2$ are units, $\varepsilon_1\varepsilon_2\gg 0$ and $\beta\in O_K$.

Proof. - Set

$$q = e^{2i\pi(z_1+z_2)/2} \qquad \tau = e^{2i\pi(z_1-z_2)/2\sqrt{D}} .$$

Writing $\nu = (s+n\sqrt{D})/2\sqrt{D}$, it is clear that we have :

$$E_f^K(z_1,z_2) = \sum_{n\geq 1} q^n \sum_{d\mid n} d^{k-1}\chi(d)\left(\frac{4D}{d}\right)\sum_{s\in\mathbb{Z}} \tau^{sd}\, a\left(\frac{(n/d)^2 D - s^2}{4}\right) .$$

Note that the inner sum is finite since by convention $a(x)=0$ for $x\notin\mathbb{N}$. It is now easy to see that :

$$\varphi_m(E_f^K)(z) = (2i\pi)^m \sum_{n\geq 1} q^n \sum_{d\mid n} d^{k+m-1}\chi(d)\left(\frac{4D}{d}\right)\sum_{s\in\mathbb{Z}} P_m(n/d,s)\, a\left(\frac{(n/d)^2 D - s^2}{4}\right)$$

and so

$$\varphi_m(E_f^K) = (2i\pi)^m\, C_m^K(f)$$

so by lemma 4.2, $\varphi_m(E_f^K)\in S_{2k+2m}(\Gamma_0(N_1),\chi^2)$ for all $m\geq 0$, and we deduce from theorem 2.2.b) that for every $\gamma = \begin{pmatrix}a & b\\ c & d\end{pmatrix}\in\Gamma_0(N_1)$:

$$E_f^K\left(\frac{az_1+b}{cz_1+d},\frac{az_2+b}{cz_2+d}\right) = \chi^2(d)(cz_1+d)^k(cz_2+d)^k E_f^K(z_1,z_2) .$$

On the other hand it is clear from the definition of E_f^K that if $\varepsilon_1,\varepsilon_2$ are units such that $\varepsilon_1\varepsilon_2\gg 0$ and $\beta\in O_K$, then :

$$(*) \qquad E_f^K\left(\frac{\alpha z_1+\beta}{\gamma z_1+\delta},\frac{\alpha' z_2+\beta'}{\gamma' z_2+\delta'}\right) = \chi(\delta\delta')(\gamma z_1+\delta)^k(\gamma' z_2+\delta')^k E_f^K(z_1,z_2)$$

is true for $\begin{pmatrix}\alpha & \beta\\ \gamma & \delta\end{pmatrix} = \begin{pmatrix}\varepsilon_1 & \beta\\ 0 & \varepsilon_2\end{pmatrix}$. Thus (*) is valid for the group Γ'_{N_1} generated by $\Gamma_0(N_1)$ and the matrices $\begin{pmatrix}\varepsilon_1 & \beta\\ 0 & \varepsilon_2\end{pmatrix}$. This group is described in detail in [4]. Let us simply show why it is a congruence subgroup.

Let $(1,\xi)$ be a $\mathbb{Z}$-basis of O_K, and let $u = u_1+u_2\xi$ be a totally positive unit. Then $\begin{pmatrix}1 & 0\\ N_1u & 1\end{pmatrix}\in\Gamma'_{N_1}$ for $\begin{pmatrix}u^{-1} & 0\\ 0 & 1\end{pmatrix}\begin{pmatrix}1 & 0\\ N_1 & 1\end{pmatrix}\begin{pmatrix}u & 0\\ 0 & 1\end{pmatrix} = \begin{pmatrix}1 & 0\\ N_1u & 1\end{pmatrix}$.

Thus for any pair $(c_1,c_2)\in\mathbb{Z}\times\mathbb{Z}$ we have

$$\begin{pmatrix}1 & 0\\ N_1u_2(c_1+c_2\xi) & 1\end{pmatrix} = \begin{pmatrix}1 & 0\\ N_1c_2u & 1\end{pmatrix}\begin{pmatrix}1 & 0\\ N_1(c_1q_2-c_2q_1) & 1\end{pmatrix}\in\Gamma'_{N_1}$$

so for any $\gamma \in N_1 u_2 O_K$ we have $\begin{pmatrix} 1 & 0 \\ \gamma & 1 \end{pmatrix} \in \Gamma'_{N_1}$. Since for any $\beta \in O_K$ we have $\begin{pmatrix} 1 & \beta \\ 0 & 1 \end{pmatrix} \in \Gamma'_{N_1}$, we deduce from a deep theorem of Vašerstein [12] that :

$$\Gamma'_{N_1} \supset \{ \begin{pmatrix} \alpha & \beta \\ \gamma & \delta \end{pmatrix} \in SL_2(O_K) \,,\ \alpha, \delta \in 1 + N_1 u_2 O_K \,,\ \gamma \in N_1 u_2 O_K \}$$

and so Γ'_{N_1} is a congruence subgroup of level dividing $N_1 u_2$. Much more is shown in [4].

However the group we want is a bit larger, and so we have to look back at what we have done. We have essentially used the infinitesimal information around the diagonal $z_1 = z_2 = z$ with the help of theorem 2.2.b). But one can also use the "twisted diagonals" $z_1 = \mu z$, $z_2 = \mu' z$, where μ is a totally positive integer of K, so we can hope to enlarge our group (I am indebted to N. Katz for this suggestion).

§. 5. - Twisted theorems

Let us fix a totally positive integer $\mu \in O_K$, and write $M = N_{K/\mathbb{Q}}\, \mu = \mu\mu'$. We shall suppose without loss of generality that μ is primitive, i.e. $\mu/e \in O_K$ and $e \in \mathbb{Z}$ implies $e = \pm 1$.

Set $F(z_1, z_2) = E_f^K(\mu z_1, \mu' z_2)$ and $\mu = \frac{a + b\sqrt{D}}{2}$ with $a \equiv bD \pmod 2$. If $e = (a, b)$, the primitivity condition is equivalent to $e = 1$ or 2, and $e = 2 \Rightarrow \frac{a}{2} \not\equiv \frac{b}{2} D \pmod 2$. It is easy to see that this is equivalent to the existence of $A, B \in \mathbb{Z}$ such that

$$Aa + Bb = e$$
$$A^2 D - B^2 \equiv 0 \pmod{e^2}.$$

Set $\lambda = A\frac{b}{e} D + B\frac{a}{e}$. Then we have

$$\lambda^2 = D - \left(\frac{A^2 D - B^2}{e^2}\right) 4M \equiv D \pmod{4M}.$$

LEMMA 5.1. - Setting again $q = e^{2i\pi(z_1+z_2)/2}$, $\tau = e^{2i\pi(z_1 - z_2)/2\sqrt{D}}$ we have :

$$F(z_1, z_2) = \sum_{n \geq 1} q^n \sum_{d|n} d^{k-1} \chi(d) \left(\frac{4D}{d}\right) \sum_{s \equiv \lambda \frac{n}{d}\ (2M)} \tau^{ds}\, a\left(\frac{(N/d)^2 D - s^2}{4M}\right).$$

Proof. - Writing $\mu\nu = (s + n\sqrt{D})/2\sqrt{D}$, it is clear that :

$$F(z_1, z_2) = \sum_{\substack{n\geq 1 \\ |s|\leq n\sqrt{D}}} q^n \tau^s \sum_{d \,|\, \frac{(s+n\sqrt{D})/2}{(a+b\sqrt{D})/2}} d^{k-1} \chi(d)\left(\frac{4D}{d}\right) a\left(\frac{(n/d)^2 D - (s/d)^2}{4M}\right)$$

$$= \sum_{n\geq 1} q^n \sum_{d|n} d^{k-1} \chi(d)\left(\frac{4D}{d}\right) \sum_{\frac{a+b\sqrt{D}}{2} \,|\, \frac{s+(n/d)\sqrt{D}}{2}} \tau^{ds}\, a\left(\frac{(n/d)^2 D - s^2}{4M}\right).$$

One checks immediately that the condition $\frac{a+b\sqrt{D}}{2} \,|\, \frac{s+(n/d)\sqrt{D}}{2}$ is equivalent to the system of congruences :

$$as - bD(n/d) \equiv 0 \pmod{2M}$$
$$bs - a(n/d) \equiv 0 \pmod{2M}$$
$$(n/d)^2 D - s^2 \equiv 0 \pmod{4M}.$$

From the first two congruences we obtain

$$s \equiv \left(A\frac{b}{e} D + B\frac{a}{e}\right)\left(\frac{n}{d}\right) \pmod{\frac{2M}{e}}$$

that is $s \equiv \lambda \frac{n}{d} \pmod{\frac{2M}{e}}$.

Write $s = \lambda \frac{n}{d} + \frac{2}{e} M\rho$. We have

$$s^2 - (n/d)^2 D = \left(\frac{B^2 - A^2 D}{e^2}\right) 4M \left(\frac{n}{d}\right)^2 + \frac{4M}{e} \rho \lambda \frac{n}{d} + \frac{4}{e^2} M^2 \rho^2 .$$

If $e = 1$, this is indeed congruent to $0 \mod 4M$.
If $e = 2$, $s^2 - (n/d)^2 D \equiv 2M\rho\lambda(n/d) + M^2\rho^2 \pmod{4M}$ so the third congruence above implies

$$2\rho\lambda(n/d) + M\rho^2 \equiv 0 \pmod 4 .$$

But since μ is primitive, $a/2 \not\equiv bD/2 \pmod 2$ and so M is odd. Thus we must have ρ even, i. e. $s \equiv \lambda \frac{n}{d} \pmod{2M}$. Conversely if this is true the three congruences above are evidently satisfied, so lemma 5. 1. follows.

COROLLARY 5. 2. -

$$\varphi_m(F)(z) = (2i\pi)^m \sum_{n\geq 1} q^n \sum_{d|n} d^{k+m-1} \chi(d)\left(\frac{4D}{d}\right) \sum_{s \equiv \lambda\frac{n}{d} \pmod{2M}} P_m(n/d, s)\, a\left(\frac{(n/d)^2 D - s^2}{4M}\right).$$

This is immediate from lemma 5. 1 by definition of φ_m and P_m .

Our goal is now to prove that $\varphi_m(F)$ is a cusp form on some group. The difficulty lies in the condition $s \equiv \lambda(n/d) \pmod{2M}$. We shall deal with this congruence by the usual method of summing over characters (mod 2M).

For any integer m write $\mathfrak{X}_m$ for the group of characters modulo m . Furthermore set

$$e_1 = (\lambda, M) \ , \ \lambda' = \lambda/e_1 \ , \ M' = M/e_1 .$$

Then we have :

PROPOSITION 5.3. - Suppose that M is odd when $D \equiv 1 \pmod 4$. Then :

$$\varphi_m(F)(z) = (2i\pi)^m \sum_{e_2 | M'} (e_2^m / \varphi(M'/e_2)) \sum_{\chi_1 \in \mathfrak{X}_{M'/e_2}} \overline{\chi}_1(\lambda') \, \varphi(F;\chi_1)(e_2 z)$$

where

$$\varphi(F;\chi_1)(z) = \sum_{n \geq 1} q^n \sum_{d|n} d^{k+m-1} \chi(d)\left(\frac{4D}{d}\right) \overline{\chi}_1\left(\frac{n}{d}\right) \times$$

$$\times \sum_{s \in \mathbb{Z}} \chi_1(s) P_m(n/d, e_1 s) \, a\left(e_2 \cdot \frac{(n/d)^2 D - e_1^2 s^2}{4M/e_2}\right) .$$

Proof. - We shall first show the following lemma :

LEMMA 5.4. - Suppose that M is odd if $D \equiv 1 \pmod 4$.

Then if $(n/d)^2 D - s^2 \equiv 0 \pmod{4M}$:

$$s \equiv \lambda(n/d) \pmod{2M} \Leftrightarrow s \equiv \lambda(n/d) \pmod M .$$

Indeed suppose $s \equiv \lambda(n/d) \pmod M$, i.e. $s = \lambda(n/d) + \rho M$. Then

$$s^2 - (n/d)^2 D = \left(\frac{B^2 - A^2 D}{e^2}\right)\left(\frac{n}{d}\right)^2 4M + 2\rho\lambda\frac{n}{d} M + \rho^2 M^2 .$$

If $D \equiv 1 \pmod 4$, M is odd and so $s^2 - (n/d)^2 D \equiv 0 \pmod{4M}$ implies $2\rho\lambda(n/d) + \rho^2 M \equiv 0 \pmod 4$ so $\rho^2 \equiv 0 \pmod 2$ hence $s \equiv \lambda(n/d) \pmod{2M}$.
If $D \equiv 0 \pmod 4$, then since $\lambda^2 \equiv D \pmod{4M}$, λ is even and so $s^2 - (n/d)^2 D \equiv 0 \pmod{4M}$ is equivalent to

$$\rho^2 M \equiv 0 \pmod 4 .$$

But $D/4 \equiv 2$ or $3 \pmod 4$ and so one checks that

$$(a/2)^2 - b^2(D/4) \equiv 0 \pmod 4 \Rightarrow a/2 \equiv b \equiv 0 \pmod 2 \Rightarrow \mu/2 \in O_K$$

in contradiction with the primitivity of μ. So we must have $M \not\equiv 0 \pmod 4$, and so it follows again that ρ must be even, whence lemma 5.4.

Thus we have :

$$(2i\pi)^{-m}\varphi_m(F)(z)=\sum_{n\geq 1} q^n \sum_{d|n} d^{k+m-1}\chi(d)\left(\frac{4D}{d}\right)\times$$

$$\times \sum_{s\equiv\lambda'(n/d)\ (\mathrm{mod}\ M')} P_m(n/d, e_1 s)\, a\left(\frac{(n/d)^2 D - e_1^2 s^2}{4M}\right)$$

$$=\sum_{e_2|M'}\sum_{n\geq 1} q^n \sum_{\substack{d|n \\ (n/d, M')=e_2}} d^{k+m-1}\chi(d)\left(\frac{4D}{d}\right)\times$$

$$\times \sum_{s\equiv\lambda'(n/d)\ (\mathrm{mod}\ M')} P_m(n/d, e_1 s)\, a\left(\frac{(n/d)^2 D - e_1^2 s^2}{4M}\right)$$

$$=\sum_{e_2|M'} e_2^m \sum_{n\geq 1} q^{e_2 n} \sum_{\substack{d|n \\ (n/d, M'/e_2)=1}} d^{k+m-1}\chi(d)\left(\frac{4D}{d}\right)\times$$

$$\times \sum_{s\equiv\lambda'(n/d)\ (\mathrm{mod}\ M/e_2)} P_m(n/d, e_1 s)\, a\left(e_2\frac{(n/d)^2 D - e_1^2 s^2}{4M/e_2}\right)$$

using the homogeneity in n/d and s^2 of P_m.

Since $(n/d, M'/e_2)=1$ and $(\lambda', M'/e_2)=1$ by definition of e_1, it is clear that

$$\sum_{\chi_1\in\mathfrak{X}_{M'/e_2}} \overline{\chi}_1(\lambda')\overline{\chi}_1(n/d)\chi_1(s)=\begin{cases} 0 & \text{if } s\not\equiv\lambda'(n/d) \quad (\mathrm{mod}\ M'/e_2) \\ \varphi(M'/e_2) & \text{if } s\equiv\lambda'(n/d) \ (\mathrm{mod}\ M'/e_2)\ . \end{cases}$$

Proposition 5.3. now follows immediately.

Our aim is now to prove :

PROPOSITION 5.4. -

$$\varphi(F;\chi_1)\in S_{2k+2m}(\Gamma_o(N_1 M^3/e_1 e_2^2), \chi^2)\ .$$

Proof. - We can clearly suppose $\chi_1(-1)=(-1)^m$ since otherwise the terms s and $-s$ cancel and so $\varphi(F;\chi_1)=0$.

Set $\theta_{\chi_1}(z)=\sum_{s\in\mathbb{Z}}\chi_1(s)q^{s^2}$ when $\chi_1(-1)=1$

$\theta_{\chi_1}(z)=\sum_{s\in\mathbb{Z}}\chi_1(s)\,s\,q^{s^2}$ when $\chi_1(-1)=-1$.

Then it is well known (see for instance [11]) that

$$\theta_{\chi_1} \in M_{1/2}(\Gamma_o(4M'^2/e_2^2), \chi_1) \quad \text{when} \quad \chi_1(-1) = 1$$

$$\theta_{\chi_1} \in M_{3/2}(\Gamma_o(4M'^2/e_2^2), \chi_1(\tfrac{-4}{\cdot})) \quad \text{when} \quad \chi_1(-1) = -1 .$$

Furthermore we have the following lemma :

LEMMA 5.5. - Let $g(z) = \sum_{n\geq 1} b(n)q^n \in S_k(\Gamma_o(N), \chi)$. Then

$$\sum_{n\geq 1} b(e_2 n)q^n \in S_k(\Gamma_o(e_2 N), \chi) .$$

Proof. - Recall that the Hecke operator T_m acts on Fourier coefficients by

$$T_m g(z) = \sum_{n\geq 1} q^n \Big(\sum_{d|(m,n)} \chi(d)\, d^{k-1}\, b(nm/d^2)\Big)$$

$$= \sum_{d|m} \chi(d)\, d^{k-1} \sum_{n\geq 1} b(\tfrac{m}{d}\cdot n)\, q^{dn}$$

Thus :

$$\sum_{d'|e_2} \chi(d')\,d'^{k-1} \mu(d')(T_{(e_2/d')}(g))(d'z) = \sum_{d_1|e_2} \chi(d_1)\, d_1^{k-1} \sum_{n\geq 1} b(\frac{e_2 n}{d_1})\, q^{d_1 n} \sum_{d'|d_1} \mu(d')$$

where we have written $d_1 = dd'$, and so this is equal to $\sum b(e_2 n)q^n$, since the last sum is zero if $d_1 \neq 1$.

Lemma 5.5. follows immediately.

We distinguish again the cases $D \equiv 0$ or $1 \pmod 4$. Suppose $D \equiv 0 \pmod 4$. In lemma 4.1, take

$$f_1(z) = \theta_{\chi_1}(e_1^2 z) \qquad \text{and} \qquad f_2(z) = \sum_{n\geq 1} a(e_2 \frac{n}{M/e_2})q^n .$$

Then by lemma 5.5. $f_2 \in S_k(\Gamma_o(MN), \chi)$, and

$$\theta_{\chi_1}(e_1^2 z) \in M_{\frac{1}{2}+\nu}(\Gamma_o(4M^2/e_2^2), \chi_1(\tfrac{-4}{\cdot})^\nu) \quad \text{if} \quad \chi_1(-1) = (-1)^\nu \quad \text{and} \quad \nu = 0 \text{ or } 1 .$$

So lemma 4.1 shows that

$$F_{(m-\nu)/2}(f_1, f_2) \in S_{k+m+\frac{1}{2}}(\Gamma_o(4M^2 N/e_2(4,N)), \chi\chi_1(\tfrac{-4}{\cdot})^{\nu+k})$$

and furthermore that

$$(2i\pi)^{-(\frac{m-\nu}{2})} e_1^{\nu} F_{(m-\nu)/2}(f_1, f_2) = \frac{m!\, 2^{-m}}{(\frac{m+\nu-3}{2}+k)!} \sum_{n\geq 1} q^n \times$$

$$\times \sum_{s\in\mathbb{Z}} \chi_1(s)\, Q_m(n, e_1 s)\, a(e_2 \frac{n-e_1^2 s^2}{M/e_2}).$$

As in the proof of lemma 4. 2, we would like now to use Shimura-Niwa's theorem 2.1. However this will not exactly give us our function $\varphi(F;\chi_1)$ since there is a $\bar{\chi}_1(n/d)$ in its definition. So we need a twisted version of theorem 2. 1.

This is as follows :

THEOREM 5. 6. - Let $g = \sum_{n\geq 1} b(n) q^n \in S_{k+\frac{1}{2}}(\Gamma_o(4N), \chi)$. Suppose that $\chi = \chi_1\chi_2(\frac{-4}{\cdot})^k$ on $(\mathbb{Z}/4N\mathbb{Z})^*$, where $\chi_1 \in \mathfrak{X}_{N_1}$, $\chi_2 \in \mathfrak{X}_{N_2}$, where we assume $N_1N_2|4N$ and $N_2^2|N$. Write F_2 = conductor of χ_2. Then for any squarefree t, if we set :

$$S(g;t;\chi_1,\chi_2)(z) = \sum_{n\geq 1} q^n \sum_{d|n} d^{k-1}\chi_1(d)(\frac{(-1)^k 4t}{d})\,\bar{\chi}_2(\frac{n}{d})\, b((n/d)^2 t)$$

we have for $k\geq 3$: $S(g;t;\chi_1,\chi_2) \in S_{2k}(\Gamma_o(2N(N_2/F_2)), \chi_1^2)$.

Assuming this theorem for the moment, we deduce from above that :

$$S(\chi_1(2).(2i\pi)^{-(\frac{m-\nu}{2})} e_1^{\nu} F_{(m-\nu)/2}(f_1, f_2); \frac{D}{4}; \chi, \chi_1) = \frac{m!\, 2^{-2m}(k-3/2)!}{(\frac{m+\nu-3}{2}+k)!}\, \varphi(F;\chi_1)$$

and proposition 5. 4 follows in the case $D\equiv 0 \pmod 4$.

In case $D\equiv 1 \pmod 4$ we must replace $f_2(z)$ by

$$f_2(z) = \sum_{n\geq 1} a(e_2 \frac{n}{4M/e_2})\, q^n$$

and take $t = D$ instead of $D/4$ in theorem 5. 6. The proof is entirely similar. This shows proposition 5. 4.

It remains to prove theorem 5. 6. We shall do this by a simple modification of Niwa's proof [8]. We sketch these modifications, using his notations.

We take the lattice $L' = \mathbb{Z} \oplus \frac{N}{N_2}\mathbb{Z} \oplus \frac{N}{4}\mathbb{Z}$ and in the definition of $\theta(z,g)$ we replace $\bar{\chi}_1(x_1)$ by

$$\bar{\chi}_1(x_1)\,\bar{\chi}_2(x_2 N_2/N).$$

We must also change the definition of $\theta_{1,\varepsilon}(z)$:

$$\theta_{1,\varepsilon}(z) = \sum_{h\in L'/L} \chi_2(hN_2/N)\,\theta(z, f_{1,\varepsilon}, h)$$

where $L = N\mathbb{Z}$, $L' = (N/N_2)\mathbb{Z}$.

The transformation formulae can then be easily proved, with some characters χ_2 popping up, and the rest of the proof is practically identical.

However the result obtained is not quite theorem 5.6, but the theorem obtained by replacing $\overline{\chi}_2(n/d)$ with

$$\tau(\chi_2, n/d) = \sum_{h\in \mathbb{Z}/N_2\mathbb{Z}} \chi_2(h)\, e^{2i\pi hk/N_2} .$$

This Gauss sum comes up in the Fourier expansion of the function $\theta_m(z)$.

Since $\tau(\chi_2, n/d) = \overline{\chi}_2(n/d)\,\tau(\chi_2)$ where $\tau(\chi_2)$ is a non zero constant when χ_2 is primitive, theorem 5.6 is proved for χ_2 primitive.

If χ_2 is not primitive, call ψ_2 the primitive character mod F_2 equivalent to χ_2. Then :

$$\sum_{n\geq 1} q^n \sum_{d|n} d^{k-1}\chi_1(d)\Big(\frac{(-1)^k 4t}{d}\Big)\overline{\chi}_2(n/d)\, b((n/d)^2 t) =$$

$$= \sum_{n\geq 1} q^n \sum_{\substack{d|n\\ (n/d,\, N_2/F_2)=1}} d^{k-1}\chi_1(d)\Big(\frac{(-1)^k 4t}{d}\Big)\overline{\psi}_2(n/d)\, b((n/d)^2 t)$$

and using the usual trick of introducing the Möbius function to get rid of conditions of the type $(n,m)=1$, this is equal to

$$\sum_{e|(N_2/F_2)} \mu(e)\overline{\psi}_2(e) \sum_{n\geq 1} q^{en} \sum_{d|n} d^{k-1}\chi_1(d)\Big(\frac{(-1)^k 4t}{d}\Big)\overline{\psi}_2\Big(\frac{n}{d}\Big)\, b(e^2(n/d)^2 t) .$$

Now according to [11] proposition 1.5, we still have

$$\sum_{n\geq 1} b(e^2 n)q^n \in S_{k+\frac{1}{2}}(\Gamma_o(4N), \chi) \qquad \text{since } e^2 \mid N .$$

We deduce from theorem 5.6 applied to (χ_1, ψ_2), where ψ_2 is now primitive, that

$$\sum_{n\geq 1} q^n \sum_{d|n} d^{k-1}\chi_1(d)\Big(\frac{(-1)^k 4t}{d}\Big)\overline{\psi}_2\Big(\frac{n}{d}\Big)\, b(e^2(n/d)^2 t) \in S_{2k}(\Gamma_o(2N), \chi_1^2)$$

and so theorem 5.6 follows in general.

Remark. - It is possible that the extra factor N_2/F_2 in the level can be removed. However for our purposes it does not matter.

§. 6. - End of the proof of theorem 1.2.

Recall first the following group-theoretical theorem :

THEOREM 6.1. - *Let* q_2, N *be fixed integers. Then the group generated by* $\Gamma_o(N)$ *and all matrices* $\begin{pmatrix} 1 & \beta \\ 0 & 1 \end{pmatrix}$ *and* $\begin{pmatrix} 1 & 0 \\ Nq_2\gamma & 1 \end{pmatrix}$ $(\beta, \gamma \in O_K)$ *is the set of matrices* $\begin{pmatrix} \alpha & \beta \\ \gamma & \delta \end{pmatrix} \in SL_2(O_K)$ *such that* :

i) $\alpha, \delta \in \mathbb{Z} + NO_K$

ii) $\gamma \in N(\mathbb{Z}+(q_2,N)O_K)$

iii) *Let* $v=2$ *if* $1 \leq v_2(N) < v_2(q_2)$, $v=1$ *otherwise. Then if* $(1, \theta)$ *is a* $\mathbb{Z}$*-basis of* O_K *and if one writes* :

$$\gamma = N(c_1 + c_2(q_2, N)\theta) \qquad \delta = d_1 + d_2 N\theta$$

we must have

$$d_2(c_1+1) \equiv c_2 \pmod v .$$

(This condition is of course empty if $v = 1$.)

This theorem is proved in [4] and depends on Vašerstein's theorem cited above [12]. The proof being long and very technical will not be reproduced here. However in the particular case $q_2 = 1$ it can be proved rather shortly.

Let us now put together the results of proposition 5.3 and 5.4. We obtain

$$\varphi_m(F) \in S_{2k+2m}(\Gamma_o(N_1 M^3), \chi^2) .$$

Hence, using theorem 2.2.b) we deduce that for every $\gamma = \begin{pmatrix} a & b \\ c & d \end{pmatrix} \in \Gamma_o(N_1 M^3)$:

$$F|_{(k,k)}(\gamma, \gamma) = \chi^2(d) F .$$

Coming back to $E_f^K(z_1, z_2) = F(z_1/\mu, z_2/\mu')$, we see that the equality

$$(*) \quad E_f^K\left(\frac{\alpha z_1 + \beta}{\gamma z_1 + \delta}, \frac{\alpha' z_2 + \beta'}{\gamma' z_2 + \delta'}\right) = \chi(\delta\delta')(\gamma z_1 + \delta)^k (\gamma' z_2 + \delta')^k E_f^K(z_1, z_2)$$

is valid for $\begin{pmatrix} \alpha & \beta \\ \gamma & \delta \end{pmatrix} = \begin{pmatrix} a & b\mu \\ c/\mu & d \end{pmatrix}$ where $\begin{pmatrix} a & b \\ c & d \end{pmatrix} \in \Gamma_o(N_1 M^3)$.

In particular (*) is valid for the matrix $\begin{pmatrix} 1 & 0 \\ N_1 M^2 \mu' & 1 \end{pmatrix}$.

If we write $\mu = \mu_1 - \mu_2 \theta$, where $(1, \theta)$ is a $\mathbb{Z}$-basis of O_K, we deduce, taking successively 1 and μ that (*) is valid for $\begin{pmatrix} 1 & 0 \\ N_1 M^2 \mu_2 \theta & 1 \end{pmatrix}$.

Now if $D \equiv 0 \pmod 4$, we choose $\theta = \sqrt{D/4}$ and we take all μ of the form $a-\theta$ with $a > \theta$, $a \in \mathbb{N}$. We have $M = a^2 - D/4$, and one checks easily that $((a+1)^2 - D/4, a^2 - D/4) = (2a+1, 1-D)$ and so is equal to 1 for a suitable value of $a_{\cdot}$.

Thus (*) is valid for $\begin{pmatrix} 1 & 0 \\ N_1\theta & 1 \end{pmatrix}$ and hence for $\begin{pmatrix} 1 & 0 \\ N_1\gamma & 1 \end{pmatrix}$ for any $\gamma \in O_K$. We can apply theorem 6.1 with $N = N_1$ and $q_2 = 1$ and theorem 1.2 follows in case $D \equiv 0 \pmod 4$. Suppose now $D \equiv 1 \pmod 4$. We choose $\theta = \frac{1+\sqrt{D}}{2}$. If $D \equiv 5 \pmod 8$, we take again all μ of the form $a-\theta$, and we see that $\begin{pmatrix} 1 & 0 \\ N_1 \frac{(2a+1)^2 - D}{4}\theta & 1 \end{pmatrix}$ satisfies (*).

Since $\left(\frac{(2a+1)^2 - D}{4}, \frac{(2a+3)^2 - D}{4}\right) = \left(a+1, \frac{1-D}{4}\right) = 1$ for a suitable value of a, we conclude again that (*) is valid for $\begin{pmatrix} 1 & 0 \\ N_1\theta & 1 \end{pmatrix}$ and theorem 1.2 follows again from theorem 6.1.

Finally if $D \equiv 1 \pmod 8$, then if $\mu = \mu_1 - \mu_2\theta$ is of odd norm, one sees easily that μ_2 must be even. Taking all μ of the form $2a+1-2\theta$ $(a > \theta,\ a \in \mathbb{N})$, we see that (*) is valid for all matrices $\begin{pmatrix} 1 & 0 \\ N_1(4a^2 - D)2\theta & 1 \end{pmatrix}$ and since $(4a^2 - D, 4(a+1)^2 - D) = (2a+1, 1-D) = 1$ for a suitable choice of a, we conclude that (*) is valid for $\begin{pmatrix} 1 & 0 \\ 2N_1\theta & 1 \end{pmatrix}$ and thus also for $\begin{pmatrix} 1 & 0 \\ 2N_1\gamma & 1 \end{pmatrix}$ for any $\gamma \in O_K$.

We must now apply theorem 6.1 with $q_2 = 2$. We have $v = 1$ and $(q_2, N_1) = 2$ so we get the extra factor $w = 2$ in this case. This ends the proof of theorem 1.2.

§. 7. - Concluding remarks

7.1. - Let us first consider possible improvements to theorem 1.2. First of all one could hope to improve on the group on which E_f^K is modular. However, inherent in our proof is the fact that χ can be multiplied by an even character $\psi \in \mathfrak{X}_N$ of order 2 without changing the resulting character $\chi \circ N_{K/\mathbb{Q}}$. This implies that for any such character ψ and any $\delta \in O_K$ such that E_f^K is modular for $\begin{pmatrix} \alpha & \beta \\ \gamma & \delta \end{pmatrix}$, we must have $\psi(\delta\delta') = 1$.

This is of course satisfied for $\delta \in \mathbb{Z} + N_1 O_K$, but it shows that a similar proof could not allow $\delta \in O_K$ arbitrary, so the group we have found, maybe not the best, cannot be too much enlarged by our method.

Second, note that our original aim of finding a suitable generalization to cusp forms of proposition 1.1 is satisfied, but with $\left(\frac{D}{d}\right)$ replaced by $\left(\frac{4D}{d}\right)$. This means simply that one must exclude even d's. It is well possible that the replacement in

$c(\nu\mathfrak{d})$ of $(\frac{4D}{d})$ by $(\frac{D}{d})$ will lower the level, for example transforming N_1 into N. I think it should be possible to do this by finding a still wider generalization of Shimura's theorem than theorem 5.6, where the annoying $(\frac{-4}{\cdot})$ is taken care of.

Lastly one could hope to generalize theorem 1.2 to non cusp forms. It would be enough to prove it for Eisenstein series. For this I see two (related) ways : either generalize Shimura's theorem to non cusp forms, or find directly what should be the image of the initial Eisenstein series. In both cases the calculations lead to complicated Gaussian sums in the field K, and I have been unable to finish the proofs except in some cases. However I conjecture that the constant terme one should add to the definition of E_f^K when $f = \sum_{n\geq 0} a(n)q^n$ is not assumed to be a cusp form, is equal to :

$$\frac{a(0)}{2} L(1-k, \chi(\frac{D}{\cdot})).$$

7.2.- Our theorem 1.2 is closely related to the Doi-Naganuma map defined in [5]. In a sense it is at the same time weaker and stronger. It is weaker mainly because of the group Γ_{N_1} obtained, which is not as large as one could hope for. It is stronger because it is of much greater generality, since Doi-Naganuma [5] and Naganuma [7] restrict themselves to $S_k(\Gamma_o(D), (\frac{D}{\cdot}))$ and to $S_k(SL_2(\mathbb{Z}))$. Furthermore one deduces Doi-Naganuma's mapping from theorem 1.2 by taking a suitable linear combination of $E_{f_s}^K$ where $f_s = f|_k \gamma_s$ and γ_s sends $i\infty$ to the cusp s of $\Gamma_o(D)$ (see [13]). One could say (in the case of $S_k(\Gamma_o(D), (\frac{D}{\cdot}))$) that theorem 1.2 is an **asymmetrical form** of the Doi-Naganuma mapping. It is easy to see on examples that it does not map Hecke eigenforms to Hecke eigenforms.

7.3.- Recent work of Kudla [6] suggests that, as one can interpret the Shimura map by a Petersson product with a certain theta function corresponding to an indefinite quadratic form of type (2,1) (see Niwa [8]), one should be able to interpret Doi-Naganuma's or our mapping as a Petersson product with a theta function of an indefinite quadratic form of type (2,2). This would have the advantage of giving certainly a "good" group on which E_f^K is modular, in the same way that Niwa's proof improved the level given in Shimura's paper. The possibility of using quadratic forms of type (2,2) was in fact suggested by Niwa himself at the end of his paper, and carried out explicitly in the case of the Doi-Naganuma map by Zagier [13].

7.4. - One can think about two further generalizations of theorem 1.2. The first one would be to extend it to fields K other than quadratic fields. This has been done by Saito [9] for the Doi-Naganuma map in case $K/\mathbb{Q}$ is cyclic of prime degree, tamely ramified and $h(K)=1$. Saito's lifting being also very much Hecke-invariant, one would need again an asymmetrical form to generalize theorem 1.2.

The other generalization would be to find a lifting to Siegel modular forms. Work of Kudla [6] mentioned above suggests there should be a map from modular forms of half integral weight to Siegel modular forms of genus 2 given by a Petersson product with a theta series of an indefinite quadratic form of type (3,2).

-:-:-:-

BIBLIOGRAPHY

[1] H. COHEN, Variations sur un thème de Siegel et Hecke, Acta Arith., 30, (1976), p. 63-93.

[2] H. COHEN, Sums involving the values at negative integers of L functions of quadratic characters, Math. Ann. 217, (1975), p. 271-285.

[3] H. COHEN, Formes modulaires à deux variables associées à une forme à une variable, C. R Acad. Sci. Paris, 281, (1975), p. 753-755.

[4] H. COHEN, Formes modulaires à une et deux variables, Thèse, Université de Bordeaux I (1976).

[5] K. DOI and H. NAGANUMA, On the functional equation of certain Dirichlet series, Invent. Math. 9, (1969), p. 1-14.

[6] S. KUDLA, Theta functions and Hilbert modular forms, to appear.

[7] H. NAGANUMA, On the coïncidence of two Dirichlet series associated with cusp forms of Hecke's Neben-type and Hilbert modular forms over a real quadratic field, J. Math. Soc. Japan, 25, (1973), p. 547-555.

[8] S. NIWA, Modular forms of half integral weight and the integral of certain theta functions, Nagoya Math. J. 56, (1974), p. 147-161.

[9] H. SAITO, Algebraic extensions of number fields and automorphic forms, Kyoto Univ. Lectures in Math. 8, Tokyo : Kinokuniya (1973).

[10] B. SCHOENEBERG, Elliptic modular functions, Springer Verlag, (1974).

[11] G. SHIMURA, Modular forms of half integral weight, Ann. of Math. 97 (1973), p. 440-481.

[12] L.W. VASERŠTEIN, On the group SL_2 over Dedeking rings of arithmetic type, Mat. Sbornik 89 (1972) = Math. USSR Sbornik 18, (1972), p. 321-332.

[13] D. ZAGIER, Modular forms associated to real quadratic fields, Invent. Math 30, (1975), p. 1-46.

-:-:-:-

Henri COHEN
Laboratoire de Mathématiques
et d'Informatique dépendant de
l'Université de Bordeaux I
associé au C. N. R. S.
351, cours de la Libération
33405 TALENCE CEDEX

THETA FUNCTIONS OVER Q AND OVER $\mathbf{Q}(\sqrt{q})$

by M. Eichler

CONTENTS

International Summer School on Modular Functions
Bonn 1976

INTRODUCTION.

In the following we present a new variation of a tune which has been played in the last 7 years on different instruments [1], [8], [10], [12], namely the lifting of modular forms of one variable to Hilbert modular forms in two (or more) variables. Our instrument is the application of similarities of quadratic forms. We have seen in 1952 [2], that the Hecke operators T(n) on theta series of quadratic forms can be realized by families of similarities on these forms, whose "similarity norm" is n, provided that such similarities exist. This is only the case if n is the norm of an element of the field $\mathbf{Q}(\sqrt{q})$ where q is the discriminant of the form.

This is no restriction if q is a rational square. Then the idea leads to a close correspondence between Hecke operators for modular forms for $\Gamma_0(q)$ and principal character and the Brandt matrices attached to rational quaternion algebras, see [5], [7].

Now we treat definite quaternary quadratic forms of prime discriminant q. Their similarities can be represented by elements of the second Clifford algebra attached to the forms. It is a quaternion algebra K over $k = \mathbf{Q}(\sqrt{q})$ which is ramified only at the two infinite places. The relationship is explained in §2, while in §1 the necessary preparations on Brandt matrices and Hecke operators in $\mathbf{Q}(\sqrt{q})$ are collected. Further preparations, namely on the representations of the similarities by spherical polynomials follow in §3.

In §4 we adapt the results of [2, chap. IV] to our present situation. In §5 we translate the findings of §4 into function theoretic language. Theorem 2 in §5 exhibits the Naganuma lift of theta series of such quadratic forms to theta series over k.

We had vainly hoped to prove or to refute the following statement : all modular forms for $\Gamma_0(q)$ of Hecke's real "Nebentype" are linear combinations of theta series for quadratic forms F of discriminant q and for the adjoint forms $\tilde{F}$. Of course we mean generalized theta series with spherical harmonics. The question seems to be much more difficult than that for modular forms of character 1, see [5], [7]. Its solution would certainly imply some knowledge on linear relations between such theta series.

§1. THETA SERIES AND HILBERT MODULAR FORMS.

Let $k = \mathbf{Q}(\sqrt{q})$ be a real quadratic number field, o its maximal order, and d its different. We assume $q \equiv 1 \bmod 4$, a prime number. Under this condition the fundamental unit of k has norm -1. Elements of k will be denoted by small Greek letters. The canonical automorphism of k is $\mu \to \mu^\sigma$.

We consider the totally definite quaternion algebra K/k which is unramified at all finite places. Elements of K will be written with big Roman letters. There exists an involutorial automorphism $M \to M^\sigma$ which extends that of k.

Let $M_1,\ldots,M_h$ represent all left ideal classes with the same maximal left order O. The right orders O_i represent all types of maximal orders; if k has ideal class number 1, these O_i are different. For a given integral ideal m of k we form all integral ideals

(1) $$M_{ij} = M_i^{-1}M_j\, M_{ij} \qquad\qquad M_{ij} \in K$$

of **reduced** norms $N(M_{ij}) = N_{K/k}(M_{ij}) = m$, which are right equivalent with $M_i^{-1}M_j$.

The multiplicative group of the Hamilton quaternion algebra over $\mathbf{R}$ has exactly one irreducible representation $r_l(M)$ of $l+1$ rows ($l = 0,1,\ldots$) We form the $(l+1)^2$-rowed representation

$$R_l(M) = r_l(M) \times r_l(M^\sigma).$$

Its class is unchanged under $M \to M^\sigma$. With a unit ε of k we have

(2) $$R_l(\varepsilon M) = n(\varepsilon)^l R_l(M) = n_{k/\mathbf{Q}}(\varepsilon)^l R_l(M).$$

Unless the opposite is stated we will always assume that l is even.

Then

(3) $$R_l(\varepsilon M) = R_l(M).$$

With these prepartions the Brandt matrices of $h(l+1)^2$ rows are defined by

(4) $$\begin{cases} \Pi_{ij}(m) = \sum R_l(M_{ij}^{\kappa})e_j^{-1} & (i,j = 1,\dots,h) \\ B_l(m) = (\Pi_{ij}(m)) \end{cases}$$

where e_j is the index of the unit group of o in that of O_j. The superscript κ means the canonical antiautomorphism of K/k. It implies that $M \to R_l(M^{\kappa})$ is an inverse representation. The sum includes all M_{ij} such that (1) is an integral ideal of the properties described, but of a set εM_{ij} with units $\varepsilon \in o$ only one element is taken. Which one is taken does not matter because of (3).

The Kloosterman-Schoeneberg theta series are

(5) $$\Theta_{ij}(z) = \frac{1}{e_j} \sum R_l(M_{ij}^{\kappa})e^{i\pi tr(zN(M_i^{-1}M_j)N(M_{ij}))}$$

with $N = N_{K/k}$. Here $z = (z_1,z_2)$ with z_1 and z_2 in the upper half plane and $tr(z\lambda)$ means $\lambda z_1 + \lambda^{\sigma}z_2$. The sum is extended over all $M_{ij} \in M_j^{-1}M_i$. It has been shown [6 : Theorem 1] that they are Hilbert modular forms of weight $k = l+2$ with respect to the group

(6) $$\Gamma = \left\{\begin{pmatrix} \alpha & \beta \\ \gamma & \delta \end{pmatrix} : \alpha\delta - \beta\gamma = 1, \alpha \in o, \beta \in o^{-1}, \gamma \in o, \delta \in o\right\}.$$

If $l > 0$, which will always be assumed, they are cusp forms. These series are arranged in a $h(l+1)^2$-rowed matrix

(7) $$\Theta_l(z) = \sum_{m \gg o} B_l(m)e^{2\pi i(z_1\mu+z_2\mu^{\sigma})}$$

with $(\mu) = \underline{m}$ and $B_l(\mu) = B_l(\underline{m})$, z_1 and z_2 being two complex variables with positive imaginary part. (In the exceptional case $l = 0$ we would have to introduce also the terms $B_o(0)$.)

The Brandt matrices satisfy the same equations as the Hecke operators [6 : Theorem 2 and 5]

$$(8)\qquad \begin{cases} B_l(m_1)B_l(m_2) = B_l(m_1m_2) & \text{for } (m_1,m_2) = 1 \\ B_l(p^\rho)B_l(p^\sigma) = \displaystyle\sum_{\tau=\sigma}^{\min(\rho,\sigma)} n(p)^{(l+1)\tau} B_l(p^{\rho+\sigma-2\tau}) & \end{cases}$$

for prime ideals p. The second formula is slightly modified if p is not principal (see [6]). They are connected with the Hecke operators T(m) by the equations [6 : Theorem 7]

$$(9)\qquad C^{-1}B_l(m)C\Theta_l(z) = \Theta_l(z)|_{l+2}\, T(m)$$

with a certain matrix C independent of m which does not interest us here. (9) means that the Brandt matrices represent the Hecke operators in the space of the theta functions which are the coefficients of the matrix $\Theta_l(z)$.

PROPOSITION 1. If $k > 2$ (or $l > 0$) all modular cusp forms with respect to the group (6) are linear combinations of the Kloosterman-Schoeneberg theta series (5) provided that the ideal class number of the field k is one.

We prove that the representation $T(m) \to B_l(m)$ is the same (up to equivalence) as that of $T(m)$ in the space of cusp forms. The latter respresentation is semisimple which is shown by using the Petersson metric [6 : Theorem 6]. It remains to compare the traces of both representations.

The traces of the Brandt matrices have been determined in 3 former papers, viz. tr $B_l(m) = 0$ if m is not principal in the narrow sense. Otherwise

$$\operatorname{tr}(B_l(m)) = \frac{1}{2}\sum \frac{h((\mu-4\sigma^2)\phi^{-2})}{w((\mu-4\sigma^2)\phi^{-2})}\,\frac{\rho^{l+1}-\bar\rho^{l+1}}{\rho-\bar\rho}\,\frac{\rho'^{l+1}-\bar\rho'^{l+1}}{\rho'-\bar\rho'} + \dots$$

where μ is such that $\mu \gg 0$ and $(\mu) = m$. The sum is extended over all σ, ϕ such that $(\mu - 4\sigma^2)\phi^{-2}$ is the discriminant of an order J in a totally imaginary quadratic extension of k, and h(...), w(...) mean the number of ideal classes and the index of the unit group of O in that of J. ρ and ρ' are the roots of the equations

$$\rho^2 - \rho\sigma + \mu = 0, \quad \rho'^2 - \rho'\sigma^\sigma + \mu^\sigma = 0$$

and the bar means the complex conjugate. Finally the dots indicate a correction term for the case that $m = m_1^2$, the square of an indeal in k.

The expression is independent of the number μ taken. Indeed, if $m = (\mu_1)$, $\mu_1 \gg 0$, we have $\mu_1 = \varepsilon\mu$ with a totally positive unit ε. Since the basic unit has norm -1 in our case, $\varepsilon = \varepsilon_1^2$ with another unit ε_1, and in the sum we can substitute μ, σ, ϕ, ρ, ρ' by $\mu\varepsilon_1^2$, $\sigma\varepsilon_1$, $\phi\varepsilon_1$, $\rho\varepsilon_1$, $\rho'\varepsilon_1^\sigma = \pm\, \rho'\varepsilon_1^{-1}$ which leaves it unchanged.

The proof is in principle contained in [3], [5]. It consists of the determination of all $M_{ii} = O_i M_{ii}$ in (1). Instead of mere counting their number one has to enumerate them with the weights $\mathrm{tr}(R_l(M_{ii}))$ which are

$$\frac{\rho^{l+1} - \bar{\rho}^{l+1}}{\rho - \bar{\rho}} \quad \text{or} \quad \frac{\rho^{l+1} - \bar{\rho}^{l+1}}{\rho - \bar{\rho}} \; \frac{\rho'^{l+1} - \bar{\rho}'^{l+1}}{\rho' - \bar{\rho}'}$$

in the case of quaternion algebras K over **Q** or k. This has been carried out in [3] and [5], [7].

The trace of the representation of the Hecke operators $T(m)$ in the space of modular cusp forms of weight $k = l + 2$ has been given by Shimizu [11]. It is the same as $\mathrm{tr}(B_l(m))$, perhaps up to the correction term for $m = m_1^2$ which had been left open. Although these correction terms can also been shown to be equal, it is easier to prove that they cannot differ. This will follow from the

LEMMA 1. Both traces are symmetric with respect to the automorphism σ of k :

$$\mathrm{tr}(T(m^\sigma)) = \mathrm{tr}(T(m)), \qquad \mathrm{tr}(B_l(m^\sigma)) = \mathrm{tr}(B_l(m)).$$

Indeed, these traces are the Fourier coefficients of the following symmetric modular forms :

$$\Phi(z) = \sum_{\mu \gg \sigma} \mathrm{tr}(T(\mu))e^{2\pi i(z_1\mu+z_2\mu^\sigma)}, \quad \Psi(z) = \sum_{\mu \gg \sigma} B_l(\mu)e^{2\pi i(z_1\mu+z_2\mu^\sigma)}.$$

Their coefficients are the same for all μ with $(\mu) \neq m_1^2$. If they would differ for the rest, $\Phi(z) - \Psi(z)$ would have Fourier coefficients $\neq 0$ only for $(\mu) = m_1^2$, which is impossible.

PROOF of the LEMMA. Let D be the matrix expressing the involution $\Phi(z_1,z_2') \to \Phi(z_2',z_1)$, and understand $T(m)$ as the matrix representing the Hecke operator. Then

$$T(m^\sigma) = D^{-1}\, T(m)\, D$$

which entails the first symmetry.

What concerns the latter, we consider an extension $M \to M^\sigma$ of the non-identical automorphism σ of k. Under this σ the ideals (1) are transformed into

$$M_{ij}^\sigma = M_i^{-\sigma} M_j^\sigma\, M_{ij}^\sigma$$

which have left orders O_i^σ, norms m_σ^σ, an right orders which are isomorphic with O_j^σ. The orders O_i^σ, O_j^σ belong to types of $O_{i'}$, $O_{j'}$ where $i \to i'$, $j \to j'$ is an involutorial permutation. Furthermore

$$R_l(M) = r_l(M) \times r_l(M^\sigma) \quad \text{and} \quad R_l(M^\sigma) = r_l(M^\sigma) \times r_l(M)$$

are equivalent. Both facts result in the equivalence of the matrices $B_l(m)$ and $B_l(m^\sigma)$, which in turn implies the second symmetry.

§2. THE CLIFFORD ALGEBRA AND SIMILARITIES OF QUADRATIC FORMS.

We consider definite quaternay quadratic forms F of discriminant $q \equiv 1 \bmod 4$, a prime number. For their matrices we use the same symbols, and we suppose

$$F = (f_{ij}) \quad \text{with} \quad \frac{1}{2} f_{ii},\ f_{ij} \in \mathbf{Z}.$$

All such F belong to the same genus. As usual, we attach to them a metric space S over k such that F is the norm form of the generic vector of a maximal lattice L in S :

$$\frac{1}{2} F[x] = \frac{1}{2} \sum f_{ij} x_i x_j = n(\sum b_i x_i)$$

with a basis b_i of L. The g. c. d. of the coefficients $\frac{1}{2} f_{ii}$, f_{ij} is the norm of L.

To the space S we attach the first Clifford algebra over $\mathbf{Q}$, spanned by all formal products of r vectors $(a_1 \ldots a_r)$ subject to the relations

1) the empty product is the unit element,
2) $(a_1 \ldots a_r)(b_1 \ldots b_s) = (a_1 \ldots a_r b_1 \ldots b_s)$,
3) $(\ldots a_{i-1}(a_i x_i + b_i y_i) a_{i+1} \ldots) = x_i(\ldots a_{i-1} a_i a_{i+1} \ldots) +$
$\qquad = y_i(\ldots a_{i-1} b_i a_{i+1} \ldots)$,
4) $(ab) + (ba) = (a,b) = \frac{1}{2}(n(a+b) - n(a) - n(b))$,

where x_i, y_i mean elements in $\mathbf{Q}$. For the following see [2, §4 and §5].

The second Clifford algebra of S over $\mathbf{Q}$ is the subalgebra of the first spanned by the products $(a_1 \ldots a_r)$ with even r. This is isomorphic to the totally definite quaternion algebra K over $k = \mathbf{Q}(\sqrt{q})$ which is unramified at all finite places [9].

The inversion of the order of factors in

$$(a_1 \ldots a_r)^{\kappa} = (a_r \ldots a_1)$$

is the canonical antiautomorphism of K. The space S always contains a vector e of norm 1. The transformation of the elements of the second Clifford algebra K with e in the sense of the first Clifford algebra induces an involutorial automorphism :

$$(e)(a_1 \ldots a_r)(e) = (e)^{-1}(a_1 \ldots a_r)(e) = (a_1 \ldots a_r)^{\sigma}$$

It is easily seen that e transforms the product of 4 naturally orthogonal vectors $b_1, \ldots, b_4$ into its negative. This product has the property

$$(b_1 \ldots b_4)^2 = 2^{-4} n(b_1) \ldots n(b_4)$$

which is up to a rational square equal to q. Furthermore it generates the center of K which is k. Thus σ induces in k the canonical automorphism.

Evidently σ and κ commute :

$$M^{\sigma\kappa} = M^{\kappa\sigma}.$$

It has to be borne in mind that σ is not uniquely determined since there exist infinitely many vectors of norm 1.

PROPOSITION 2. All similarities of the metric space S, of positive determination, are given by

$$(10) \qquad (x) \to (x') = mM^{\kappa}(x)M$$

where x is the generic vector of S and (x) the corresponding element in the first Clifford algebra. $m \neq 0$ is an arbitrary rational number and $M \neq 0$ an arbitrary element in the second Clifford algebra. The norm of

<u>the similarity is</u> $m^2 n_{k/\mathbf{Q}}(N_{K/k}(M)) = m^2 n(N(M))$.

<u>Two pairs</u> m_1, M_1 <u>and</u> m_2, M_2 <u>yield the same similarity if and only if</u>

$$M_2 = \mu M_1, \quad m_2 = n(\mu)^{-1} m_1, \quad \mu \in k.$$

<u>PROOF</u>. Let b_1 be an orthogonal basis of S. The canonical antiautomorphism κ transforms an element $M = \xi_0 + \xi_1(b_2 b_3) + \ldots$ with $\xi_\nu \in k$ into $M^\kappa = \xi_0 - \xi_1(b_2 b_3) - \ldots$. Their product is the norm :

$$MM^\kappa = M^\kappa M = N(M) = N_{K/k}(M).$$

From (10) we find for two vectors

$$(x'y') = m^2 M^\kappa(x) N(M)(y) M.$$

The transformation of an element K by a one-member element (a) of the first Clifford algebra induces an isomorphism which is not trivial in k. But since k has only one non-identic automorphism, we have

$$(x'y') = m^2 N(M)^\sigma M^\kappa(xy) M.$$

The sum of this formula and that with x, y inverted yields

$$(x',y') = m^2 n(M(M)) \, (x,y)$$

which exhibits a similarity of norm $m^2 \, n(N(M))$.

The last statement on two pairs m_1, M_1 and m_2, M_2 is easily checked.

It remains to be shown that all similarities are given in this way. Since every totally positive $\mu \in k$ is the norm of some $M \in K$, all $n(\mu) > 0$ occur as norms of similarities. (10). According to [2 : Satz 11.1] all similarities have such norms.

At last we show that all proper isometries can be obtained in the way (10) with $m^2 n(N(M)) = 1$. Indeed, a proper isometry induces an automorphism in both Clifford algebras and is therefore representable as

$$(x) \to (x') = M^{-1}(x)M = N(M)^{-1}M^{\kappa}(x)M$$

which has the form (10).

From now on we consider only the special similarities

(11) $$(x) \to (x') = M^{\kappa}(x)M.$$

To a maximal lattice L in S we attach the order O in K which is generated by all symbols $n(L)^{-r}(a_1 \ldots a_{2r})$ with $a_i \in L$ [see 2 : §14]; n(L) means the norm of L. Since L is a maximal lattice, O is a maximal order [9 : §5]. Conversely, L is uniquely determined by O up to a trivial similarity $L \to mL$ with $m \in \mathbf{Q}$.

<u>THEOREM</u> 1. <u>Let</u> L <u>be a maximal lattice and</u> O <u>the maximal order attached to it</u>. <u>If</u> M <u>is an ideal with left order</u> O <u>and right order</u> O', <u>there exists a maximal lattice</u> L' <u>attached to</u> O' <u>such that each of the equations</u>

(12) $$\begin{cases} \cap_{p,M_p} M_p^{\kappa} O_p M_p = N(M)O', \\ \cap_{p,M_p} M_p^{\kappa}(L_p)M_p = (L') \end{cases}$$

<u>implies the other, and</u>

(13) $$n(N(M))n(L) = n_{k/\mathbf{Q}}(N_{K/k}(M))n(L) = n(L').$$

<u>Conversely, if</u> L <u>and</u> L' <u>are given maximal lattices, there exists such an ideal</u> M. M <u>is determined by</u> L' <u>resp.</u> O' <u>up to an ideal factor</u> m <u>of</u> k.

In (12) p runs over all prime numbers and M_p over all elements of the p-adic extensions M_p. (L_p) and (L) mean the elements in the first Clifford algebra attached to the generic vectors of these lattices.

The equations (12) can briefly be written

$$\text{(12a)} \quad \begin{cases} M^{\kappa} \mathcal{O} M = N(M)\mathcal{O}', \\ M^{\kappa}(L)M = (L'). \end{cases}$$

The proof is evident for principal ideals $M = \mathcal{O}M$, and the p-adic extension of ideals are always principal. We have to observe that ideals and lattices are the intersections of their p-adic extensions. The uniqueness of M in the sense of the theorem stems from the fact that all ambigue ideals are $\mathcal{O}m$, m an ideal of k.

COROLLARY. There is a one-to-one correspondence between the similarity classes of maximal lattices in S and the (inner) isomorphy types of maximal orders in K.

PROPOSITION 3. The group $U(L)$ of proper units of a maximal lattice L (i.e. those of determinants +1), divided by its centre $\{\pm E\}$, and the groups $U(\mathcal{O})$ and $U(\mathfrak{o})$ of its order $\mathcal{O}$ in K and of $\mathfrak{o}$ are connected by

$$U(L)/\{\pm E\} \cong U(\mathcal{O})/U(\mathfrak{o}).$$

PROOF. An isometry or a unit of L induces an isomorphism of $\mathcal{O}$. The units $\pm E$, or in other words $x \to \pm x$, yield the identical isomorphism of $\mathcal{O}$. We distinguish automorphisms of 1st kind which leave the elements of k fixed, and those of 2nd kind which do not. Because K is unramified at all finite places and $\mathcal{O}$ is maximal, all automorphisms of 1st kind are given by the transformation of $\mathcal{O}$ by units V of $\mathcal{O}$. V and αV with $\alpha \in K$ yield the same automorphism.

Let a_ν be an orthogonal basis of S. Then $\alpha = (a_1 \dots a_4)$ is an element of the second Clifford algebra which generates the subfield k. An isometry of S, as a linear transformation of the a_ν maps α on

α.det(A). So it is proper if and only if k is left fixed. This completes the proof.

§3. THE REPRESENTATION OF THE SIMILARITIES.

Now we make some preparations for the following.

PROPOSITION 4. Let b_ν be a basis of S. The representations (11), namely

$$(b_\nu) \to (b'_\nu) = M^\kappa(b_\nu)M = (\sum b_\mu q_{\mu\nu})$$

with $q_{\mu\nu} \in k$ is equivalent with $R_1(M) = r_1(M) \times r_1(M^\sigma)$.

For the proof we compare the traces. We may extend S by **R**. The space $S\mathbf{R}/\mathbf{R}$ has an orthogonal basis s_ν with $n(s_\nu) = 1$, and we replace the b_ν by the s_ν. The Clifford algebra is likewise extended to K**R** and then has a basis [2 : p.31]

$$[1, (s_2 s_3), (s_3 s_1), (s_1 s_2)] \times [1, (s_1 s_2 s_3 s_4)].$$

The $(s_2 s_3)$ etc., satisfy the same relations as the commonly known Hamilton quaternions. The second factor is isomorphic with $\mathbf{R} \oplus \mathbf{R}$, since the square of $\delta = (s_1 s_2 s_3 s_4)$ is 1, and $k\mathbf{R} = \mathbf{R} \oplus \mathbf{R}$. The automorphism σ keeps the $(s_\mu s_\nu)$ fixed and transforms δ in $-\delta$. The representations and the trace $tr(R_1(M))$ are given in the following table, where we use the Pauli matrices

$$p_0 = \begin{pmatrix}1 & 0\\ 0 & 1\end{pmatrix},\ p_1 = \begin{pmatrix}i & 0\\ 0 & -i\end{pmatrix},\ p_2 = \begin{pmatrix}0 & 1\\ -1 & 0\end{pmatrix},\ p_3 = \begin{pmatrix}0 & i\\ i & 0\end{pmatrix} :$$

M	1	δ	$(s_2 s_3)$	$\delta(s_2 s_3)$	$(s_3 s_1)$	$\delta(s_3 s_1)$	$(s_1 s_2)$	$\delta(s_1 s_2)$
$r_1(M)$	p_0	δp_0	p_1	δp_1	p_2	δp_2	p_3	δp_3
$r_1(M^\sigma)$	p_0	$-\delta p_0$	p_1	$-\delta p_1$	p_2	$-\delta p_2$	p_3	$-\delta p_3$
$tr(R_1(M))$	4	-4	0	0	0	0	0	0

It is easily verified that the similarities generated in the way (11) by these elements M have the same traces.

PROPOSITION 5. Let L be a lattice of norm $n(L)$ and F the quadratic form attached to a basis b_ν of L. Furthermore, let C be a real matrix satisfying

$$C^t\ C = F$$

and $p_\nu(x)$ a basis of the homogeneous spherical harmonic polynomials of degree l in 4 variables x_i (understood as a row vector of length $(l+1)^2$). Finally let Q(M) be the representation of $M \in K$ in the way of Proposition 4.
Then $\tilde{R}_l(M) = (\tilde{r}_{\mu\nu}(M))$, defined by

$$p_\nu(CQ(M)x) = \sum_\mu p_\mu(Cx)\tilde{r}_{\mu\nu}(M)$$

is an inverse representation of M which is equivalent to the representation $R_l(M^\kappa)$ defined in §1.

PROOF. It is evident that $M \to \tilde{R}_l(M)$ is inverse. As in the proof of Proposition 4 we may use an orthonormal basis s_ν of $S\mathbf{R}/\mathbf{R}$. Then M and M^σ become independent, and according to Proposition 4 the group of the Q(M) becomes a direct product isomorphic with $r_1(M) \times r_1(M^\sigma)$.

All homogeneous polynomials of degree l can be written as

$$f_l(x) = \sum_{\lambda=0}^{[l/2]} (x_1^2 + \ldots + x_4^2)^\lambda \, p_{l-2\lambda}(x)$$

with homogeneous spherical polynomials $p_{l-2\lambda}(x)$ of degrees $l-2\lambda$. The representation of the orthogonal group O_4 in the space of the $f_l(x)$ splits up into the representations in the spaces of the $p_{l-2\lambda}(x)$, the latter being irreducible. **(This is known for the representations of $O(3,\mathbb{R})$ and is easily extended to those of $O(4,\mathbb{R})$.)** The same is true for the similarities. There exists $\binom{l+3}{3}$ linearly independent $f_l(x)$, and $\binom{l+3}{3} - \binom{l+1}{3} = (l+1)^2$ linearly independent $p_l(x)$.

The matrix C translates $n(L)\ \frac{1}{2}\ F[x]$ into $\sum x_i^2$. As was already remarked , $Q(M)$ is isomorphic with $r_1(M) \times r_1(M^\sigma)$, and we may assume M and M^σ as independent. Then all irreducible representations of this group are equivalent with $r_{l_1}(M) \times r_{l_2}(M^\sigma)$, their degrees being $(l_1+1)(l_2+1)$. In our case the representation of M by the $p_\nu(Cx)$ is equivalent with that of M^σ, because $r_1(M) \times r_1(M^\sigma) \sim r_1(M^\sigma) \times r_1(M)$. Hence l_1 and l_2 must be equal, and equal to l. This completes the proof.

§4. SIMILARITIES AND THETA FUNCTIONS OF QUADRATIC FORMS.

Let $F_1,\ldots,F_h$ represent all classes of quadratic forms of the properties fixed in §2, and $L_1,\ldots,L_h$ the underlying lattices in S. By $b_{i,\nu}$ $(\nu = 1,\ldots,4)$ we denote bases of the L_i such that

$$n(\sum b_{i,\nu}x_\nu) = \frac{1}{2}\ F_i[x].$$

Without loss of generality we may assume the F_i primitive or, in other words, that the norms of the L_i are 1.

For a given n we let x_j be a column vector with coefficients $x_{j,\nu} \in \mathbf{Z}$ satisfying

(14) $$\frac{1}{2}\ F_j[x_j] = n.$$

Furthermore for a given m let X_{ij} be a matrix with coefficients in $\mathbf{Z}$

such that

(15) $$X_{ij}^t F_i X_{ij} = mF_j, \qquad \det(X_{ij}) > 0$$

Then

(16) $$x_i = X_{ij} x_j$$

is a column vector with the property

(17) $$\frac{1}{2} F_i[x_i] = nm.$$

We will count all situations (14)-(17), but we will attach to the x_i, x_j, X_{ij} certain weights.

For this purpose we fix a real matrix C_1 such that $C_1^t C_1 = F_1$. Then we have

(18) $$C_i^t C_i = F_i$$

with $C_1^{-1} C_i$ transforming the basis $b_{1,\nu}$ into $b_{i,\nu}$ or more generally, $C_i^{-1} C_j$ transforming $b_{i,\nu}$ into $b_{j,\nu}$. We claim that the $C_i^{-1} C_j$ can be found in such a way that the denominators of the coefficients are powers of one particular prime $p_o \neq q$. Indeed, considering the L_o as maximal lattice over the domain $\mathbf{Z}(p_o)$ of numbers whose denominators are powers of p_o, they are arithmetically indefinite, and by Meyers' theorem [13 : 104 : 5] they are isomorphic and their norm forms equivalent. This proves the contention. We take a p_o with $(\frac{q}{p_o}) = 1$. It is then the norm of an element of k.

Let $Q_1(M)$ be the representation of a similarity of S by the basis $b_{1,\nu}$ as described in Proposition 4. It acts on a basis $p_\nu(x)$ of homogeneous spherical harmonic polynomials in the way of Proposition 5

$$p_\nu(C_1 Q_1(M) C_1^{-1} x) = \sum p_\mu(x) \tilde{r}_{\mu\nu}(M), \qquad \nu = 1, \dots, (l+1)^2$$

or more briefly with $p_\nu(x)$ understood as a row vector

(19) $$p(C_1 Q_1(M) C_1^{-1} x) = p(x)\tilde{R}(M).$$

The matrices X_{ij} in (15) can be written

$$X_{ij} = C_i^{-1} C_1 \; Y \; C_1^{-1} C_j$$

with a proper similarity Y of F_1. According to proposition 2, $Y = mQ_1(M_{ij})$ with an $m \in \mathbf{Q}$ and an element $M_{ij} \in K$ attached to X_{ij}. If m is the norm of an element of k, we have $mQ_1(M_{ij}) = Q_1(M'_{ij})$ with some other M'_{ij}. Now we suppose that X_{ij} be a primitive matrix. Therefore also Y must be primitive with respect to all primes for which $C_i^{-1} C_1$ and $C_1^{-1} C_j$ are integral. Due to our construction these are all primes except one p_0. So m can at most be a power of p_0. But since p_0 is the norm of an element of k, we arrive at $Y = Q_1(M_{ij})$ and

(20) $$X_{ij} = C_i^{-1} C_1 \; Q_1(M_{ij}) C_1^{-1} C_j.$$

With these preparations we attach to each vector x_1 with components $x_{i,\nu}$ (in $\sum b_{i,\nu} x_{i,\nu}$) the vector

$$p(C_i x_i) = (p_\nu(\sum_\lambda c_{i,\mu\lambda} x_{i,\lambda})), \qquad \nu = 1,\ldots,(l+1)^2.$$

Equation (16) and (20) lead to

$$p(C_i x_i) = p(C_i X_{ij} x_j) = p(C_1 Q_1(M_{ij}) C_1^{-1} C_j x_j)$$

or more briefly

(21) $$p(C_i x_i) = p(C_j x_j)\tilde{R}(M_{ij}).$$

Now we report on some results from [2 : §18]. We quote them without repeating [2]. Only we replace the representations of the vectors and matrices by those just introduced. (In [2] representations were used which were not irreducible and which contained the present ones as summands.) We restrict ourselves to two cases :

1) $m = p_1$, a prime number with $(\frac{q}{p_1}) = 1$,

2) $m = p_{-1}^2$, the square of a prime with $(\frac{q}{p_{-1}}) = -1$, and the g. c. d. of the elements of the matrix X_{ij} in (15) is 1.

We introduce $h(l+1)^2$ rowed vectors

$$(22)\qquad m_l(n) = (\ldots, \sum p_\nu(C_j x_j), \ldots)$$

$(j = 1,\ldots,h;\ \nu = 1,\ldots,(l+1)^2)$, the sums extended over all integral solutions of (14). Similarly we form the $h(l+1)^2$-rowed matrices

$$(23)\qquad P_l(m) = (\sum \tilde{R}_l(M_{ij})e_j^{-1}),$$

for each pair i,j summed over all primitive matrices satisfying (15), (that X_{ij} be primitive has been assumed in §11.4), where e_j means the number of proper units of F_j. According to Proposition 3, e_j equals the group index $[U(\mathcal{O}_j) : U(o)]$. Each individual case of (14)-(17) yields an entry into the vector $m_l(n)$ and the matrix $P_l(m)$, and the weights of the x_i, x_j are connected by (21).

Satz 18.4 is an intermediate result which must be adapted to our present purpose. The solutions of (17) are at first distributed into certain residue classes C_ν mod p_1 or p_{-1}^2, and we have

$$(24)\qquad m_l(n)P_l(n) = \sum_\nu \rho(C_\nu)m_l(nm, C_\nu),$$

where the $m_l(nm, C_\nu)$ is the sum (22) restricted to $x_i \in C_\nu$, and the $\rho(C_\nu)$ are certain invariants of the classes. They have been calculated in §11, p. 68-71. In the first case $m = p_1$, there are two classes C_1 and C_0, consisting of the primitive and the imprimitive vectors x_i. The number of "ideals" in the sense of p.70 is $\pi = 2(p_1+1)$, and (11.13), (11.16) yield $\rho(C_1) = 2$, while (11.9) and (11.15) yield $\rho(C_0) = 2(p_1+1)$. Adding up the primitive and the imprimitive vectors gives

$$m_l(np_1) = m_l(np_1, C_1) + m_l(np_1, C_0)$$

and writing the imprimitive $x_i = p_1 x_i'$

$$m_l(np_1, C_0) = p_1^l m_l(np_1^{-1}).$$

The considerations end up with

$$(25) \qquad m_l(n)\, \tfrac{1}{2}\, P_l(p_1) = m_l(np_1) + p_1^{l+1} m_l(np_1^{-1})$$

for (24), where the second summand vanishes if n is not divisible by p_1.

In the second case $m = p_{-1}^2$ we have 4 classes (see p.71, case B). C_1 contains the primitive x_i, C_{01} and C_{00} contain the $x_i = p_{-1} x_i'$ with x_i' primitive and $\frac{1}{2} F_i[x_i'] \not\equiv 0 \bmod p_{-1}$ resp. $\frac{1}{2} F_i[x_i'] \equiv 0 \bmod p_{-1}$. Lastly the $x_i \in C_{000}$ are divisible by p_{-1}^2. The number of "ideals" is this time $\pi = p_{-1}^2 + 1$, and from (11.9) and (11.18) we collect

$$\rho(C_1) = 1,\ \rho(C_{01}) = p_{-1} + 1,\ \rho(C_{00}) = 1,\ \rho(C_{000}) = p_{-1}^2 + 1.$$

Similarly to the first case we have

$$m_l(np_{-1}^2) = m_l(np_{-1}^2, C_1) + \ldots + m_l(np_{-1}^2, C_{000})$$

and

$$m_l(np_{-1}^2, C_{01}) = \begin{cases} p_{-1}^l(n) m_l(n) & \text{if } (n, p_{-1}) = 1 \\ 0 & \text{if } (n, p_{-1}) > 1, \end{cases}$$

$$m_l(np_{-1}^2, C_{000}) = p_{-1}^{2l} m_l(np_{-1}^{-2}) \qquad \text{if } (n, p_{-1}) > 1.$$

Now (24) becomes

$$(26) \qquad m_l(n) P_l(p_{-1}^2) = \begin{cases} m_l(np_{-1}^2) + p_{-1}^{l+1} m_l(n) & \text{if } (n, p_{-1}) = 1, \\ m_l(np_{-1}^2) + p_{-1}^{2l+2} m_l(np_{-1}^{-2}) & \text{if } (n, p_{-1}) > 1. \end{cases}$$

The matrices $\frac{1}{2} P_l(p_1)$, $P_l(p_{-1}^2)$ express the action of the Hecke operators $T(p_1)$, $T(p_{-1}^2)$ on the theta functions

$$\theta_{i,\nu}(\tau) = \sum_{x \in L^4} p_\nu(C_i x) e^{\pi i \tau F_i[x]} .$$

Arranging them in $h(l+1)^2$-rowed vectors we can write them

(27)
$$\theta_l(\tau) = \sum_{n=1}^{\infty} m_l(n) e^{2\pi i n \tau}.$$

With this notation (25) becomes

$$\theta_l(\tau)\, \frac{1}{2}\, P_l(p_1) = \sum_{n=1}^{\infty} m_l(np_1) e^{2\pi i n\tau} + p_1^{l+1} \sum_{n=1}^{\infty} m_l(n) e^{2\pi i n p_1 \tau}$$

(in the second summand we have written n instead of np_1^{-1}). The right hand side is

$$p_1^{l+1}\, \theta_l(p_1\tau) + p_1^{-1} \sum_{p=0}^{p_1-1} \theta_l\!\left(\frac{\tau+r}{p_1}\right) + \theta_l(\tau) \mid_{l+2} T(p_1).$$

So we have proved

(28)
$$\theta_l(\tau)\, \frac{1}{2}\, P_l(p_1) = \theta_l(\tau) \mid_{l+2} T(p_1).$$

Equation (26) is

$$\theta_l(\tau)\, P_l(p_{-1}^2) = \sum_{m=1}^{\infty} m_l(np_{-1}^2) e^{2\pi i n\tau} + p_{-1}^{l+1}\left(\sum_{n=1}^{\infty} m_l(n) e^{2\pi i n\tau} - \sum_{n=1}^{\infty} m_l(np_1) e^{2\pi i n p_1\tau}\right)$$

$$+\, p_{-1}^{2l+2} \sum_{n=1}^{\infty} m_l(n) e^{2\pi i n p_{-1}^2 \tau}$$

$$= p_{-1}^{2l+2}\, \theta_l(p_{-1}^2\tau) - p_{-1}^{l} \sum_{r=0}^{p_{-1}-1} \theta_l\!\left(\frac{p_{-1}\tau+r}{p_{-1}}\right) + p_{-1}^{-2} \sum_{r=0}^{p_{-1}^2-1} \theta_l\!\left(\frac{\tau+r}{p_{-1}^2}\right)$$

$$+\, p_{-1}^{l+1}\, \theta_l(\tau),$$

and this is

(29)
$$\theta_l(\tau)\, P_l(p_{-1}^2) = \theta_l(\tau) \mid_{l+2} T(p_{-1}^2) + p_{-1}^{l+1}\, \theta_l(\tau).$$

With (28) and (29) we have proved

PROPOSITION 6. The space spanned by the theta series $\Theta(\tau,F,p)$ with

<u>spherical harmonic polynomials of degree</u> l <u>is closed under the Hecke operators</u> $T(p_1)$ and $T(p_{-1}^2)$.

One must be careful not to confound the matrices $\frac{1}{2}\,P(p_1)$, $P(p_{-1}^2)$ with the matrices representing the $T(p_1)$, $T(p_{-2}^2)$ since the $\theta(\tau,F,p)$ are in general not linearly independent.

§5. <u>THE NAGANUMA LIFT.</u>

In order to avoid a difficulty (however small), we assume in §5 that the ideal class number of k is 1. Then the correspondence between the F_i, L_i, O_i, and the ideals M_i is one-to-one.

If the basis $p_\nu(x)$ of the spherical polynomials is suitably chosen, the following matrices are equal :

$$\tilde{R}_l(M_{ij}) = R_l(M_{ij}^\kappa)$$

according to Proposition 5, and the matrices (23) can be written

$$P_l(m) = \sum B_l(\mu),$$

summed over all ideals (μ) of k with $\mu \gg 0$ and $n(\mu)\ m$. Here we have used the one-to-one correspondence between the ideals M_i and the classes F_i. In our two special cases this is

$$(30)\qquad \begin{cases} P_l(p_1) = B_l(\pi) + B_l(\pi^\sigma) & (p_1 = (\pi)(\pi^\sigma)). \\ P_l(p_{-1}^2) = B_l(p_{-1}) & \end{cases}$$

The matrices $B_l(\mu)$ can be transformed simultaneously into diagonal form. This follows from Proposition 1 and the analogue property of the representations of the Hecke operators (but it can also be proved independently). Let

$$(31)\qquad A^{-1}B_l(\mu)A = \mathrm{diag}(\beta_i(\mu))$$

and

$$(32)\qquad \begin{cases} A^{-1}\ \Theta_l(z)A = \mathrm{diag}(\Phi_i(z)), \\ \Phi_i(z) = \sum\limits_{\mu} \beta_i(\mu)e^{2\pi i(z_1\mu+z_2\mu^\sigma)}. \end{cases}$$

With the same matrix we also transform the theta series (27) :

$$(33)\qquad \begin{cases} \theta_l(\tau)A = (\ldots,\phi_i(\tau),\ldots) \\ \phi_i(\tau) = \sum\limits_{m} \lambda_i(m)e^{2\pi i m\tau}, \end{cases}$$

$(i = 1,\ldots,h(l+1)^2)$. Some of the $\phi_i(\tau)$ will in general vanish. For those which do not we may even assume without loss of generality $\lambda_i(1) = 1$. Then $\lambda_i(m)$ for such m which are norms from k, are eigenvalues of T(n).

Now we restrict ourselves to the indices i for which

$$(34)\qquad \beta_i(\mu) = \beta_i(\mu^\sigma)$$

holds. For these the equations (28) (together with (30)) and (29) imply

$$(35)\qquad \begin{cases} \beta_i(\pi) = \lambda_i(p_1) & (p_1 = (\pi)(\pi^\sigma)) \\ \beta_i(p_{-1}) = \lambda_i(p_{-1}^2) + p_{-1}^{l+1}. \end{cases}$$

The equations (35) can be interpreted as a relation between the zeta functions

$$Z_i(s) = \sum \beta_i(\mu)n(\mu)^{-s},$$

$$\zeta_i(s) = \sum \lambda_i(m)m^{-s}$$

attached to the $\Phi_i(z)$, $\phi_i(\tau)$. According to Hecke's theory they are Euler products

$$Z_i(s) = \Pi(1-\beta_i(\pi)n(\pi)^{-s} + n(\pi)^{l+1)2s})^{-1},$$

$$\zeta_i(s) = \Pi(1-\lambda_i(p)p^{-s} + (\tfrac{q}{p})\, p^{l+1-2s})^{-1}.$$

We need yet other functions, namely

(36) $$\widetilde{\phi}_i(\tau) = \phi_i(\tau)\,\Big|_{l+2}\begin{pmatrix}0 & -1\\ q & 0\end{pmatrix} = \phi_i(\tfrac{-1}{q\tau})(\sqrt{q}\tau)^{-l-2}.$$

They are also modular forms with respect to $\Gamma_0(q)$ and, by the way, they are obtained by the analogue linear combinations as (33) from the

(37) $$\Theta_l(\frac{-1}{q\tau},\ \widetilde{F},\ p)\ \frac{-1}{q^2\tau^{l+2}}$$

where $\widetilde{F}$ means the quadratic form adjoint to F. Let

(38) $$\widetilde{\phi}_i(\tau) = \sum \widetilde{\lambda}_i(m)e^{2\pi im\tau}.$$

From the well known fact for $p \neq q$,

(39) $$T(p)\begin{pmatrix}0 & -1\\ q & 0\end{pmatrix} = (\tfrac{q}{p})\begin{pmatrix}0 & -1\\ q & 0\end{pmatrix}\ T(p)$$

we conclude that the $\widetilde{\phi}_i(\tau)$ are also eigenfunctions of the T(p), and that

(40) $$\widetilde{\lambda}_i(p) = (\tfrac{q}{p})\ \lambda_i(p).$$

The zeta functions attached to the $\widetilde{\theta}_i(\tau)$ are also Euler products :

$$\widetilde{\zeta}_i(s) = \Pi(1-\widetilde{\lambda}_i(p)p^{-s} + (\tfrac{q}{p})p^{l+1-2s})^{-1}.$$

Now (35) and (40) imply

(41) $$Z_i(s) = \zeta_i(s)\widetilde{\zeta}_i(s),$$

perhaps up to the Euler factors for q.

But we can also show that the Euler factors for q are equal on both sides of (41). For this we apply the functional equations. $\Phi_i(z)$ is a modular forn with respect to the group (6). Especially it satisfies

$$\Phi_i\left(\frac{-1}{qz_1}, \frac{-1}{qz_2}\right)\left(\frac{1}{qz_1z_2}\right)^{l+2} = \Phi_i(z_1,z_2)$$

whence the functional equation

$$(42) \qquad \Phi_i(s) = q^s \int_0^\infty \Phi_i(it_1,it_2)(t_1t_2)^{s-1}dt_1dt_2$$

$$= q^s(2\pi)^{-s}\Gamma(s)^2 Z_i(s) = \Psi_i(l+2-s).$$

From (36) follows in the same way the functional equation

$$(43) \qquad \psi_i(s) = q^{s/2}(2\pi)^s\Gamma(s)\zeta_i(s)$$

$$= q^{s/2}\int_0^\infty \phi_i(it)t^{s-1}dt = \tilde{\psi}_i(k+2-s)$$

where $\tilde{\psi}_i(s)$ is formed in the same way with $\tilde{\zeta}_i(s)$ instead of $\zeta_i(s)$. (43) implies that $\psi_i(s)\tilde{\psi}_i(s)$ satisfies the same functional equation (42) as $\Psi_i(s)$. The quotient of both sides of (41) (we omit the subscript i) satisfies the same functional equation

$$\frac{(1-\lambda(q)q^{-s})(1-\tilde{\lambda}(q)q^{-s})}{1-\beta(\sqrt{q})q^{-s}+q^{l+1-2s}} = \frac{(1-\lambda(q)q^{l+2-s})(1-\tilde{\lambda}(q)q^{l+2-s})}{1-\beta(\sqrt{q})\theta q^{s-l-2}+q^{2s-l-1}}$$

An easy computation concludes from this

$$\lambda(q) + \tilde{\lambda}(q) = \beta(\sqrt{q}),\ \lambda(q)\tilde{\lambda}(q) = q^{l+1},$$

which indeed implies (41).

<u>THEOREM</u> 2. <u>We assume that the ideal class number of</u> k <u>is</u> 1. <u>Let</u> $\Phi_i(z) = \Phi_i(z_1,z_2)$ <u>be the modular forms with respect to the group (6) which are eigenfunctions of the Hecke operators</u>, <u>and whose eigenvalues are symmetric</u> : $\beta_i(\mu) = \beta_i(\mu^\sigma)$. <u>They are linear combinations of the theta functions</u> $\Theta_l(z)$ <u>introduced in §1</u>. <u>Lastly we assume that the analogue linear combinations (33) of the theta functions</u> $\theta_l(\tau)$ <u>introduced in §4 (see (27)) do not vanish</u>.
<u>Then the zeta functions attached to</u> $\Phi_i(z)$, $\phi_i(\tau)$, <u>and the "adjoint" functions</u> $\tilde{\phi}_i(\tau)$ <u>defined in (36) are connected by (41)</u>.

The map $\phi_i(\tau) \to \Phi_i(z)$ is the Naganuma lift, restricted to these $\phi_i(\tau)$. It is left open here whether the $\phi_i(\tau)$ span all modular forms for $\Gamma_0(q)$ and of "Nebentype". Saito has shown [10] that all $\Phi_i(z)$ with symmetric coefficients $\beta_i(\mu) = \beta_i(\mu^\sigma)$ are either obtained by the Naganuma lift from modular forms for $\Gamma_0(q)$ and of "Nebentype", or by the Doi-Naganuma lift from modular forms for the full modular group $SL(2,\mathbf{Z})$. Both sorts of $\Phi_i(z)$ are different. Thus the existence of the latter $\Phi_i(z)$ implies the existence of linear dependencies between $\theta_l(\tau)$ (which can be written $\phi_i(\tau) = 0$).

The $\phi_i(\tau)$ corresponding in the way (32), (33) to those $\Phi_i(z)$ whose coefficients are not symmetric ($\beta_i(\mu) \neq \beta_i(\mu^\sigma)$) vanish also. This is an easy consequence of Naganuma's theorem and our Proposition 1.

Lastly even some of the $\Phi_i(z)$ may vanish, and then the corresponding $\phi_i(\tau)$ must vanish, too.

REMARK. Proposition 1 has only been proved under the assumption that the ideal class number of the field k is one. Without this assumption further operators $V(m)$ in the Hecke ring and matrices $A(m)$, introduced in [6], and their products with the $T(m)$ and Brandt matrices $B(m)$ have to be considered and their traces compared.

LITERATURE

[1] COHEN H., Formes modulaires à une et deux variables, Thèse 1976, Université de Bordeaux I

[2] EICHLER M., Quadratische Formen und orthogonale Gruppen, 2. Aufl. Springer Verlag, Berlin-Heidelberg-New York 1974.

[3] EICHLER M., Zur Zahlentheorie der Quaternionen-Algebren, Journal reine und angew. Math. 195 (1956), 127-151.

[4] EICHLER M., Quadratische Formen und Modulfunktionen, Acta Arithmetica 4 (1958), 217-239.

[5] EICHLER M., The basis problem for modular forms and the traces of Hecke operators, in Modular Forms of One Variable I, Lecture Notes in Mathematics nr. 320, Springer Verlag, Berlin-Heidelberg-New York 1973.

[6] EICHLER M., On theta functions in real algebraic number fields, to appear in Acta Arithmetica.

[7] HIJIKATA H. and SAITO H., On the representability of modular forms by theta series, p. 13-21 in Number Theory, Algebraic Geometry, and Commutative Algebra, a volume published in honor of Y. Akizuki, Kinokuniya, Tokyo 1973.

[8] NAGANUMA H., On the coincidence of two Dirichlet series associated with cusp forms of Hecke's "Nebentype" and Hilbert modular forms over real quadratic fields, Journ. Math. Soc. Japan (4) 25 (1973), 547-555.

[9] PETERS M., Ternäre und quaternäre quadratische Formen und Quaternionenalgebren, Acta Arithmetica 15 (1969), 319-365.

[10] SAITO H., Automorphic forms and algebraic extensions of number fields, Lecture Note in Mathematics nr. 8, Kyoto Univ. 1975.

[11] SHIMIZU H., On traces of Hecke operators,
Journ. Fac. Sci. Univ. Tokyo 10 (1963), 1-19.

[12] ZAGIER D., Modular forms associated to real quadratic fields,
Inventiones Math. 30 (1975), 1-46.

[13] O'MEARA O.T., Introduction to Quadratic Forms,
Springer Verlag, Berlin-Heidelberg-New York, 1963.

International Summer School on Modular Functions
BONN 1976

SÉRIES THÊTA DES FORMES QUADRATIQUES INDÉFINIES

Marie-France VIGNÉRAS

Introduction.

Le but de cet exposé est de décrire un critère simple pour la construction générale de formes modulaires associées à une forme quadratique indéfinie, qui est l'analogue des résultats classiques sur la construction des séries thêta associées à des formes quadratiques définies positives.

Théorème 1. Soit $q(x)$ une forme quadratique sur $\mathbb{R}^n$, de signature quelconque (s,t), soit $L \subset \mathbb{R}^n$ un réseau sur lequel $q(x)$ prend des valeurs entières et soit $p(x) : \mathbb{R}^n \to \mathbb{C}$ une fonction avec les propriétés suivantes :

*) La fonction $f(x) = p(x)\, e^{-2\pi q(x)}$, ainsi que $D(x)f(x)$ et $R(x)f(x)$ pour toute dérivation $D(x)$ d'ordre ≤ 2 et tout polynôme $R(x)$ de degré ≤ 2 sont définies et appartiennent à $L^2(\mathbb{R}^n) \cap L^1(\mathbb{R}^n)$.

**) $p(x)$ satisfait l'équation différentielle $(E-\lambda)p(x) = \frac{\Delta}{4\pi}\, p(x)$, pour un entier λ, où E est l'opérateur d'Euler et Δ le laplacien associé à $q(x)$.

Alors la série thêta

$$\theta(\tau) = v^{-\lambda/2} \sum_{x \in L} p(x\sqrt{v})\, e^{2i\pi q(x)\tau} \qquad [\tau \in \mathbb{C}\ ,\ v = \mathrm{Im}(\tau) > 0]$$

est une forme modulaire non holomorphe de poids $k = \lambda + n/2$; son niveau et son caractère sont ceux de $q(x)$.

Pour faire mieux comprendre ce théorème, je voudrais donner quelques exemples d'applications. Un cas très important est celui où $p(x)$ est homogène, de poids λ , et sphérique pour $q(x)$, c'est-à-dire où **) est remplacé par les deux conditions : $Ep(x) = \lambda p(x)$, $\Delta p(x) = 0$. Alors la fonction $\theta(\tau)$ égale à

$$\sum_{x \in L} p(x)\, e^{2i\pi q(x)\tau}$$

est holomorphe. Si la forme quadratique $q(x)$ est définie positive, on peut montrer que les fonctions sphériques et homogènes sont des polynômes, et on retrouve ainsi les séries thêta classiques (Hecke [4], Ogg [9], Schoeneberg [11], Shimura [14]).

Si $q(x)$ est de signature $(2,n-2)$, soit $\langle x,y\rangle = q(x+y) - q(x) - q(y)$ la forme bilinéaire associée à $q(x)$ et soit $z \in \mathbb{C}^n$, tel que $q(z) = 0$, $\langle z,\bar{z}\rangle > 0$. Alors la fonction

$$\sum_{x \in L} \frac{q(x)^{k-1}}{\langle x,z\rangle^{k+n-2}}\, e^{2i\pi q(x)\tau} \qquad [k > n/2]$$

est une forme modulaire (holomorphe) de poids k .

Ce corollaire qui m'a été suggéré par Eichler, Zagier et Deligne redémontre que les fonctions $\Omega(z_1,z_2,\tau)$ définies par Zagier [18], sont modulaires (elles correspondent au cas $n = 2$).

Si l'on rajoute la condition

$$***)\quad (E + \lambda + n - 2 - 4\pi q(x))(E - \lambda)p(x) = 4r\, p(x) \ ,\quad r \in \mathbb{R}$$

la fonction $\theta(\tau)$ est alors une forme de Maass [7].

De telles formes permettent de construire les séries thêta des corps quadratiques réels, associées aux Grössencharakter (Hecke [4], Maass [6], Gelbart [3]) et ont été utilisées dans des situations diverses (correspondances entre formes modulaires à une variable et à deux variables ([2], [5], [18]) ou entre formes modulaires de poids ½ entier et de poids entier ([8], [10], [15])).

Mentionnons enfin que ce théorème peut être généralisé pour donner des formes modulaires de Hilbert et de Siegel.

Ce théorème admet une démonstration élémentaire reposant sur la formule de Poisson et les propriétés des fonctions d'Hermite, qui donne en plus toutes les fonctions $p(x)$ vérifiant *) et **). Mais, il existe une autre démonstration [16] consistant à remarquer que les séries thêta construites dans ce théorème sont des séries thêta de Weil [17] et que **) est une condition sur le comportement de $f(x)$ pour le groupe des rotations opérant sur $L^2(R^n)$ par la représentation de Weil, ce qui permet d'obtenir le théorème en utilisant un résultat de Shintani ([15], proposition 1-7).

1. Définition des séries thêta.

On suppose donnés :

- un espace vectoriel réel quadratique, non dégénéré, V, de dimension n, muni d'une forme quadratique $q(x)$, à laquelle on associe la forme bilinéaire $\langle x,y\rangle = q(x+y) - q(x) - q(y)$;

- un réseau L, sur lequel $q(x)$ est entière, dont on note L^* le dual

$$L^* = \{x \in V \mid \langle x,y\rangle \in \mathbb{Z}, \forall y \in L\} .$$

On rappelle que le groupe modulaire est engendré par les deux applications $\tau \to \tau+1$, $\tau \to -1/\tau$. On désire déterminer un ensemble de fonctions $p : V\times\mathbb{R}^+ \to \mathbb{C}$, telles que les séries de Fourier (appelées séries thêta)

$$\theta_h(\tau) = \sum_{x\in L+h} p(x,v)\, e^{2i\pi q(x)\tau} , \quad h \in L^* , \ \tau = u+iv \in \mathbb{C} , \ v>0$$

convergent et se transforment simplement sous l'action du groupe modulaire. Le comportement sous l'action de $\tau \to \tau+1$ est déterminé par la forme de $\theta_h(\tau)$, on a :

(1) $$\theta_h(\tau+1) = e^{2i\pi q(h)}\,\theta_h(\tau) .$$

Aussi, on est ramené à la question suivante : déterminer un ensemble de fonctions $p(x,v)$ telles que $\theta_h(-1/\tau)$ ait une valeur convenable.

2. La formule de Poisson.

Il s'avère très naturellement que la formule de Poisson est l'outil adapté au calcul de $\theta_h(-1/\tau)$. Très brièvement, la formule de Poisson peut se résumer ainsi (Bochner [1] th. 67).

Etant donnés :

- une fonction $f(x)$ sur V , pour laquelle l'intégrale suivante (appelée transformée de Fourier de $f(x)$) :

$$f^*(x) = \int_V f(y)\, e^{-2i\pi\langle x,y\rangle}\, dy$$

a un sens (dy est une mesure de Haar quelconque sur V) ;

- un réseau L de V dont le volume d'un domaine fondamental dans V , pour la mesure dy , est noté $\mathrm{vol}(L)$.

La formule de Poisson s'écrit :

$$\sum_{x\in L} f(x) = \mathrm{vol}(L)^{-1} \sum_{x\in L^*} f^*(x) .$$

Elle est indépendante du choix de dy et est valable si $f(x)$ vérifie certaines hypothèses. Elle est vraie pour la fonction $f_{-1/\tau}(x)$ définie par :

$$f_\tau(x) = p(x,v)\, e^{2i\pi\tau q(x)}$$

où elle donne :

$$\theta_h(-1/\tau) = \mathrm{vol}(L)^{-1} \sum_{\substack{h\in L^*\\ h(\mathrm{mod}\ L)}} e^{2i\pi\langle h,k\rangle} \sum_{x\in k+L} f^*_{-1/\tau}(x) .$$

Dans l'hypothèse où

(2) $$f^*_{-1/\tau}(x) = (-i)^a\, \tau^{\lambda+n/2}\, f_\tau(x) , \quad a\in\mathbb{R}$$

la valeur de $\theta_h(-1/\tau)$ est égale à :

(3) $$\theta_h(-1/\tau) = \mathrm{vol}(L)^{-1}\ (-i)^a\ \tau^{\lambda+n/2} \sum_{k\in L^*/L} \theta_k(\tau)\ .$$

Une relation telle que (3) s'introduit assez naturellement si on suppose d'abord que $\mathrm{Re}(\tau) = 0$ à cause de la règle suivante : $[f(x/\sqrt{v})]^*(y) = v^{n/2} f^*(y\sqrt{v})$, $v>0$. La définition de $(\tau/i)^{\lambda+n/2}$ est précisée de la sorte : Arg w étant la valeur de l'argument d'un nombre complexe w vérifiant $-\pi < \mathrm{Arg}\, w \leq \pi$, on pose $\mathrm{Log}\, w = \mathrm{Log}\, |w| + i\, \mathrm{Arg}\, w$ et $w^{\frac{1}{2}} = e^{\frac{1}{2}\mathrm{Log}\, w}$. On vérifie que $(w_1 w_2)^{\frac{1}{2}} = \varepsilon\, w_1^{\frac{1}{2}} w_2^{\frac{1}{2}}$ où $\varepsilon = 1$ si $-\pi < \mathrm{Arg}\, w_1 + \mathrm{Arg}\, w_2 \leq \pi$ et $\varepsilon = -1$ sinon.

3. Formule de transformation de thêta.

Que la forme quadratique $q(x)$ soit définie positive ou non, les formules (1) et (3) impliquent des formules de transformations de $\theta_h(\tau)$ pour l'action du groupe modulaire. Je réfère à Shimura ([14], p. 454) pour la démonstration dans le cas positif ; elle s'applique aussi au cas quelconque.

On note :

- A la matrice de $q(x)$ sur une base de L (si $x = (x_1,\dots,x_n)$ sur cette base, $q(x) = \frac{1}{2}\, x\, A\, {}^t x)$;

- D le déterminant de A , Δ son discriminant $\Delta = (-1)^{n/2}D$ si n est pair et $2D$ si n est impair ;

- N le niveau de A , c'est-à-dire le plus petit entier tel que la matrice NA^{-1} soit paire ;

- χ le caractère de A défini par $\chi(p) = (\frac{\Delta}{p})$, le symbole $(\frac{c}{d})$, c et $d \in \mathbb{Z}$, étant défini comme dans [14]. En fait, D , Δ , N , ne dépendent pas du choix de A , mais de $q(x)$ et de L .

Théorème 2. Sous l'hypothèse (2), on a :

* pour tout $\gamma = \begin{pmatrix} a & b \\ c & d \end{pmatrix} \in SL_2(\mathbb{Z})$, $c \equiv 0 \pmod N$

$$\theta_h\left(\frac{a\tau+b}{c\tau+d}\right) = \left(\frac{\Delta}{d}\right) v(\gamma)\ (c\tau+d)^{\lambda+n/2}\ \theta_{ah}(\tau)\ ,$$

où $v(\gamma) = 1$ si n est pair et $v(\gamma) = (\frac{c}{d})(\frac{-4}{d})^{-n/2}$ si n est impair.

** <u>la limite de</u> $\theta_h(\eta(\tau))(c\tau+d)^{-\lambda-n/2}$ <u>est finie quand</u> τ <u>tend vers</u> $i\infty$, <u>pour tout</u> $\eta = \begin{pmatrix} a & b \\ c & d \end{pmatrix} \in SL_2(\mathbb{Z})$.

4. <u>Les fonctions d'Hermite</u>.

Une solution $f_\tau(x)$ de (3) doit être pour $\tau = i$ une fonction propre pour la transformation de Fourier ; on aura :

(4) $f^*(x) = (-i)^a f(x)$ $\qquad (f(x) = f_i(x))$.

Mais inversement, la relation (4) entraîne que les fonctions $f_\tau(x)$ définies par

(5) $f_\tau(x) = v^{-\lambda/2} f(x\sqrt{v})\, e^{2i\pi uq(x)}$ $\qquad \tau = u+iv$, $v > 0$

sont des solutions de (3) lorsque $u = \mathrm{Re}(\tau) = 0$ (d'après la règle donnant $f(x\sqrt{v})^*$). L'idée est de trouver toutes les fonctions $f(x)$, propres pour Fourier, telles que les $f_\tau(x)$ définies par (5) soient des solutions de (3) pour τ quelconque.

L'espace $L^2(V)$ possède une base orthonormale, formée de fonctions propres pour la transformation de Fourier, appelées les fonctions d'Hermite (Appel et Kampé de Fériet [O], 3e partie).

On note

- $(x_1,\ldots,x_n)$ les composantes de $x \in V$ sur une base telle que $q(x) = \frac{1}{2}(x_1^2 + \ldots x_s^2 - x_{s+1}^2 - \ldots - x_n^2)$, $n = s+t$
- $q_+(x) = \frac{1}{2}(x_1^2 + \ldots + x_n^2)$
- pour $m = (m_1,\ldots,m_n) \in \mathbb{N}^n$, $\lambda \in \mathbb{R}$ et $h = (h_1,\ldots,h_n) \in \mathbb{R}^n$, $m! = m_1!\ldots m_n!$, $\lambda^m = \lambda^{m_1+\ldots+m_n}$, $H^m = h_1^{m_1}\ldots h_n^{m_n}$, $\varepsilon(m) = m_1 + \ldots m_s - m_{s+1} - \ldots - m_n$, et

$$c_m = (-1)^m \sqrt{m!}\; 2^{m-1/4}\, \pi^{m/2} .$$

Les fonctions d'Hermite sont définies par :

$$e^{-4\pi q_+(x+h)} = e^{-2\pi q_+(x)} \sum_{m \in \mathbb{N}^n} c_m \mathcal{H}_m(x) \frac{h^m}{m!} .$$

Elles sont propres pour la transformation de Fourier et sont solution d'une équation différentielle du 2$^{\underline{e}}$ ordre :

(6) $\mathcal{H}^*_m(x) = (-i)^{\varepsilon(m)}\,\mathcal{H}_m(x)$

(7) $(\Delta - 8\pi^2 q(x))\,\mathcal{H}_m(x) = -4\pi(\varepsilon(m) + \frac{s-t}{2})\,\mathcal{H}_m(x)$

où $\Delta = \sum_{i=1}^{s} \frac{\partial^2}{\partial x_i^2} - \sum_{i=s+1}^{n} \frac{\partial^2}{\partial x_i^2}$ est le laplacien de $q(x)$.

La référence donnée [O] convient pour une variable. On généralise (6) et (7) à un nombre quelconque de variables et à une forme quadratique de signature (s,t) en utilisant que $\mathcal{H}_m(x)$ est le produit $\mathcal{H}_{m_1}(x_1)\ldots\mathcal{H}_{m_n}(x_n)$, que $\mathcal{H}^*_m(x) = \mathcal{H}^*_{m_1}(x_1)\ldots\mathcal{H}^*_{m_s}(x_s)\mathcal{H}^*_{m_s+1}(-x_{s+1})\ldots\mathcal{H}^*_{m_n}(-x_n)$ et que $\mathcal{H}_{m_i}(x_i) = (-1)^{m_i}\,\mathcal{H}_{m_i}(-x_i)$.

Si l'on considère la fonction $p_m(x) = \mathcal{H}_m(x)\,e^{2\pi q(x)}$, l'équation différentielle (7) devient :

(8) $(E - \varepsilon(m) + q)\,p_m(x) = \frac{\Delta}{4\pi}\,p_m(x)$

où E est l'opérateur d'Euler, $E = \sum_{i=1}^{n} x_i \frac{\partial}{\partial x_i}$.

5. Calcul d'intégrales.

Lemme 1. $\int_{\mathbb{R}} e^{i\pi\tau y^2 - 2i\pi xy}\,dy = (\tau/i)^{-\frac{1}{2}}\,e^{-i\pi x^2/\tau}$.

C'est bien connu. On généralise à un plus grand nombre de variables.

Soit

$$g_\tau(x) = e^{2i\pi uq(x) - 2\pi vq_+(x)} = \prod_{j=1}^{s} e^{i\pi\tau x_j^2} \prod_{j=s+1}^{s+t} e^{i\pi\bar{\tau}x_j^2}.$$

Lemme 2. $g_\tau(x)^* = (\tau/i)^{-s/2}\,(\bar{\tau}/i)^{-t/2}\,g_{-1/\tau}(x)$.

En considérant $v^{t/2}\,g_\tau(x)$, on élimine le terme en $\bar{\tau}$; autrement dit, la transformée de Fourier de la fonction

$$\mathcal{H}_{0,\tau}(x) = v^{t/2}\,\mathcal{H}_0(x\sqrt{v})\,e^{2i\pi uq(x)}$$

est donnée par :

Lemme 3. $\mathcal{H}^*_{0,\tau}(x) = (-i)^{(t-s)/2}\,\tau^{(t-s)/2}\,\mathcal{H}_{0,-1/\tau}(x)$.

On généralise à m quelconque en utilisant les propriétés remarquables des fonctions d'Hermite (on applique le fait que $\frac{\partial}{\partial x}$ et $-2\pi ix$ sont duales par rapport à Fourier, et les relations

$$c_{m+1}\,\mathcal{H}_{m+1}(x) + 4\pi c_m\, x\,\mathcal{H}_m(x) + 4\pi c_{m-1}\, m\,\mathcal{H}_{m-1}(x) = 0\ ,$$
$$\mathcal{H}'_m(x) - 2\pi x\,\mathcal{H}_m(x) = c_{m+1}/c_m\ \mathcal{H}_m(x)\ ,\ \text{pour}\ x\in\mathbb{R}).$$

<u>Lemme</u> 4. $\mathcal{H}^*_{m,\tau}(x) = (-i)^{(t-s)/2}\,\tau^{-(\lambda+n/2)}\,\mathcal{H}_{m,-1/\tau}(x)$ <u>avec</u> $\lambda = \varepsilon(m) - q$.

6. Le lemme 4 nous donne une base dans $L^2(V)$ des fonctions $f(x)$ cherchées. Elle est formée des fonctions $\mathcal{H}_m(x)$, telles que $\lambda = \varepsilon(m)-q$. Les fonctions $f(x)$ sont (avec l'hypothèse *) supplémentaire du théorème de l'introduction) les solutions de l'équation différentielle (7), équivalente pour $p(x) = f(x)\,e^{2\pi q(x)}$ à l'équation différentielle de **). Le théorème 1 est démontré.

Si la forme $q(x)$ est définie positive, les fonctions $p_m(x)$ sont des polynômes. Les polynômes $p_m(x)$ tels que $m_1+\ldots+m_n = \lambda$ forment une base (finie) des fonctions $p(x)$ vérifiant les deux conditions du théorème. Si $q(x)$ est indéfinie, la base est infinie.

7. <u>Dérivées et formes de Maass</u>.

Parmi les fonctions du demi-plan supérieur, se transformant sous l'action du groupe modulaire comme dans le théorème 2, se trouvent les formes de Maass, qui sont les valeurs propres de l'opérateur de Beltrami (généralisé) du demi-plan supérieur. Cet opérateur au moyen de thêta se relève sur les fonctions $p(x)$. On note :

- P_λ l'ensemble des solutions $p(x)$ de l'équation differentielle $(E-\lambda)p(x) = \Delta/4\pi\ p(x)$ telles que $f(x) = p(x)\,e^{-2\pi q(x)}$ appartienne à l'espace de Schwarz de V .
- $\theta(\tau,p)$ la série thêta associée à $p(x)\in P_\lambda$, dans le théorème 1.
- Θ_k , l'image de P_λ par l'application $\theta(\tau,p)$, $k = \lambda + n/2$.

Sur P_λ, on définit deux opérateurs $\widetilde{K}_\lambda$ et $\widetilde{\Lambda}_\lambda$ qui sont les relèvements des opérateurs usuels de dérivation K_k et Λ_k sur θ_k. On démontre la proposition suivante :

<u>Proposition</u>. <u>Si</u> $p(x) \in P_\lambda$ <u>satisfait l'équation différentielle</u>

$$(E+\lambda+n-2-4\pi q(x))(E-\lambda)p(x) = 4r\, p(x) \quad , \quad r \in \mathbb{R}$$

<u>les séries</u> $\theta(\tau,p)$ <u>sont des formes de Maass</u>.

<u>Démonstration</u> : $K_k = \frac{k}{v} + i\,\frac{\partial}{\partial\tau} \qquad \Lambda_k = iv^2\,\frac{\partial}{\partial\bar\tau}$

(on pose $\frac{\partial}{\partial\tau} = \frac{\partial}{\partial u} - i\,\frac{\partial}{\partial v}$, $\frac{\partial}{\partial\bar\tau} = \frac{\partial}{\partial u} + i\,\frac{\partial}{\partial v}$) .

<u>Définition</u>. $\widetilde{K}_\lambda = \frac{E+\lambda}{2} + \frac{n}{2} - 2\pi q(x) \qquad \widetilde{\Lambda}_\lambda = -\frac{E-\lambda}{2}$.

On vérifie que $\widetilde{K}_\lambda$ et $\widetilde{\Lambda}_\lambda$ sont des applications de P_λ , respectivement dans $P_{\lambda+2}$ et $P_{\lambda-2}$, c'est-à-dire que les diagrammes suivants sont commutatifs :

$$\begin{array}{ccc} P_\lambda & \xrightarrow{\widetilde{K}_\lambda} & P_{\lambda+2} \\ {\scriptstyle\theta(\tau,p)}\downarrow & & \downarrow{\scriptstyle\theta(\tau,p)} \\ \theta_k & \xrightarrow{K_k} & \theta_{k+2} \end{array} \qquad\qquad \begin{array}{ccc} P_\lambda & \xrightarrow{\widetilde{\Lambda}_\lambda} & P_{\lambda-2} \\ {\scriptstyle\theta(\tau,p)}\downarrow & & \downarrow{\scriptstyle\theta(\tau,p)} \\ \theta_k & \xrightarrow{\Lambda_k} & \theta_{k-2} \end{array}$$

Les égalités $\theta(\tau,\widetilde{K}_\lambda p) = K_k\,\theta(\tau,p)$ et $\theta(\tau,\widetilde{\Lambda}_\lambda p) = \Lambda_k\,\theta(\tau,p)$ montrent que les images de séries thêta par les opérateurs de dérivation sont des séries thêta.

Dans θ_k l'opérateur $K_{k-2}\Lambda_k$

$$\Delta_k = -v^2\,\frac{\partial^2}{\partial\tau\,\partial\bar\tau} + ikv\,\frac{\partial}{\partial\bar\tau} \qquad \text{(opérateur de Beltrami généralisé)}$$

admet des valeurs propres, qui sont des formes de Maass ([7]

le relèvement de Δ_k à P_λ étant

$$\widetilde{\Delta}_\lambda = -((E+\lambda)/2 + n/2 - 1 - 2\pi q(x))(E-\lambda)/2$$

on en déduit la proposition.

<u>Exemples</u> : <u>Les séries thêta des corps quadratiques réels</u>.

On note :

$V = \mathbb{R}^2$.

L = un idéal d'un corps quadratique réel K , de discriminant Δ , d'élément générique μ , dont le conjugué est μ', plongé dans $\mathbb{R}^2$ par $\mu \to (\mu,\mu')$; $n(L)$ est la norme de L .

$h \in \sqrt{\Delta^{-1}}\, L$, $h \neq 0$.

$q(x,y) = \frac{xy}{n(L)}$.

On peut construire des séries thêta de poids $\lambda \in \mathbb{Z}$, de niveau Δ , qui sont des formes de Maass avec $4r = \lambda^2 + t^2$, où t est de la forme $t = 2\pi n/\mathrm{Log}|\varepsilon|$, $\varepsilon >> 0$ unité de K .

Le théorème 2 nous permet de construire des séries thêta de poids λ , de niveau N :

$$\theta_h(\tau) = v^{-\lambda/2} \sum_{\mu \in L+h} f_{m,n}(\mu\sqrt{v}, \mu'\sqrt{v})\, e^{2i\pi u \frac{\mu\mu'}{n(L)}}$$

avec $m-n = \lambda+1$ et

$$f_{m,n}(x,y) = \mathcal{H}_m((x+y)\sqrt{\tfrac{n(L)}{2}})\ \mathcal{H}_n((x-y)\sqrt{\tfrac{n(L)}{2}}) .$$

Remarquons que pour $a \in \mathbb{R}^+$, l'application $(x,y) \to (ax,a^{-1}y)$ est une unité de $q(x,y)$ et par conséquent $f_{m,n}(ax,a^{-1}y)$ vérifie (7). D'autre part, $f_{m,n}(a\mu\sqrt{v}, a^{-1}\mu'\sqrt{v})$ est le produit d'un terme polynômial en $a\mu\sqrt{v}$ et $a^{-1}\mu'\sqrt{v}$ par une exponentielle $e^{-\pi v(a^2\mu^2 + a^{-2}\mu'^2)}$.

Pour tout caractère $a \to a^{it}$ fixant globalement $L+h$, donc $t = 2\pi n/\mathrm{Log}|\varepsilon|$, ε unité de k , $n \in \mathbb{Z}$,

$$\overline{\theta}_h(\tau) = \int_0^\infty a^{it}\, \theta_h(\tau,a)\, \frac{da}{a}$$

où l'on a noté $\theta_h(\tau,a)$ la série thêta correspondante à $f_{m,n}(ax,a^{-1}y)$. Les séries $\overline{\theta}_h(\tau)$ sont des séries thêta de poids λ de niveau N ; elles s'écrivent

$$\overline{\theta}_h(\tau) = \sum_{\substack{\mu \in L+h \\ \mu \bmod (\varepsilon^n)_{n\in\mathbb{Z}}}} v^{-\lambda/2}\, g_{m,n}(\mu\sqrt{v}, \mu'\sqrt{v})\, e^{2i\pi u \frac{\mu\mu'}{n(L)}}$$

où

$$g_{m,n}(x,y) = \int_0^\infty a^{it}\, f_{m,n}(x,y)\, \frac{da}{a} \qquad (xy \neq 0)$$

est bien définie pour $xy \neq 0$. Si l'on pose :

$$p(x,y) = g_{m,n}(x,y)\, e^{-2\pi \frac{xy}{n(L)}}$$

on obtient des fonctions $p(x,y)$ vérifiant (8) ainsi qu'une autre équation différentielle

$$x\, p'_x(x,y) - y\, p'_y(x,y) = -it\, p(x,y)$$

que l'on obtient en dérivant par rapport à b l'égalité

$$p(bx, b^{-1}y) = b^{-it}\, p(x,y) .$$

Il est aisé d'en déduire que $p(x,y)$ vérifie $\tilde{\Delta}_\lambda\, p(x,y) = 4r\, p(x,y)$ avec $4r = \lambda^2 + t^2$, donc les séries $\bar{\theta}_h(\tau)$ sont des formes modulaires de Maass.

Bibliographie

[0] APPEL P. et KAMPÉ DE FÉRIET J.- Fonctions hypergéométriques et hypersphériques. Polynômes d'Hermite - Gauthier-Villars, (1926) (3e partie).

[1] BOCHNER S.- Lectures on Fourier Integrals. Ann. of Math. Studies, N° 42, Princeton University Press (1959).

[2] DOI K., NAGANUMA H.- On the coincidence of two Dirichlet series associated with cusp forms of Hecke's "Neben"-type and Hilbert modular forms over a real quadratic field. J. Math. Soc. Japan 25 (1973), 547-555.

[3] GELBART S.- Automorphic forms on adele groups. Ann. of Math. Studies, N° 83, Princeton University Press 1975.

[4] HECKE E.- Mathematische Werke, Vandenhoeck und Ruprecht, Göttingen (1959).

[5] KUDLA Theta Functions and Hilbert Modular Forms. (Preprint).

[6] MAASS H.- Über eine neue Art von nichtanalytischen automorphen Funktionen und die Bestimmung Dirichletscher Reihen durch Funktional Gleichungen. Math. Annalen 121 (1949), 141-183.

[7] MAASS H.- Lectures on Modular Functions of One Complex Variable. Tata Institute (1964), chap. IV et V.

[8] NIWA S.- Modular Forms of half integral weight and the integral of certain theta functions. Nagoya Math. Journal 56 (1975), 147-163.

[9] OGG A.- Modular Forms and Dirichlet Series. Benjamin (1969), chap. VI.

[10] RALLIS S. et SCHIFFMANN G.- Automorphic forms constructed from the Weil Representation, holomorphic case (1976) (preprint).

[11] SCHOENEBERG B.- Das Verhalten von mehrfachen Thetareihen bei Modulsubstitutionen. Math. Annalen 116 (1939), 511-523.

[12] SHALIKA I., TANAKA S.- On an explicit construction of a certain class of automorphic forms. Amer. J. of Math. 91 (1969), 1049-1076.

[13] SHIMIZU H.- Theta series and automorphic forms on GL(2). Journ. Math. Soc. Japan, 24 (1972), 638-683.

[14] SHIMURA G.- On modular forms of half-integral weight. Ann. of Math. 97 (1973), 440-481.

[15] SHINTANI T.- On construction of holomorphic cusp forms of half integral weight. Nagoya Math. Journal 58 (1975), 83-126.

[16] VIGNÉRAS M.-F.- Séries thêta des formes quadratiques indéfinies. Séminaire de théorie des nombres Delange-Pisot-Poitou (1976-1977).

[17] WEIL A.- Sur certains groupes d'opérateurs unitaires. Acta Math. 111 (1964), 143-211.

[18] ZAGIER D.- Modular Forms associated to real quadratic fields. Inventiones Mathematicae 30 (1975), 1-48.

International Summer School on Modular Functions
Bonn 1976

AUTOMORPHIC FORMS AND ARTIN'S CONJECTURE

by Stephen Gelbart[1)]

1) Work supported by a grant from the National Science Foundation

CONTENTS

INTRODUCTION

Various objects in algebraic geometry and number theory have associated to them L-functions with Euler product. Examples include the L-functions attached to algebraic varieties, complex Galois representations of a number field, and strictly compatible systems of ℓ-adic representations. The conjecture is that all these L-functions should possess analytic continuations and functional equations. In fact, according to R.P. LANGLANDS and A. WEIL, these functions should coincide with the Euler products naturally associated to automorphic forms on groups such as GL(n).

In [13] Langlands described this philosophy in a sequence of dazzling conjectures. The particular conjecture which is of interest to us here relates Artin L-functions of degree n to cusp forms on GL(n). When $n = 1$ this reduces to the fundamental reciprocity law of abelian class field theory. Thus we refer to it as "Langlands' reciprocity conjecture". It asserts that there is a natural map between n-dimencional irreducible Galois representations and cusp forms on GL(n). Since this map preserves L-functions, its existence implies the truth of Artin's conjecture.

Recently Langlands has been able to prove his reciprocity conjecture -and hence Artin's conjecture- for a wide class of 2-dimensional Galois representations. A brief sketch of the proof appears in [14] and [15]. The modular ingredients include :

(i) the theory of base change for GL(2) as developed by Saito [17], Shintani [19] and Langlands [14];

(ii) the theory of "lifting" from GL(2) to GL(3) developed by myself and Jacquet (Gelbart-Jacquet [7]);

(iii) the work in progress of Jacquet, Piateckii-Shapiro and Shalika on the "converse theorem for GL(3)" and L-functions on GL(3) × GL(3) (cf. [12]).

The purpose of this paper is to describe the significance of these modular results in more detail. Particular emphasis is on the role these results play in Langlands' proof of Artin's conjecture for the so-called tetrahedral representations of the Galois group of an arbitrary number field.

Some of the work described here is still in progress, and still more is not yet published. Thus I am grateful to all those concerned for allowing me to make this report. I am also grateful to P. CARTIER, J-P. SERRE and J. TUNNELL for helpful remarks on the material of Sections II and III.

I. ARTIN'S CONJECTURE

Fix an integer n and a number field F.

1.1. Artin's Original Conjecture.

Suppose K is any Galois extension of F and

$$\sigma : \mathrm{Gal}(K/F) \to GL(n,\mathbb{C})$$

is an n-dimensional representation of Gal(K/F). For each place v of F let σ_v denote the restriction of σ to the decomposition group of Gal(K/F) at v. The Artin L-function attached to σ is then given by an infinite Euler product

$$L(s,\sigma) = \prod_v L(s,\sigma_v)$$

extending over all the places of F. If v is unramified in K, and Fr_v denotes a Frobenius element over v, then

$$L(s,\sigma_v) = [\det(I-\sigma(Fr_v)N_v^{-s})]^{-1}.$$

Artin's Conjecture. Suppose σ is irreducible and non-trivial. Then $L(s,\sigma)$, originally defined only in some right half-plane, extends to an entire function of $\mathbb{C}$.

Non-trivial results in the direction of Artin's Conjecture were first obtained by E. Artin and R. Brauer. Artin proved his conjecture for monomial representations -those induced from one-dimensional representations of a subgroup. In fact, for such σ Artin proved that $L(s,\sigma)$ is $L(s,\chi)$, a Hecke L-series with character χ. Thus Artin proved a dual form of the fundamental reciprocity law of abelian class-field theory.

For arbitrary σ, Brauer proved that L(s,σ) is at least <u>meromorphic</u> in ℂ. Until recently, however, very little was known in general about the entirety of L(s,σ).

As already indicated, one purpose of this paper is to report on recent progress mady by R.P. Langlands on the proof of Artin's conjecture for two-dimensional σ. For irreducible such σ, it is known (cf. [18], §2.5.) that the image of σ(Gal(K/F)) in PGL(2,ℂ) is either :

(i) dihedral, in which case σ is <u>monomial</u>;

(ii) isomorphic to A_4 - the <u>tetrahedral</u> case;

(iii) isomorphic to S_4 - the <u>octahedral</u> case; or

(iv) isomorphic to A_5 - the <u>icosahedral</u> (or <u>non-solvable</u>) case.

Langlands' work concerns cases (ii) and (iii) but not (iv). Before describing his results, we need to reformulate Artin's Conjecture in terms of the so-called Hecke theory for GL(n).

1.2. <u>Hecke Theory for GL(n)</u>.

Let G denote the algebraic group GL(n). In this section we shall describe the automorphic representations of G. To save space, we shall often sacrifice precision for the sake of speed. For a more detailed development of the theory, see [1], [5] or [6].

For each place v of F let F_v denote the completion of F at v. Let $\mathbb{A}_F$ denote the adele ring of F, and $G_{\mathbb{A}}$ the adele group

$$GL(n,\mathbb{A}) = \prod_v GL(n,F_v) \quad \text{(a restricted direct product)}.$$

For each algebraic subgroup H of G, $H_{\mathbb{A}}$ will denote the group of adelic points of H.

An <u>automorphic form on GL(n) (over F)</u> is a slowly increasing left GL(n,F)-invariant function on $G_{\mathbb{A}}$ which is right invariant by an appropriate compact subgroup; for the definition of slowly increasing, see [5]. An automorphic form is <u>cuspidal</u> if

$$\int_{U_F \backslash U_{\mathbb{A}}} \phi(ux)du = 0$$

for each unipotent radical U of a proper F-parabolic subgroup of GL(n).

<u>Examples (n = 1,2).</u>

(a) Examples of automorphic forms on GL(1) are provided by <u>grössencharakters</u> of F, i.e., characters of the idele class group $F^x \backslash \mathbb{A}_F^x$. Since GL(1) has no proper parabolic subgroups, every automorphic form on GL(1) is (automatically) cuspidal.

(b) Suppose F = $\mathbb{Q}$. Then an automorphic form ϕ on GL(2) extends the notion of a modular form

$$f(z) = \sum_{n=0}^{\infty} a_n e^{2\pi inz}$$

defined classically in $H = \{z : Im(z) > 0\}$. The left invariance of ϕ with respect to GL(2,F) corresponds to the fact that f is "automorphic" in the classical sense for some congruence subgroup of SL(2,$\mathbb{Z}$). The cuspidal condition on ϕ corresponds to the fact that f is a cusp form in the usual sense, i.e. f vanishes at each cusp; in particular $a_0 = 0$.

(c) Suppose F is a real quadratic field. Then <u>automorphic forms on GL(2) over F</u> generalize the classical notions of <u>Hilbert modular form</u>.

Now suppose π is any irreducible unitary representation of GL(n,$\mathbb{A}$). If π can be realized by right translation operators in the space of automorphic (resp. cuspidal automorphic) forms on GL(n) we call π an automorphic (respectively cuspidal) representation of GL(n). To attach to π an L-function with Euler product we need to factor π as a product of local representations.

According to [8] and [11] there is associated to π a family of local representations π_v -uniquely determined by π- such that

(i) for every v, π_v is irreducible;

(ii) for almost every v, π_v is "unramified"; and

(iii) in a sense to be made precise below,

$$\pi = \bigotimes_v \pi_v. \tag{1.1}$$

This result is really only true for "admissible" π. However, it is easy to show that every cuspidal π is admissible.

When n = 1 the decomposition (1.1) corresponds to the fact that every grössencharakter χ can be written as a product of local characters χ_v. Condition (ii) corresponds to the fact that the restriction of χ to the group of units in $F_v^\times$ is trivial for almost every v, i.e., for almost every v, χ_v is "unramified" in the usual sense.

In general, π_v unramified means that the restriction of π_v to the standard maximal compact subgroup K_v of $GL(n,F_v)$ contains at least one fixed vector. In this case, the theory of spherical functions for GL(n) shows that :

(a) K_v has exactly one fixed vector in the space of π_v; and

(b) π_v corresponds canonically to a semi-simple conjugacy class A_v in $GL(n,\mathbb{C})$.

The significance of (a) is that (1.1) is meaningful when interpreted as a restricted direct product with respect to these K_v-fixed vectors. The significance of (b) is that one can canonically attach to most π_v an "Euler factor" of degree n.

Examples (n = 1,2).

(a) Suppose n = 1 and π_v is an unramified character of F_v^x. Then A_v is $\pi_v(\tilde{\omega})$, the value of π_v at any local uniformizing variable of F_v.

(b) Suppose n = 2, F = $\mathbb{Q}$, and $\pi_f = \otimes \pi_p$ is generated by the classical modular form

$$f(z) = \sum_{n=1}^{\infty} a_n e^{2\pi i n z}$$

on SL(2,$\mathbb{Z}$). The decomposition $\pi_f = \otimes \pi_p$ corresponds to the fact that f is an eigenfunction for all the Hecke operators T_p, i.e., $T_p f = a_p f$ for all p. The unramified representation π_p then corresponds to the conjucagy class

$$A_p = \begin{pmatrix} \alpha_p & 0 \\ 0 & \beta_p \end{pmatrix}$$

if and only if $\det(A_p) = 1$ and $\mathrm{tr}(A_p) = p^{-(\frac{k-1}{2})} a_p$.

As already indicated, Hecke theory for GL(n) starts with the notion of an L-function attached to each irreducible unitary reperesentation

$$\pi = \otimes \pi_v$$

of GL(n,$\mathbb{A}$). Given π, let S denote the finite set of places v of F outside of which π_v is unramified. For each v $\notin$ S let A_v denote the semi-simple conjugacy class in GL(n,$\mathbb{C}$) corresponding to π_v. Then consider the infinite Euler product

$$(1.2.)\qquad \prod_{v \notin S} L(s,\pi_v) = \prod_{v \notin S} [\det(I-\{A_v\}(N_v)^{-s})]^{-1}$$

extending over (most of) the non-archimedean places of F. This is an <u>Euler product of degree n</u> in the sense that each <u>Euler factor</u> $L(s,\pi_v)$ is of the form $P^{-1}((N_v)^{-s})$ with P a polynomial of degree n such that $P(0) = 1$. The infinite product can be shown to converge for Re(s) sufficiently large.

In many known cases the function defined by (1.2) actually extends to a meromorphic function in $\mathbb{C}$ with simple functional equation. In the context of Example (a) this amounts to Hecke's fundamental result for the Euler product

$$L(s,\chi) = \prod_{v \notin S} (1-\chi(\tilde{\omega}_v)N_v^{-s})^{-1},$$

a Hecke L-series with grössencharakter χ. In the context of Example (b) it amounts to Hecke's classical theory of Dirichlet series associated to modular forms. In this case

$$\prod_{v \notin S} L(s,\pi_v) = \prod_{p<\infty} (1-\alpha_p p^{-s})^{-1}(1-\beta_p p^{-s})^{-1}$$

$$= \prod_p (1-a_p p^{-s'} + p^{k-1-2s'})^{-1}$$

$$= \sum a_n n^{-s'} = D(s',f),$$

with $s' = s + (\frac{k-1}{2})$. Thus the analytic continuation of $\prod_{v \notin S} L(s,\pi_v)$ results from the known holomorphy of $D(s',f)$, and the functional equation results from the invariance (up to sign) of

$$(2\pi)^{-s'}\Gamma(s')D(s',f)$$

with respect to the change of variable $s' \to k-s'$.

In general, one has the following result :

THEOREM 1. (Jacquet-Langlands [11], Godement-Jacquet [9]).
Suppose $\pi = \bigotimes_v \pi_v$ *is an irreducible unitary representation of* $GL(n,\mathbb{A})$ *with* S *and* A_v *as above. Then one can define an infinite Euler product*

$$L(s,\pi) = \prod_v L(s,\pi_v),$$

and local factors $\varepsilon(s,\pi_v)$, *so that for* $v \notin S$, $L(s,\pi_v) = [\det(I-\{A_v\}(N_v)^{-s})]^{-1}$, *and* $\varepsilon(s,\pi_v) = 1$. *Moreover, if* π *is an automorphic representation of* GL(n), *then :*

(i) $L(s,\pi)$, *initially defined only in some right half-plane, extends to a meromorphic function of* s *with only finitely many poles in* $\mathbb{C}$;

(ii) $L(s,\pi) = (\prod_v \varepsilon(s,\pi_v))L(1-s,\tilde{\pi})$ *with* $\tilde{\pi}$ *the representation contragredient to* π; *and*

(iii) if π *is cuspidal (and non-trivial when* n = 1*), then* $L(s,\pi)$ *is* *entire* *and of finite order.*

Remarks.

(a) Properties (i)-(iii) above essentially characterize the irreducible unitary representations of $GL(2,\mathbb{A})$ wich are cuspidal. This is Theorem 11.3 of [11]; it generalizes earlier special results of Hecke and Weil. Similar results for GL(3) have been obtained in [12]; together these results constitute a converse theorem to Hecke theory for GL(2) and GL(3).

(b) Langlands has conjectured in [13] that Theorem 1 should hold in a much more general context. More precisely, fix a finite-dimensional semi-simple analytic representation

$$r : GL(n,\mathbb{C}) \to GL(m,\mathbb{C})$$

with $n \leq m$, and define, for $v \notin S$, an Euler factor of the form

$$L(s,\pi_v,r) = [\det(I-r(A_v)(N_v)^{-s})]^{-1}.$$

<u>Conjectured generalization of Theorem 1</u>. Suppose $\pi = \otimes \pi_v$ is an automorphic representation of GL(n). Then one can define $L(s,\pi_v,r)$ for $v \in S$ so that the function

$$L(s,\pi,r) = \prod_v L(s,\pi_v,r)$$

satisfies conditions (i) and (ii) of Theorem 1 with some $\varepsilon(s,\pi_v,r)$ in place of $\varepsilon(s,\pi_v)$.

Note that the assertions of this conjecture reduce to those of Theorem 1 when r is the standard representation of $GL(n,\mathbb{C})$ by itself. Indeed if we denote this representation by

$$\rho_n : GL(n,\mathbb{C}) \to GL(n,\mathbb{C}),$$

then

$$L(s,\pi,\rho_n) = L(s,\pi).$$

In general, $L(s,\pi,r)$ is defined so that

$$L(s,\pi,r_1 \oplus r_2) = L(s,\pi,r_1)L(s,\pi,r_2).$$

(c) <u>Functoriality of Automorphic Forms with respect to the Associate Group</u>.

Suppose $G' = GL(m)$ and $p : GL(n,\mathbb{C}) \to GL(m,\mathbb{C})$ is a homomorphism. Then there should be a map p_* taking automorphic representations of G to automorphic representations of G' so that for each semi-simple representation r of $GL(m,\mathbb{C})$,

$$L(s,\pi,r \circ \rho) = L(s,\rho_*(\pi),r).$$

But $GL(n,\mathbb{C})$ (resp. $GL(m,\mathbb{C})$) is (essentially) the associate group of $GL(n)$ (resp. $GL(m)$) (cf. [13] and [14]). Thus this assertion amounts to Langlands' principle of "functoriality of automorphic forms with respect to the associate group".

1.3. Langland's Reciprocity Conjecture.

Theorem 1 suggests we ask if every Artin L-function of degree n is the Hecke L-function of an automorphic representation of $GL(n)$? In other words, fix a Galois representation

$$\sigma : Gal(K/F) \to GL(n,\mathbb{C})$$

and consider the corresponding collection of conjugacy classes $\{\sigma(Fr_v)\}$ in $GL(n,\mathbb{C})$. Does there exists an automorphic representation $\otimes_v \pi_v$ of $GL(n,\mathbb{A})$ such that A_v almost always coincides with the class of $\sigma(Fr_v)$? The first precise response to this question was conjectured by Langlands in [13].

Langlands' Reciprocity Conjecture. For each Galois representation

$$\sigma : Gal(K/F) \to GL(n,\mathbb{C})$$

there exists an automorphic representation $\pi(\sigma)$ of $GL(n,\mathbb{A}_F)$ such that

$$L(s,\sigma) = L(s,\pi(\sigma)).$$

Moreover, if σ is irreducible and non-trivial, then $\pi(\sigma)$ is cuspidal.

Remarks.

(a) This conjecture implies Artin's conjecture. In fact, as already mentioned, this is how Artin proved his conjecture for one-dimensional σ.

(b) In [16] (see also [2]) Langlands showed that the factor $\varepsilon(s,\sigma)$ arising in the functional equation of $L(s,\sigma)$ can be factored as a product of appropriate local factors $\varepsilon(s,\sigma_v)$. Thus, by the converse theorems to Hecke theory already discussed, Artin's conjecture can be shown to be <u>equivalent</u> to Langlands' reciprocity conjecture <u>when n = 2 or 3</u>.

(c) These same converse theorems also imply the truth of the reciprocity conjecture for two and three dimensional irreducible <u>monomial</u> representations σ, i.e., irreducible representations of Gal(K/F) induced from grössencharakters of quadratic (resp. cubic) extensions of F. In either case the resulting automorphic representations $\pi(\sigma)$ will be called <u>monomial</u>. The case $n = 2$ is due to Jacquet-Langlands [11] (generalizing earlier classical constructions of Hecke and Maass); the case $n = 3$ is due to Jacquet, Piateckii-Shapiro, and Shalika [12].

The classical content of Langlands' reciprocity law is this. Suppose $F = \mathbb{Q}$, and σ is an irreducible two-dimensional representation of $\mathrm{Gal}(\bar{\mathbb{Q}}/\mathbb{Q})$ taking complex conjugation to $\begin{bmatrix} 1 & 0 \\ 0 & -1 \end{bmatrix}$. Then the (hypothetical) representation $\pi(\sigma)$ corresponds to a classical cusp form of weight 1.

In [3] Deligne and Serre prove that <u>all</u> forms of weight 1 are so obtained. More precisely, suppose f is a cusp form of weight 1 and a "primitive form" of odd character in the sense of [3]. Then there exists an irreducible two-dimensional representation σ of $\mathrm{Gal}(\bar{\mathbb{Q}}/\mathbb{Q})$ with odd determinant such that $L(s,\sigma) = D(s,f)$. In other words, modulo Artin's conjecture, there is a 1-to-1 correspondence between such representations of $\mathrm{Gal}(\bar{\mathbb{Q}}/\mathbb{Q})$ and appropriate cusp forms of weight 1.

We close this Section by describing Langlands' reciprocity law in a local setting. This law will be particularly useful in Section II.

For the moment, let F denote a <u>local</u> field and σ an n-dimensional representation of Gal(K/F). From the global theory we expect to be able to attach to σ an irreducible (admissible) representation $\pi(\sigma)$ of GL(n,F) whose L and ε factors coincide with those of σ. We want to assert more, however. Thus we consider representations not just of Gal(K/F) but also of W_F -the absolute <u>Weil group</u> of F. Our assertion then is that there is a natural correspondence

$$\sigma \leftrightarrow \pi(\sigma)$$

which is <u>bijective</u>. For precise definitions, and a discussion of the present state of this conjecture, see [1] and [6]. The case n = 1 is equivalent to the local reciprocity law of abelian class field theory. In general, every representation of Gal(K/F) may be regarded as a representation of W_F but not conversely.

<u>Concluding remark</u>. Although the correspondence $\sigma \leftrightarrow \pi(\sigma)$ should preserve L and ε factors, it is not expected that these factors should always determine the representation. In particular, in the statement of Langlands' reciprocity conjecture, it is not asserted that $\pi(\sigma)$ is uniquely determined by the condition $L(s,\pi(\sigma)) = L(s,\sigma)$. For n = 2 or 3, however, the L and ε factors do locally determine π.

II. AUTOMORPHIC FORMS ON GL(2) AND GL(3)

Although the results of this Section are included primarily because of the role they play in the proof of Artin's conjecture, all are improtant in their own right.

2.1. Base Change for GL(2).

Suppose F is a global field, and E is intermediate between K and galois over F. If σ is a two-dimensional representation of Gal(K/F), restriction to Gal(K/E) gives rise to a representation Σ. Thus Langlands' reciprocity law suggests that the map

$$\sigma \to \Sigma$$

should index a "base change" map

$$\pi(\sigma) \to \Pi(\Sigma)$$

between automorphic representations of $GL(2,\mathbb{A}_F)$ and $GL(2,\mathbb{A}_E)$.

To be more precise, suppose $\pi = \otimes \pi_v$ is a cuspidal representation of $GL(2,\mathbb{A}_F)$. Let w denote an arbitrary place of E and suppose $\pi' = \otimes \pi'_w$ is a cuspidal representation of $GL(2,\mathbb{A}_E)$. If v is such that π_v is unramified, then $\pi_v = \pi_v(\sigma_v)$ for some two-dimensional representation σ_v of the Weil group W_{F_v}. In particular, $L(s,\sigma_v) = L(s,\pi_v)$. If w is a place of E dividing v, we write Σ_w for restriction of σ_v to the Weil group of E_w. We say π' is a <u>base change lift</u> of π if

$$\pi'_w = \pi'_w(\Sigma_w)$$

for almost every w.

Since the elements of Gal(E/F) act on $GL(2,E) \setminus GL(2,\mathbb{A}_E)$ they also act on automorphic representations of $GL(2,\mathbb{A}_E)$. If π' is such that $(\pi')^{\tau} = \pi'$ for all $\tau \in Gal(E/F)$, we say that π' is Galois invariant.

THEOREM 2. (Langlands [14]). *Suppose* E/F *is cyclic of prime degree. Then :*

(a) every cuspidal representation of $GL(2,\mathbb{A}_F)$ *has a base change lift to* $GL(2,\mathbb{A}_E)$;

(b) a cuspidal representation of $GL(2,\mathbb{A}_E)$ *is the base change lift of some* π *on* $GL(2,\mathbb{A}_F)$ *if and only if it is Galois invariant;*

(c) if π *and* π' *have the same base change lift to* $GL(2,\mathbb{A}_E)$ *then there exists a character* ω *of* $F^{\times}N_{E/F}(\mathbb{A}_E^{\times}) \setminus \mathbb{A}_F^{\times}$ *such that* $\pi' = \pi \otimes \omega$. *(Here* $N_{E/F}$ *denotes the norm map from* E *to* F *and* $\pi \otimes \omega$ *denotes the representation* $\pi(g) \otimes \omega(\det g)$.*)*

Langlands proves this theorem in [14] using the Selberg trace formula. Special cases of it had been proved earlier by Doi and Naganuma ([4]), Jacquet ([9]), Shintani ([19]), and Saito ([17]).

Remark. Because E/F is assumed to be cyclic of prime degree, say q, the relation bewteen the functions $L(s,\pi)$ and $L(s,\pi')$ is particularly simple to describe. In general, almost every component π_v of π is unramified. Thus π_v corresponds to some semi-simple cojucagy class

$$\begin{pmatrix} \alpha_v & 0 \\ 0 & \beta_v \end{pmatrix}$$

and $L(s,\pi_v) = (1-\alpha_v(N_v)^{-s})^{-1}(1-\beta_v(N_v)^{-s})^{-1}$. But since q is prime, almost every v either splits completely in E or else remains prime. If v splits, and w denotes a prime of E lying over v, then $L(s,\pi'_w) = L(s,\pi_v)$. On the other hand, if v is inert, and w lies over v, then

$$L(s,\pi_W') = (1-\alpha_v^q(N_v)^{-qs})^{-1}(1-\beta_v^q(N_v)^{-qs})^{-1}.$$

<u>Example</u>. Suppose $E = Q(\sqrt{d})$ is a real quadratic field with discriminant D and class number 1. Let $f(z) = \sum a_n e^{2\pi i n z}$ be a cusp form of weight k on $SL(2,\mathbb{Z})$ and assume

$$\sum_{n=1}^{\infty} a_n n^{-s} = \prod_p (1-a_p p^{-s}+p^{k-1-2s})^{-1} = \prod_p (1-\alpha_p p^{-s})^{-1}(1-\beta_p p^{-s})^{-1}.$$

Then by Theorem 2, and the above remark, f lifts to a Hilbert modular form F over E whose L-series is described as follows. If p splits in E, and v divides p, put a_v equal to a_p. On the other hand, if p is inert, and v divides p, put a_v equal to $a_p^2-2p^{k-1}$. Then

$$L(s,F) = \prod_v (1-a_v(N_v)^{-s} + (N_v)^{k-1-2s})^{-1}$$

$$= \prod_{p \text{ splits}} (1-a_p p^{-s} + p^{k-1-2s})^2 \prod_{p \text{ inert}} (1-\alpha_p^2 p^{-2s})^{-1}(1-\beta_p^2 p^{-s})^{-1}.$$

This generalizes the set-up of [4].

2.2. <u>Lifting Forms from GL(2) to GL(3)</u>.

Let A denote the three-dimensional representation of $PGL(2,\mathbb{C})$ determined by the adjoint action of $PGL(2,\mathbb{C})$ on the Lie algebra of $SL(2,\mathbb{C})$. Denote the resulting three-dimensional representation

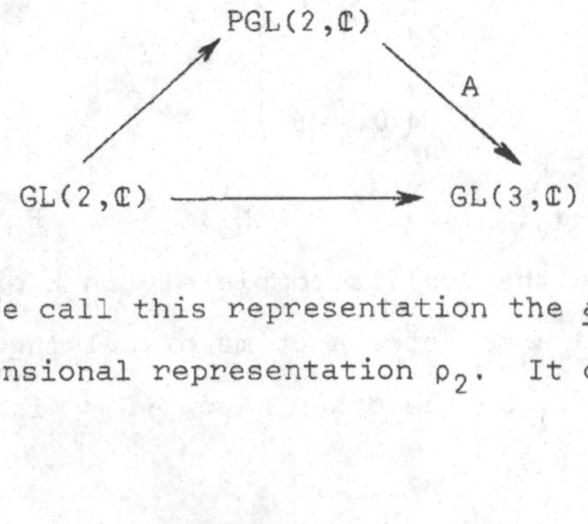

of $GL(2,\mathbb{C})$ by $A^2(\rho_2)$. We call this representation the <u>adjoint square</u> of the standard two-dimensional representation ρ_2. It differs from the usual symmetric square $Sym^2(\rho_2)$ by a power of the determinant. In particular, $A^2(\rho_2)$ is trivial on scalar matrices, whereas $Sym^2(\rho_2)$ is not.

Suppose σ is any two-dimensional representation of the Galois (or Weil) group of a local (or global) field. Then we let $A^2(\sigma)$ denote the composition of σ with $A^2(\rho_2)$. This is the adjoint square of σ.

Now suppose $\pi = \otimes \pi_v$ is an automorphic representation of $GL(2,\mathbb{A}_F)$ and $\Pi = \otimes \Pi_v$ is an automorphic representation of $GL(3,\mathbb{A}_F)$. Using the notation of the local reciprocity correspondence, we say Π is a GL(3)-lift of π if, for almost every v,

$$\Pi_v = \Pi_v(A^2(\sigma_v))$$

whenever $\pi_v = \pi_v(\sigma_v)$.

THEOREM 3. (Gelbart-Jacquet [7]).

(i) Every automorphic cuspidal representation π *of* $GL(2,\mathbb{A}_F)$ *has a lift to* $GL(3,\mathbb{A}_F)$;

(ii) the lift of π *is cuspidal if and only if* π *is cuspidal but not monomial, i.e., not of the form* $\pi(\sigma)$ *for some monomial representation* σ *of* Gal(K/F).

THEOREM 4. ([7]). *The conjectured Generalization of Theorem 1 is true with* r *equal to* $A^2(\rho_2)$. *Moreover,*

$$L(s,\pi,A^2(\rho_2)) = L(s,\Pi,\rho_3)$$

is entire.

Example. Suppose $F = \mathbb{Q}$. If π is generated by the (normalized) "new" form

$$f(z) = \sum_{n=1}^{\infty} a_n e^{2\pi i n z}$$

on $SL(2,\mathbb{Z})$ then $T_p f = a_p f$ for each p. If $\sum a_n n^{-s} = \prod_p (1-\alpha_p p^{-s})^{-1}(1-\beta_p p^{-s})^{-1}$, define an Euler product $L^2(s,f)$ <u>of degree</u> 3 by the formula

$$L^2(s,f) = \pi^{-3s/2}\Gamma(\tfrac{s}{2})\Gamma(\tfrac{s+1}{2})\Gamma(\tfrac{s-k+2}{2})\prod_p (1-\alpha_p\beta_p^{-1}p^{-s})^{-1}(1-\alpha_p^{-1}\beta_p p^{-s})^{-1}(1-p^{-s})^{-1}.$$

Then

$$L(s,\pi,A^2(\rho_2)) = L^2(s+k-1,f).$$

Thus the Corollary above implies $L^2(s,f)$ is entire (cf. Shimura [20]). What Theorem 3 implies is that

$$L(s,\Pi) = L(s,\pi,A^2(\rho_2))$$

for some cuspidal representation Π of $GL(3,\mathbb{A}_{\mathbb{Q}})$.

<u>Remark</u>. The lifting

$$\pi \to \Pi$$

from GL(2) to GL(3) is a special case of the type of correspondence predicted by Langlands' "principle of functoriality of automorphic forms with respect to the associate group". Indeed the "associate group" of $GL(n)$ is just $GL(n,\mathbb{C}) \times Gal(K/F)$. Thus the correspondence

$$\pi \to \Pi$$

taking automorphic representation of GL(2) to GL(3), corresponds naturally to the homomorphism of associate groups determined by the map

$$A^2(\rho_2) : GL(2,\mathbb{C}) \to GL(3,\mathbb{C}).$$

Cf. Remark (c) at the end of Section 1.2.; the correspondence $\pi \to \Pi$ is "natural" in that it preserves L-functions, i.e., $L(s,\Pi,\rho_3) = L(s,\pi,A^2(\rho_2))$.

Similarly, the base change lifting

$$\pi \to \pi'$$

described in 2.1. corresponds to an appropriate homomorphism of associate groups. More precisely, let E denote a Galois extension of F contained in K. Let G' denote the algebraic group over F obtained from GL(2) by restriction of scalars from E to F. Then the associate group of G' is

$$\Big(\prod_{\mathrm{Gal}(K/E)\backslash \mathrm{Gal}(K/F)} GL(2,\mathbb{C})\Big) \times \mathrm{Gal}(K/F)$$

with Gal(K/F) acting on $\Pi\, GL(2,\mathbb{C})$ via its action on coordinates. The base change lifting

$$\pi \to \pi'$$

taking automorphic representation of $GL(2,\mathbb{A}_F)$ to $GL(2,\mathbb{A}_E)$ corresponds to the homomorphism of associate groups which maps $GL(2,\mathbb{C})$ onto the diagonal and operates as the identity on Gal(K/F).

For a more detailed discussion of Langlands' functoriality principle see [1], [6], and [14]. In [1] the "associate group" is renamed the "L-group".

2.3. Automorphic Forms on GL(n) × GL(m).

Suppose $2 \leq n \leq m \leq 3$. If π_v and π_v' are unramified representations of $GL(n,F_v)$ and $GL(m,F_v)$ respectively, let A_v and A_v' denote the corresponding conjugacy classes in $GL(n,\mathbb{C})$ and $GL(m,\mathbb{C})$. A natural Euler factor attached to the pair (π_v, π'_v) is given by the formula

$$L(s,\pi_v \times \pi'_v) = [\det(I-(A_v \otimes A'_v)(N_v)^{-s}]^{-1}.$$

This is an Euler factor of degree nm.

Now suppose $\pi = \otimes \pi_v$ is an automorphic cuspidal representation of $GL(n,\mathbb{A}_F)$ and $\pi = \otimes \pi'_v$ is one of $GL(m,\mathbb{A}_F)$. According to work still in progress of Jacquet, Piateckii-Shapiro, and Shalika one can define an infinite Euler product

$$L(s,\pi \times \pi') = \prod_v L(s,\pi_v \times \pi'_v)$$

which extends over <u>all</u> the places of F.

<u>THEOREM</u> 5. *The function* $L(s,\pi \times \pi')$ *, originally only in some right half-plane, extends to a meromorphic function of* $\mathbb{C}$ *with functional equation*

$$L(s,\pi \times \pi') = (\prod_v \varepsilon(s,\pi_v \times \pi'_v))L(1-s,\pi \times \pi').$$

<u>THEOREM</u> 6. *Suppose* $n = m = 3$*, and* π *and* π' *are self-contragredient. Then :*

(i) $L(s,\pi' \times \pi)$ *has a pole at* $s = 1$ *if and only if* π *is equivalent to* π'*;*

(ii) for each place v*,* $L(s,\pi_v \times \pi'_v)$ *is non-zero at* $s = 1$*;*

(iii) for each place v*,* $L(s,\pi_v \times \pi_v)$ *is pole-free in the <u>closed</u> half-space* $\mathrm{Re}(s) \geq 1$.

<u>Remarks</u>.

(i) Theorem 6, as stated above, has not yet actually been proved; however, a weaker form of it, already sufficient for application to Artin's conjecture, will appear in a forthcoming note of Jacquet and Shalika.

(ii) The proof of Theorem 5 for $n = m = 2$ already appears in [10]; for function fields, the general global theory is discussed in [12].

(iii) By combining Theorems 3, 4 and 5 one can obtain parts of the Conjectured Generalization of Theorem 1 for the third and fourth symmetric squares of ρ_2. More precisely, if $r = \mathrm{Sym}^3(\rho_2)$ or $\mathrm{Sym}^4(\rho_2)$, then $L(s,\pi,r)$ should have a meromorphic continuation and functional equation. The argument below, due to Deligne, was shown to me by Serre. It does not prove that $L(s,\pi,r)$ has only finitely many poles.

Assume π is not monomial. By Theorem 3 we can lift π to a cuspidal representation Π of GL(3) with the property that $L(s,\pi,A^2(\rho_2)) = L(s,\Pi)$. Thus it follows from Theorem 5 that the functions $L(s,\pi \times \Pi)$ and $L(s,\Pi \times \Pi)$ have analytic continuations and functional equations. But $L(s,\pi \times \Pi) = L(s,\pi,\rho_2 \otimes \mathrm{Sym}^2(\rho_2))$, and

$$L(s,\Pi \times \Pi) = L(s,\pi,\mathrm{Sym}^2(\rho_2) \times \mathrm{Sym}^2(\rho_2)).$$

Moreover,

$$\mathrm{Sym}^2(\rho_2) \otimes \rho_2 = \mathrm{Sym}^3(\rho_2) \oplus (\rho_2 \otimes \Lambda^2\rho_2),$$

$$\mathrm{Sym}^2(\rho_2) \otimes \mathrm{Sym}^2(\rho_2) = \mathrm{Sym}^4(\rho_2) \oplus (\mathrm{Sym}^2(\rho_2) \otimes \Lambda^2(\rho_2)) \oplus (\Lambda^2\rho_2)^{\otimes 2}.$$

Now apply $L(s,\pi,.)$ to both sides of the equations above and solve for $L(s,\pi,\mathrm{Sym}^3(\rho_2))$ and $L(s,\pi,\mathrm{Sym}^4(\rho_2))$. From this we obtain the desired analytic continuations and functional equations of these functions.

Concluding Remark. As already indicated, Theorem 6 will be used in the proof of Artin's conjectue given in Section III.

III. LANGLANDS' PROOF OF ARTIN'S CONJECTURE FOR TETRAHEDRAL σ

As always, F is an arbitrary number field and K is a finite Galois extension of F. If σ is a two-dimensional representation of Gal(K/F), $\pi(\sigma)$ denotes an automorphic representation of $GL(2,\mathbb{A}_F)$ with the property that

$$L(s,\pi(\sigma)) = L(s,\sigma)$$

(at least in the sense that the local factors of these L-functions agree almost everywhere).

The assertion that $\pi(\sigma)$ actually exists comprises the content of Langlands' reciprocity conjecture. In particular, if σ is irreducible, $\pi(\sigma)$ should be cuspidal. Thus Artin's conjecture is a consequence of Langlands' conjecture. Since both are theorems when F is a function field, we consider only number fields in this paper.

The purpose of this Section is to describe Langlands' proof of his reciprocity conjecture -and hence Artin's conjecture- for tetrahedral σ. The proof breaks up naturally into two parts. The first part produces a natural candidate for $\pi(\sigma)$. This is the cuspidal representation Langlands calls $\pi_{pseudo}(\sigma)$. The second part establishes that $\pi_{pseudo}(\sigma)$ actually equals $\pi(\sigma)$.

The construction of $\pi_{pseudo}(\sigma)$ was first outlined in [14]. The primary tool used is Langlands' theory of base change (Theorem 2). The proof that $\pi_{pseudo}(\sigma) = \pi(\sigma)$ can be carried out in two different ways. The first uses the result of Deligne-Serre quoted earlier; thus it works only over $\mathbb{Q}$. The second works for arbitrary F but uses all the modular results described in Section 2.

3.1. Construction of $\pi_{pseudo}(\sigma)$.

We assume that

$$\sigma : Gal(K/F) \to GL(2,\mathbb{C}) \xrightarrow{p} PGL(2,\mathbb{C})$$

is such that $p \circ \sigma(Gal(K/F))$ is isomorphic to A_4 embedded in $PGL(2,\mathbb{C}) \approx SO(3,\mathbb{C})$ in the geometric way. If we let E denote the intermediate field

$$F \subset E \subset K$$

corresponding to the pull-back of $(\mathbb{Z}/2\mathbb{Z}) \times (\mathbb{Z}/2\mathbb{Z})$ in A_4 then E is cubic over F.

Let Σ denote the restriction of σ to Gal(K/E). Since Σ must be monomial, we can apply to it the reciprocity law for Galois representations of E. The result is that $\pi(\Sigma)$ exists as an automorphic cuspidal representation of $GL(2,\mathbb{A}_E)$.

On the other hand, Σ is invariant under conjugation by Gal(K/F). Thus it follows that $\pi(\Sigma)$ is Galois invariant (in the sense of Section 2.1) and we can apply to it the theory of base change for GL(2).

According to Theorem 2, $\pi(\Sigma)$ is also the lift of $\pi \otimes \omega$ whenever ω is a character of $F^x N_{E/F}(\mathbb{A}_E^x) \backslash \mathbb{A}_F^x$. Since only one of these "twists" is what we want to call $\pi_{pseudo}(\sigma)$, we proceed as follows. Recall that class field theory implies that $F^x N_{E/F}(\mathbb{A}_E^x) \backslash \mathbb{A}_F^x$ is isomorphic to Gal(E/F) and hence cyclic of order 3.

The central character of π is defined by the formula

$$\pi\begin{pmatrix} a & 0 \\ 0 & a \end{pmatrix} = \omega_\pi(a)I \quad (a \in \mathbb{A}_F^\times).$$

Thus the central character of $\pi \otimes \omega$ is $\omega_\pi\omega^2$. In particular, if $\pi \otimes \omega$ is to equal $\pi(\sigma)$ (or $\pi_{\text{pseudo}}(\sigma)$), we must have

$$\omega_\pi\omega^2 = \det \sigma.$$

Thus we choose ω so that $\omega_\pi\omega^2 = \det \sigma$. Having done so, we write $\pi_{\text{pseudo}}(\sigma)$ for $\pi \otimes \omega$. (Since ω is defined up to a character of order 3, and its square is known, it is uniquely determined.)

Our task now is to prove that

$$\pi_{\text{pseudo}}(\sigma) = \pi(\sigma).$$

What we know <u>a priori</u> is that $\pi_{\text{pseudo}}(\sigma)$ is at least the best possible candidate for $\pi(\sigma)$. In other words, <u>if</u> $\pi(\sigma)$ exists at all, it must be $\pi_{\text{pseudo}}(\sigma)$. Indeed, suppose $\pi_{\text{pseudo}}(\sigma) = \otimes\, \pi_v$, and $\pi_v = \pi_v(\sigma_v)$ for almost every v. If we regard σ_v' and σ_v as two-dimensional representations of W_{F_v}, it follows from the definition of base change lift that σ_v and σ_v' agree <u>on</u> W_{E_w}. What remains to be shown is that σ_v and σ_v' agree <u>on</u> W_{F_v}. This is what is done in Sections 3.2. and 3.3.

<u>Remark</u>. We can also use the theory of base change to construct $\pi_{\text{pseudo}}(\sigma)$ when σ is octahedral. In this case $p \circ \sigma(\text{Gal}(K/F))$ is isomorphic to S_4. Thus the pull-back of the normal subgroup A_4 determines a quadratic extension L of F and the restriction of σ to Gal(K/L) is tetrahedral. This means we can construct $\pi_{\text{pseudo}}(\sigma)$ as follows.

Let Σ denote the restriction of σ to Gal(K/L). In Sections 3.2. and

3.3. we shall prove that Langlands' reciprocity law <u>is</u> true for tetrahedral Galois representations. Thus we know that $\pi(\Sigma)$ exists as an automorphic cuspidal representation of $GL(2,\mathbb{A}_L)$. But $\pi(\Sigma)$ is again seen to Galois invariant. Thus by base change we know that $\pi(\Sigma)$ is the lift of some cuspidal representation π of $GL(2,\mathbb{A}_F)$.

The next step is to define $\pi_{pseudo}(\sigma)$ as an appropriate "twist" of π. In this context, however, $\pi \otimes \omega$ lifts to $\pi(\Sigma)$ whenever ω is a character of $F^x N_{L/F}(\mathbb{A}_L^x) \setminus \mathbb{A}_F^x$. But L is quadratic. Thus, there is <u>no</u> natural way to distinguish between these twists. Indeed the square of any character ω of $F^x N_{L/F}(\mathbb{A}_L^x) \setminus \mathbb{A}_F^x$ is trivial. I.e., the twist of π by a nontrivial ω has the same central character as π. Thus we can no longer uniquely specify $\pi_{pseudo}(\sigma)$ by requiring that $\omega_{\pi_{pseudo}(\sigma)} = \det \sigma$.

<u>Concluding remark</u>. When σ is icosahedral, the image of $\sigma(Gal(K/F))$ in $PGL(2,\mathbb{C})$ is <u>not</u> solvable. Thus the theory of base change can <u>not</u> be used to construct $\pi_{pseudo}(\sigma)$ <u>or</u> $\pi(\sigma)$.

3.2. <u>Proof that</u> $\pi_{pseudo}(\sigma) = \pi(\sigma)$ $(F = \mathbb{Q})$.

Fix $\sigma : Gal(K/F) \to GL(2,\mathbb{C})$ and assume that :

(i) σ is irreducible;

(ii) $F = \mathbb{Q}$;

(iii) σ takes complex conjugation to $\begin{bmatrix} 1 & 0 \\ 0 & -1 \end{bmatrix}$. Let f denote the classical cusp form which corresponds (canonically) to the hypothetical representation $\pi(\sigma)$. Then f is a normalized primitive cusp form of weight 1 and odd character. Thus we call σ classical.

In [3] Deligne and Serre prove that every such cusp form has associated to it some σ of this type. More precisely, the Mellin transform of any such cusp form is the Artin L-series of some σ satisfying (i)-(iii) above. Thus we have a map

$$f \to \sigma_f$$

which inverts the hypothetical reciprocity map

$$\sigma \to f_\sigma.$$

The purpose of this paragraph is to prove that this last map exists whenever σ is <u>tetrahedral</u>. In terms of the theory of base change, this amounts to proving that

$$\pi_{\text{pseudo}}(\sigma) = \pi(\sigma)$$

for classical tetrahedral σ. Using the result of Deligne-Serre, however, this result is simple to establish.

Note that our assumptions on σ imply that the field E used to define $\pi(\Sigma)$ is totally real. Now let ρ_∞^0 denote the two-dimensional representation of $\mathrm{Gal}(\mathbb{C}/\mathbb{R})$ which takes complex conjugation to $\begin{bmatrix} 1 & 0 \\ 0 & -1 \end{bmatrix}$. In the sense of the local Galois classification of representations of $GL(2,\mathbb{R})$,

$$\pi(\Sigma_w) = \pi(\rho_\infty^0)$$

for each infinite place w of E; for details, see [6] and [14]. But $\pi_{\text{pseudo}}(\sigma) = \otimes\, \pi_p$ lifts to $\pi(\Sigma)$. Thus the definition of base change lifting implies that π_∞ must also be of the form $\pi(\rho_\infty^0)$. Hence $\pi_{\text{pseudo}}(\sigma)$ corresponds to a classical cusp form of weight 1 to which the theorem of Deligne-Serre applies.

Applying Deligne-Serre, we conclude that $\pi_{\text{pseudo}}(\sigma) = \pi(\sigma')$ for some "classical" Galois representation σ'. To conclude that $\sigma = \sigma'$ we note that σ and σ' have the same restriction to W_E. Moreover, $\sigma' = \sigma \otimes \omega$ with ω some character of $F^\times N_{E/F}(\mathbb{A}_E^\times) \setminus \mathbb{A}_F^\times$. So since det σ = det σ', we must have $\omega \equiv 1$. This proves $\pi(\sigma) = \pi(\sigma') = \pi_{\text{pseudo}}(\sigma)$.

Remark. A similar argument proves Artin's conjecture for octahedral classical σ provided the quadratic field used to define $\pi_{\text{pseudo}}(\sigma)$ is real; cf. the remark towards the end of 3.1. For more details, see [14].

3.3. Proof that $\pi_{\text{pseudo}}(\sigma) = \pi(\sigma)$: σ arbitrary tetrahedral.

In this paragraph we describe a more natural proof of the fact that

$$\pi_{\text{pseudo}}(\sigma) = \pi(\sigma)$$

for tetrahedral σ. This proof is also due to Langlands and was sketched by him in [15]. It is "natural" in the sense that it makes no appeal to the result of Deligne and Serre, hence no appeal to the algebraic geometry underlying their result. Instead it draws only on basic (albeit non-trivial) results in the theory of modular forms. It also works for arbitrary F and arbitrary tetrahedral σ.

Recall that $\pi_{\text{pseudo}}(\sigma) = \otimes\, \pi_v$ with $\pi_v = \pi_v(\sigma_v')$ for every unramified π_v. What we want to prove is that

(3.1.) $$\sigma_v' = \sigma_v$$

for almost every v. Here σ_v' is a two-dimensional representation of W_{F_v} whose restriction to W_{E_w} agrees with the restriction of σ_v to W_{E_w}. We denote these restrictions by Σ_w and Σ_w' respectively.

If v splits in E, then (3.1) is immediate. Thus we assume henceforth that E_w is unramified and cubic over F_v. If Fr_v denotes a Frobenius element of $Gal(E_w/F_v)$ we suppose

$$\sigma_v(Fr_v) = \begin{pmatrix} a_v & 0 \\ 0 & b_v \end{pmatrix}$$

and

$$\sigma_v'(Fr_v) = \begin{pmatrix} c_v & 0 \\ 0 & d_v \end{pmatrix}.$$

To prove (3.1) we have to prove that

$$\begin{pmatrix} a_v & 0 \\ 0 & b_v \end{pmatrix}$$

is conjugate to $\begin{pmatrix} c_v & 0 \\ 0 & d_v \end{pmatrix}$.

But the fact that $\Sigma_W = \Sigma_W'$ implies that $\sigma_v(Fr_v^3) = \sigma_v'(Fr_v^3)$ (since $(Fr_v)^3$ belongs to W_{E_W}). Thus

$$\begin{pmatrix} a_v^3 & 0 \\ 0 & b_v^3 \end{pmatrix}$$

is conjugate to $\begin{pmatrix} c_v^3 & 0 \\ 0 & d_v^3 \end{pmatrix}$.

In particular, for some pair of cube roots of 1, say ξ and ξ',

$$c_v = \xi\, a_v$$

and

$$d_v = \xi' b_v.$$

We claim now that $\xi' = \xi^2$. Indeed $\pi_{\text{pseudo}}(\sigma)$ is chosen so that $\omega_{\pi_{\text{pseudo}}}(\sigma) = \det \sigma$. Since this implies $\det \sigma_v' = \det \sigma_v$, we must have $\xi\xi' = 1$, i.e., $\xi' = \xi^2$. To prove (3.1) it suffices to prove

(3.2) $$\xi = 1.$$

To continue, consider the three-dimensional adjoint-square represent ation

$$A^2(\rho_2) : GL(2,\mathbb{C}) \to GL(3,\mathbb{C}),$$

described in Section 2.2. If we <u>assume</u> that

(3.3) $$A^2(\rho_2) \circ \sigma'_v = A^2(\rho_2) \circ \sigma_v,$$

then $\sigma_v(Fr_v)$ and $\sigma'_v(Fr_v)$ must differ by some scalar. Indeed the kernel of the homomorphism $A^2(\rho_2)$ is precisely the group of scalar matrices $\{\begin{bmatrix}\lambda & 0\\ 0 & \lambda\end{bmatrix}\}$. Thus for some $\lambda \neq 0$,

$$\begin{pmatrix}\xi a_v & 0\\ 0 & \xi^2 b_v\end{pmatrix} \text{ is conjugate to } \begin{pmatrix}\lambda a_v & 0\\ 0 & \lambda b_v\end{pmatrix}.$$

To prove (3.1) it suffices to prove $\lambda = 1$.

Since the last two matrices above are conjugate, either

(i) $$\begin{Bmatrix}\lambda a_v = \xi a_v\\ \lambda b_v = \xi^2 b_v\end{Bmatrix}$$ or

(ii) $$\begin{Bmatrix}\lambda a_v = \xi^2 b_v\\ \lambda b_v = \xi a_v\end{Bmatrix}$$

But case (i) implies $\lambda = \xi = \xi^2$, i.e. $\lambda = 1$ (since ξ is a cube root of 1). Thus $\sigma_v(Fr_v)$ is conjugate to $\sigma'_v(Fr_v)$ and there is nothing left to prove.

In case (ii), all we can deduce immediately is that

$$\sigma_v(Fr_v) = \begin{pmatrix}a_v & 0\\ 0 & a_v\end{pmatrix}\begin{pmatrix}1 & 0\\ 0 & \lambda\xi\end{pmatrix}$$

with $\lambda^2 = 1$. But <u>if</u> $\lambda = -1$, this means $A^2(\rho_2) \circ \sigma_v$ takes Fr_v to

$$\begin{pmatrix} \lambda\xi & 0 & 0 \\ 0 & \lambda^{-1}\xi^{-1} & 0 \\ 0 & 0 & 1 \end{pmatrix}$$

with $\xi\lambda$ of order 6. This is a contradiction since A_4 has no element of order 6. Therefore $\lambda = 1$ and we are done.

We have now shown that Artin's conjecture for tetrahedral σ is true once we prove (3.3) for almost all v. To prove (3.3) we introduce the following cuspidal representations of $GL(3,\mathbb{A}_F)$ (not $GL(2,\mathbb{A}_F)$).

Let π_1 denote the cuspidal representation of $GL(3,\mathbb{A}_F)$ associated to the three-dimensional Galois representation $A^2(\rho_2) \circ \sigma$. As observed by Serre, $A^2(\rho_2) \circ \sigma$ is actually monomial. Thus by the known case of Langlands' reciprocity conjecture, π_1 exists as a cuspidal representation.

Now consider $\pi_{pseudo}(\sigma)$ as a cupsidal representation of $GL(2,\mathbb{A}_F)$. Recall that if $\pi_{pseudo}(\sigma) = \pi(\sigma')$, then σ' must be σ. In particular, $\pi_{pseudo}(\sigma)$ cannot be monomial in the sense of Section 1.3. Therefore $\pi_{pseudo}(\sigma)$ has a cuspidal lift to $GL(3,\mathbb{A}_F)$ by Theorem 3. Call this lift π_1^*. To prove (3.3) for almost every v it suffices to prove

(3.4) $$\pi_1 = \pi_1^* .$$

Indeed, the left hand side of (3.4) corresponds almost everywhere to $A^2(\rho_2) \circ \sigma_v$ and the right hand side corresponds almost everywhere to $A^2(\rho_2) \circ \sigma'_v$.

To prove (3.4), we argue analytically. Let $L(s,\pi_1 \times \check{\pi}_1)$ and $L(s,\pi_1^* \times \check{\pi}_1)$ denote the L-functions on $GL(3) \times GL(3)$ described in Sectio 2.3. By definition,

$$L(s,\pi_1 \times \check{\pi}_1) = \prod_v L(s,(\pi_1)_v \times (\check{\pi}_1)_v)$$

and

$$L(s,\pi_1^* \times \check{\pi}_1) = \prod L(s,(\pi_1^*)_v \times (\check{\pi}_1)_v).$$

Moreover, for almost every v,

$$L(s,(\pi_1)_v \times (\check{\pi}_1)_v) = L(s,A^2(\sigma_v) \otimes A^2(\check{\sigma}_v))$$

and

$$L(s,(\pi_1^*)_v \times (\check{\pi}_1)_v) = L(s,A^2(\sigma_v') \otimes A^2(\check{\sigma}_v)).$$

Keeping in mind that $A^2(\sigma)$ is monomial it is easy to check that

$$(3.5) \qquad L(s,(\pi_1)_v \times (\check{\pi}_1)_v) = L(s,(\pi_1^*)_v \times (\check{\pi}_1)_v)$$

for almost every v. Indeed if $A^2(\sigma)$ is induced from the grössencharakter θ of the extension E, then

$$A^2(\check{\sigma}_v) = \bigoplus_{w|v} \mathrm{Ind}(W_{F_v},W_{E_w},\theta_w).$$

From this it follows that

$$A^2(\sigma_v) \otimes A^2(\check{\sigma}_v) = \bigoplus_{w|v} \mathrm{Ind}(W_{F_v},W_{E_w},\Sigma_w \otimes \theta_w^{-1}).$$

On the other hand,

$$A^2(\sigma_v) \otimes A^2(\check{\sigma}_v) = \bigoplus_{w|v} \mathrm{Ind}(W_{F_v},W_{E_w},\Sigma_w' \otimes \theta_w^{-1}).$$

So since $\Sigma_w = \Sigma_w'$ almost everywhere,

$$\begin{aligned} L(s,(\pi_1)_v \otimes (\check{\pi}_1)_v) &= L(s,A^2(\sigma_v) \otimes A^2(\check{\sigma}_v)) \\ &= L(s,A^2(\sigma_v') \otimes A^2(\check{\sigma}_v)) \\ &= L(s,(\pi_1^*)_v \times (\check{\pi}_1)_v) \end{aligned}$$

for almost all v.

Using (3.5), it remains now to show that $\pi_1^* = \pi_1$. So suppose (3.5) holds for all v outside the finite set S. Then

$$L(s,\pi_1^* \times \check{\pi}_1) = \Big(\prod_{v\in S} \frac{L(s,(\pi_1^*)_v \times (\check{\pi}_1)_v)}{L(s,(\pi_1)_v \times (\check{\pi}_1)_v)}\Big)L(s,\pi_1 \times \check{\pi}_1).$$

But by Theorem 6 of Section 2.3, $L(s,\pi_1 \times \check{\pi}_1)$ has a pole at $s = 1$. Moreover, the expression in parentheses is non-zero at $s = 1$. Therefore, $L(s,\pi_1^* \times \check{\pi}_1)$ has a pole at $s = 1$, and this implies (by the same Theorem) that $\pi_1^* = \pi_1$. (Actually, since the complete story on Theorem 6 has not yet been worked out, a slight modification of the above argument is needed; we suppress this subtlety for the sake of exposition.)

This completes our discussion of Artin's conjecture for tetrahedral σ.

BIBLIOGRAPHY

[1] BOREL A., "Formes automorphes et séries de Dirichlet", Sém. Bourbaki, No. 466, Juin 1975.

[2] DELIGNE P., "Les constantes des équations **fonctionnelles** des fonctions L", Lecture Notes in Mathematics, Nr. 349, Springer-Verlag, Berlin-New-York, 1973, pp. 501-597.

[3] DELIGNE P. and SERRE J-P., "Formes modulaires de poids 1", Ann. Sci. Ecole Norm. Sup., 4e série (1974), pp. 507-530.

[4] DOI K. and NAGANUMA H., "On the functional equation of certain Dirichlet series", Inv. Math. 9 (1969), pp. 1-14.

[5] GELBART S., "Automorphic Forms on Adele Groups", Annals of Math. Studies 83 (1975), Princeton Univ. Press.

[6] GELBART S., "Elliptic curves and automorphic representations", Advances in Math. 21 (1976).

[7] GELBART S. and JACQUET H., "A relation between automorphic forms on GL(2) and GL(3)", Proc. Nat. Acad. Sci., U.S.A., October 1976; also paper in preparation.

[8] GELFAND I.M., GRAEV M.I. and PIATECKII-SHAPIRO II., "Representation Theory and Automorphic Functions", Saunders, Philadelphia, 1969.

[9] GODEMENT R. and JACQUET H., "Zeta Functions of Simple Algebras", Lecture Notes in Mathematics, Nr. 260, Springer-Verlag, Berlin-New York, 1972.

[10] JACQUET H., "Automorphic Forms on GL(2); Part II", Lecture Notes in Mathematics, Nr. 278, Springer-Verlag, Berlin-New York, 1972.

[11] JACQUET H. and LANGLANDS R.P., "Automorphic Forms on GL(2)", Lecture Notes in Mathematics, Nr. 114, Springer-Verlag, Berlin-New York, 1970.

[12] JACQUET H., PIATECKII-SHAPIRO I.I. and SHALIKA J., "Construction of Cusp Forms on GL(3)", Lecture Note Nr. 16, Department of Math., Univ. of Maryland, College Park, Md.

[13] LANGLANDS R.P., "Problems in the theory of automorphic forms", Lecture Notes in Mathematics, Nr. 170, Springer-Verlag, Berlin-New York, 1970, pp.18-86.

[14] LANGLANDS R.P., "Base change for GL(2)", The theory of Saito-Shintani with applications", mimeographed notes, Institute for Advanced Study, Princeton, N.J. 1975.

[15] LANGLANDS R.P., "Letter to J-P. Serre", December 1975.

[16] LANGLANDS R.P., "On the functional equation of the Artin L-functions", mimeographed notes, Yale University, (approximately) 1965.

[17] SAITO H., "Automorphic forms and algebraic extensions of number fields", Lecture Notes in Mathematics, Nr. 8, Kinokuniya Book Store Co., Ltd., Tokyo, Japan, 1975.

[18] SERRE J-P., "Propriétés galoisiennesdes points d'ordre fini des courbes elliptiques", Inv. Math. 15 (1972), pp. 259-331.

[19] SHINTANI T., "On liftings of holomorphic automorphic forms", 1975 U.S.-Japan Seminar on Number Theory, Ann Arbor.

[20] SHIMURA G., "On the holomorphy of certain Dirichlet series", Proc. London Math. Soc. (3) 31 (1975), pp. 79-98.

RELATIONS BETWEEN AUTOMORPHIC FORMS PRODUCED BY THETA-FUNCTIONS

by
Stephen S. Kudla

1. GENERAL PHILOSOPHY

I want to begin with a few philosophical remarks on how theta-functions attached to indefinite quadratic forms can be expected to produce relations between modular forms and automorphic forms of one and several variables.

First it should be pointed out that this 'philosophy' is, in a certain sense, not new. It is certainly contained in Siegel's papers [7] on indefinite quadratic forms and function theory (1951/52). More recently, Shintani [10] and Niwa [5] have revived these ideas and have made a beautiful application to the theory of automorphic forms of $\frac{1}{2}$-integral weight.

Now suppose

$$\mathbb{Q}^n \;=\; \text{an } n\text{-dimensional vector space over } \mathbb{Q}$$

and

$$Q\colon \mathbb{Q}^n \longrightarrow \mathbb{Q} \qquad X \longmapsto Q[X] \;=\; {}^tXQX$$

is an indefinite quadratic form of signature (p,q). Take

$$L \subset \mathbb{Q}^n \quad \text{a } \mathbb{Z}\text{-lattice such that } Q[L] \subseteq 2\mathbb{Z}.$$

If we want to construct a *theta-function* attached to Q and L, we will also need a *majorant* R of Q; that is: $R \in M_n(\mathbb{R})$, $n = p + q$ such that (1) ${}^tR = R$, $R > 0$, and (2) $RQ^{-1}R = Q$. For example, if Q is in diagonal form for some basis of $\mathbb{R}^n$

International Summer School on Modular Functions
BONN 1976

$$Q = \begin{pmatrix} 1 & & & \\ & \ddots\, 1 & & \\ & & -1 & \\ & & & \ddots\, -1 \end{pmatrix} \quad \text{then} \quad R = \begin{pmatrix} 1 & & \\ & \ddots & \\ & & 1 \end{pmatrix} \quad \text{is a majorant.}$$

If we let

$$SO(Q) = \{g \in SL_n(\mathbb{R}) \text{ such that } {}^t gQg = Q[g] = Q\},$$

and if R is one majorant, then *every* majorant has the form $R[g] = {}^t gRg$ for some $g \in SO(Q)$. Let X = the space of all majorants of Q. Then, by the preceding comment, X is the symmetric space attached to $SO(Q)$, since $SO(Q) \cap SO(R)$ is a maximal compact subgroup of $SO(Q)$.

We can now make a *theta-function*:

$$\theta(z,R) = v^{q/2} \sum_{\ell \in L} e^{i\pi(uQ+ivR)[\ell]}$$

which we view as a function of *two variables*, $(z,R) \in \mathfrak{H} \times X$ where $z = u+iv \in \mathfrak{H}$ = upper half-plane.

For our purposes, the important fact about θ is that it has a *transformation law*, like that of an automorphic form, in each variable. More precisely:

(1) $$\theta(\gamma z,R) = (cz+d)^{\frac{p-q}{2}}\, \theta(z,R)$$

$\forall\ \gamma = \begin{pmatrix} a & b \\ c & d \end{pmatrix} \in \Gamma_1(N)$ for suitable N, e.g., $N = 2\det Q$ will do. If we let $\Gamma_L = \{U \in SO(Q)$ such that $UL = L\}$ = unit group of L, then

(2) $$\theta(z,R[U]) = \theta(z,R).$$

<u>Remarks</u>:

a) (2) is obvious from the definition of θ, while (1) follows from the usual Poisson-summation argument.

b) Notice that $n = p+q \equiv p-q \bmod 2$ so that, if n is even, θ

has integral "weight"; and if n is odd, θ has half-integral "weight."

c) θ is *not* holomorphic in z, and up to the $v^{q/2}$ factor it is Siegel's 1951 θ.

Now that we have θ, we can state the basic idea: If you have $\varphi(z)$ on $\mathfrak{H}$ or $\psi(R)$ on X which transform in the same way as θ in z or R, *then* you can make the integrals:

$$\Theta_{\varphi}(R) = \int_{\Gamma_1(N)\backslash\mathfrak{H}} \varphi(z)\overline{\theta(z,R)v^{\frac{p-q}{2}}}\,\frac{du\,dv}{v^2}$$

and

$$\Theta^{\psi}(z) = \int_{\Gamma_L\backslash X} \psi(R)\theta(z,R)\,dR.$$

And the functions $\Theta_{\varphi}(R)$ and $\Theta^{\psi}(Z)$ will again have a nice transformation law with respect to Γ_L or $\Gamma_1(N)$ respectively. If everything goes well, this procedure will actually carry automorphic forms to automorphic forms:

automorphic forms on $\mathfrak{H}$ for $\Gamma \subset SL_2(\mathbb{Z})$	$\rightleftarrows$	automorphic forms on X for Γ_L.

§2. SOME PARTICULAR CASES

In order to get something more in keeping with the general themes of this conference, however, we want to interpret "automorphic forms on X" as something more modular. To do this, we make use of some "accidental" isomorphisms which occur for orthogonal groups of low dimension. In particular we have:

(1) $SO(2,1) \sim SL_2(\mathbb{R})$ so $X = \mathfrak{H}$ and $\Gamma_L \subset SL_2(\mathbb{Z})$

for a certain choice of Q.

(2) $SO(2,2) \sim SL_2(\mathbb{R}) \times SL_2(\mathbb{R})$ so $X = \mathfrak{H} \times \mathfrak{H}$ and $\Gamma_L \subset SL_2(\mathcal{O}_k)$ where k = real quadratic field, $\mathcal{O}_k$ = integers of k, again for a certain choice of Q. Using these, and recalling the earlier remark about the "weight" of θ, we see that there ought to be correspondences:

(1)	modular forms of $\frac{1}{2}$-integral weight on $\mathfrak{H}$	$\rightleftarrows$	modular forms (of integral weight) on $\mathfrak{H}$
(2)	modular forms of integral weight on $\mathfrak{H}$	$\rightleftarrows$	Hilbert modular forms on $\mathfrak{H} \times \mathfrak{H}$.

The first of these relations has been worked out by Shintani ($SO(2,1) \longrightarrow SL_2$) and Niwa ($SL_2 \longrightarrow SO(2,1)$) in their Nagoya Journal papers [10], [5]. The $SL_2 \longrightarrow SO(2,1)$ relation gives an alternative proof of Shimura's result [9], on forms of $\frac{1}{2}$-integral weight. As a corollary, Niwa is able to settle Simura's conjecture about the *precise level* of the form of even integral weight. The relation $SL_2 \longrightarrow SO(2,2)$ has been investigated by Niwa and Asai and by myself [3] independently. In this case, the construction with the theta-function gives an alternate proof of the results of Doi-Naganuma [2], [4].

§3. THE SO(2,2) CASE

Let $k = \mathbb{Q}(\sqrt{\Delta})$, $\Delta > 0$, the discriminant, and let σ = the Galois automorphism of k/ℚ. Let

$$V = \left\{ X = \begin{pmatrix} x_1 & x_4 \\ x_3 & -x_1^\sigma \end{pmatrix}, \quad x_1 \in k, \quad x_3, x_4 \in \mathbb{Q} \right\} \simeq \mathbb{Q}^4$$

and define $Q: V \longrightarrow \mathbb{Q}$ by

$$Q[X] = -2 \det X = -2(x_1 x_1^\sigma + x_3 x_4).$$

Then Q has signature $(2,2)$. Let

$$\rho: SL_2(k) \longrightarrow SO(Q)$$

be the representation given by $\rho(g)X = g^\sigma X g^{-1}$. Now, over $\mathbb{R}$, we get a representation:

$$\rho: SL_2(\mathbb{R}) \times SL_2(\mathbb{R}) \longrightarrow SO(Q)_{\mathbb{R}} \simeq SO(2,2)$$

and a corresponding isomorphism:

$$\mathfrak{H} \times \mathfrak{H} \xrightarrow{\sim} X = \text{space of } \textit{majorants of } Q$$

say $\quad (z_1, z_2) \leftrightarrow R_{z_1,z_2}$.

Now for the *lattice*, take:

$$L = \left\{ X = \begin{pmatrix} x_1 & x_4 \\ x_3 & -x_1^\sigma \end{pmatrix}, \quad x_1 \in \mathcal{O}_k, \quad x_3, x_4 \in \mathbb{Z} \right\}.$$

Choose $N \in \mathbb{Z}_{>0}$, χ = character of $(\mathbb{Z}/N\mathbb{Z})^\times$, and $\nu \in \mathbb{Z}_{>0}$ such that $\chi(-1) = (-1)^\nu$. Finally take $z = u + iv \in \mathfrak{H}$ and $(z_1, z_2) \in \mathfrak{H} \times \mathfrak{H}$ with $z_1 = u_1 + iv_1$ and $z_2 = u_2 + iv_2$; and define the *theta-function*:

$$\theta(z, z_1, z_2)$$
$$= v(v_1 v_2)^{-\nu/2} \sum_{X \in L} \chi(x_3)(-x_3 z_1 z_2 + x_1 z_1 + x_1^\sigma z_2 + x_4)^\nu e^{i\pi(uQ + ivR_{z_1,z_2})[X]}.$$

Obviously this is like Siegel's theta function except that there is now a character and a 'spherical function' thrown in. The relevant *transformation law* is:

$$\theta(\gamma z, z_1, z_2) = \chi(d)\left(\frac{\Delta}{d}\right)(cz+d)^\nu \theta(z, z_1, z_2)$$

$\forall\gamma = \begin{pmatrix} a & b \\ c & d \end{pmatrix} \in \Gamma_0(N')$ with N' = l.c.m. of N and Δ.

Now take $\varphi \in S_\nu(\Gamma_0(N'),\ \chi\cdot(\frac{\Delta}{*}))$ and let

$$\psi(z_1,z_2) = \int_{F_{\Gamma_0(N')}} \varphi(z)\overline{\theta(z,z_1,z_2)}v^{\nu-2}\,du\,dv.$$

where $S_\nu(\cdots)$ is the space of cusp forms of Neben type and weight ν, and $F_{\Gamma_0(N')}$ is a fundamental domain for $\Gamma_0(N')$ in $\mathfrak{H}$. Let $\tilde{\Gamma}_0(N) = \left\{ g = \begin{pmatrix} \alpha & \beta \\ \gamma & \delta \end{pmatrix} \in SL_2(\mathcal{O}_k) \text{ such that } \gamma \in N\mathcal{O}_k \right\}$ and put $\tilde{\chi}(\delta) = \chi(\delta\delta^\sigma)$ for $\delta \in \mathcal{O}_k$.

<u>THEOREM 1</u>. *If* $\nu \geq 7$, $\psi(z_1,z_2)$ *is holomorphic on* $\mathfrak{H}\times\mathfrak{H}$; *and*

$$\psi(gz_1,g^\sigma z_2) = \tilde{\chi}(\delta)(\gamma z_1+\delta)^\nu(\gamma^\sigma z_2+\delta^\sigma)^\nu\psi(z_1,z_2)$$

for all $g \in \tilde{\Gamma}_0(N)$. *In fact* $\psi \in S_\nu(\tilde{\Gamma}_0(N),\tilde{\chi})$.

Suppose now that $N = 1$, $\chi = 1$, and Δ = prime $\equiv 1(4)$ with $h(\Delta)$ = class number of $k = 1$. Then $N' = \Delta$. Let $\varphi(z) = \sum_{n=1}^\infty a_n e^{2\pi inz}$ be the Fourier expansion of φ, and let $L(s,\varphi) = \sum_{n=1}^\infty a_n n^{-s}$ be the corresponding Dirichlet series. Set $\varphi_1(z) = \varphi(-1/\Delta z)\Delta^{\nu/2}(\Delta z)^{-\nu}$ and let

$$\psi_1(z_1,z_2) = \int_{F_{\Gamma_0(\Delta)}} \varphi_1(z)\overline{\theta(z,z_1,z_2)}v^{\nu-2}\,du\,dv$$

be the corresponding Hilbert modular form. Finally put $\psi(z_1,z_2) = (z_1z_2)^{-\nu}\psi_1(-1/z_1,-1/z_2)$. Then $\psi \in S_\nu(SL_2(\mathcal{O}_k))$, and it has a Fourier expansion:

$$\psi(z_1,z_2) = \sum_{\substack{\xi\in\mathcal{D}^{-1} \\ \xi>>0 \bmod U_k^2}} c(\xi) \sum_{n=-\infty}^{\infty} e^{2\pi i(\xi\varepsilon_0^{2n}z_1+\xi^\sigma\varepsilon_0^{-2n}z_2)}$$

where $\mathcal{D}^{-1}$ is the inverse different of k, and ε_0 is a generator for the group U_k of units of k. Let

$$D(s,\psi) = \sum_{\substack{\xi\in\mathcal{D}^{-1} \\ \xi>>0 \bmod U_k^2}} c(\xi)(\xi\xi^\sigma)^{-s}$$

be the corresponding Dirichlet series.

THEOREM 2. *With the above assumptions on* N, χ *etc., and if* φ *is an eigenfunction of all the Hecke operators* $T(n)$ *with* $a_1 = 1$, *then:*

$$D(s,\psi) = C\Delta^{\frac{2s-\nu+1}{2}} L(s,\varphi)L(s,\varphi_1).$$

This shows that the mapping $\varphi \longrightarrow \psi$ produced by the theta-function is essentially the same as the Doi-Naganuma mapping on forms of Neben-type described in [4].

Remarks:

a) The restriction $\nu \geq 7$ is the result of some bad estimates. Presumably you can do better.

b) The restrictions on N, χ and $h(k)$ in Theorem 2 can certainly be dropped.

c) It is possible, by taking a different Q, to produce a mapping from $S_\nu(\Gamma_0(*),*) \longrightarrow S_\nu(\Gamma,*)$ where $\Gamma \subset SL_2(\mathbb{R}) \times SL_2(\mathbb{R})$ is the unit group of a *division quaternion algebra* B over k. More precisely, let B_0 be an indefinite quaternion algebra over $\mathbb{Q}$ and put $B = B_0 \otimes_{\mathbb{Q}} k$. Let ι be the main involution on B, and let σ act on B via $1 \otimes \sigma$. Then take $V = \{X \in B$ such that $X^\iota = -X^\sigma\}$ and

$Q: V \longrightarrow \mathbb{Q}$, $Q[X] = -2\nu(X)$ where ν is the reduced norm from B to k.

d) Let $\varphi_n \in S_\nu(\Gamma_0(N'), \chi(\frac{\Delta}{*}))$ be the n-th Poincaré series. Then it is not difficult to show that the corresponding Hilbert modular form is the function $\omega_n(z_1,z_2)$ which was introduced by Zagier [11]. Therefore, Zagier's function $\Omega(\tau,z_1,z_2)$ is essentially the 'holomorphic part' of $\theta(\tau,z_1,z_2)$.

§4. SOME OTHER POSSIBILITIES

There are several other "accidental" isomorphisms which allow the correspondence described in Section 1 to be interpreted in a more classical way.

First, SO(3,1), the Lorentz group, is essentially the same as $SL_2(\mathbb{C})$. Therefore there is a correspondence between modular forms of integral weight on $\mathfrak{H}$ and (non-holomorphic!) automorphic forms on H = hyperbolic 3-space with respect to subgroups of $SL_2(\mathcal{O}_K)$ where $K = \mathbb{Q}(\sqrt{-\Delta})$. In fact V and Q may be taken precisely as in Section 3 with K replacing k. This case has been considered by Asai [1].

Another possibility is the relation between SO(3,2) and $Sp(2,\mathbb{R})$ which is described, for example, in part X of Siegel's Symplectic Geometry [8]. In this case we may identify X and $\mathfrak{H}_2 = \{Z \in M_2(\mathbb{C})$ such that ${}^tZ = Z$ and $Im(Z) > 0\}$, the Siegel space of genus 2. Applying the procedure of Section 1 we obtain the mapping:

automorphic forms of $\frac{1}{2}$-integral weight on $\mathfrak{H}$	$\longrightarrow$	Siegel modular forms of genus-2 on $\mathfrak{H}_2$

It should be very interesting to give a description of this mapping in terms of Dirichlet series.

If Q has signature (p,2) for *any* p, X is of Hermitian type, so that it is possible to consider holomorphic automorphic forms on X. The relation between ordinary modular forms and such forms on X has been investigated by Oda and by Rallis & Schiffmann [6] from a slightly different point of view.

Finally it should be noted that the above type of relation between automorphic forms can be formulated in the language of group representations via the Weil representation. The problem in this more general context has been considered by R. Howe, and I am indebted to him for describing his program to me several years ago.

REFERENCES

[1] T. Asai, *On the Doi-Naganuma lifting associated to imaginary quadratic fields*, (to appear).

[2] K. Doi & H. Naganuma, *On the functional equation of certain Dirichlet series*, Invent. Math. 9(1969) 1-14.

[3] S. Kudla, *Theta-functions and Hilbert modular forms*, to appear.

[4] H. Naganuma, *On the coincidence of two Dirichlet series associated with cusp forms of Hecke's "Neben"-type and Hilbert modular forms over a real quadratic field*, J. Math. Soc. Japan, 25(1973) 547-554.

[5] S. Niwa, *Modular forms of half-integral weight and the integral of certain theta-functions*, Nagoya Math. J. 56(1974) 147-161.

[6] S. Rallis & G. Schiffmann, *Automorphic forms constructed from the Weil representation, Holomorphic case*, Lecture Notes, University of Notre Dame, (1976).

[7] C.L. Siegel, *Indefinite quadratische Formen und Funktionen Theorie I & II*, Math. Ann. 124(1951/52) 17-54, 364-387.

[8] C.L. Siegel, *Symplectic Geometry*, Amer. J. Math. 65(1943) 1-86.

[9] G. Shimura, *On modular forms of half-integral weight*, Ann. of Math. 97(1973) 440-481.

[10] T. Shintani, *On construction of holomorphic cusp forms of half integral weight*, Nagoya Math. J. 58(1975) 83-126.

[11] D. Zagier, *Modular forms associated to real quadratic fields*, Invent. Math. 30(1975) 1-46.

THE RING OF HILBERT MODULAR FORMS FOR REAL QUADRATIC FIELDS OF SMALL DISCRIMINANT

F. Hirzebruch

Contents

International Summer School on Modular Functions
Bonn 1976

THE RING OF HILBERT MODULAR FORMS FOR REAL QUADRATIC FIELDS OF SMALL DISCRIMINANT

F. Hirzebruch

In this lecture we shall show how the resolution of the singularities at the cusps of the Hilbert modular surfaces [7] can be used for a detailed study of these surfaces which makes it possible in some cases to determine the structure of the ring of Hilbert modular forms.

§1. CUSP SINGULARITIES AND INVOLUTIONS.

Let K be a real quadratic field, $M \subset K$ a module (free $\mathbb{Z}$-module of rank 2) and U_M^+ the group of the totally positive units ε of K with $\varepsilon M = M$. The group U_M^+ is infinite cyclic. Let $V \subset U_M^+$ be a subgroup of finite index. The semi-direct product

$$G(M,V) = \{\begin{pmatrix}\varepsilon & \mu\\ 0 & 1\end{pmatrix} \mid \varepsilon \in V, \mu \in M\}$$

acts freely on H^2 by

$$(z_1, z_2) \mapsto (\varepsilon z_1 + \mu, \varepsilon' z_2 + \mu'),$$

where $x \mapsto x'$ is the non-trivial automorphism of K. We add a point to $H^2/G(M,V)$ and topologize $H^2/G(M,V) \cup \{\infty\}$ by taking

$$\{(z_1,z_2) \in H^2 \mid y_1 y_2 > C\}/G(M,V) \cup \{\infty\}$$

(for $C > 0$) as neighborhoods of ∞. (Notation : $z_j = x_j + i\, y_j$ with

$x_j, y_j \in \mathbb{R}$ and $y_j > 0$). Then $H^2/G(M,V) \cup \{\infty\}$ is a normal complex space with ∞ as the only singular point. This is the "cusp singularity" defined by M,V. The local ring at ∞ is denoted by $\mathcal{O}(M,V)$. It is the ring of all Fourier series f convergent in some neighborhood of ∞ of the form

$$(1) \qquad f = a_0 + \sum_{\substack{\lambda \in M^* \\ \lambda \gg 0}} a_\lambda \cdot e^{2\pi i(\lambda z_1 + \lambda' z_2)}$$

$$a_\lambda = a_{\varepsilon\lambda} \quad \text{for } \varepsilon \in V$$

where M^* is the dual module of M, i.e.

$$M^* = \{\lambda \in K \mid Tr(\lambda\mu) \in \mathbb{Z} \quad \text{for all } \mu \in M\}.$$

The singular point ∞ can be resolved [7]. Under the process of minimal desingularisation it is blown up into a cycle of r non-singular rational curves ($r \geq 2$) or into one rational curve with a double point ($r = 1$). Such a cycle is indicated by a diagram

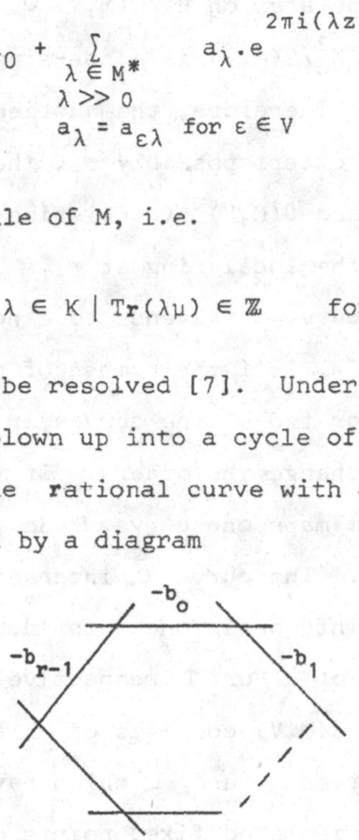

where $-b_0$, $-b_1$,... are the selfintersection-numbers (for $r \geq 2$). We have $b_i \geq 2$. This cycle of numbers is denoted by $((b_0, b_1, \ldots, b_{r-1}))$. It is determined by the denominators of a periodic continued fraction associated to M, see [7].

The non-singular surface obtained from $H^2/G(M,V) \cup \{\infty\}$ by resolving the singular point will be called X(M,V). Of course, it is not compact. For the intersection point of two consecutive curves of the cycle we have a natural coordinate system (u,v) centered at that point [7]. Any $f \in \mathcal{O}(M,V)$ can be written as a power series in u,v (this is analogous to the q-expansion in one variable.)

If $M = M'$, then the cusp is called symmetric. The involution $\tau : (z_1,z_2) \mapsto (z_2,z_1)$ operates on $H^2/G(M,V) \cup \{\infty\}$ with $\tau(\infty) = \infty$. The fixed point set of τ in $H^2/G(M,V)$ is a non-singular curve C consisting of 2,3 or 4 components. Therefore, the quotient of $H^2/G(M,V) \cup \{\infty\}$ by τ has no singular point except possibly ∞_τ, the image of ∞. The involution τ acts on $\mathcal{O}(M,V)$, and $\mathcal{O}(M,V)^\tau$ (consisting of all f in (1) with $a_\lambda = a_{\lambda'}$ for $\lambda \in M$) is the local ring at ∞_τ. The involution τ can be lifted to X(M,V). The curve C extends to a non-singular curve in X(M,V), also denoted by C. If the number of curves in the cycle is even, then τ maps none or two of the curves in the cycle, say S and T, to themselves and interchanges the others. If the number of curves in the cycle is odd, then τ maps one curve S in the cycle to itself and interchanges the others. The curve C intersects each curve S and T transversally in two points or in one point depending on whether the selfintersection number of S or T respectively is even or odd. The fixed point set of τ in X(M,V) consists of C and an isolated fixed point on each of the curves S and T which have odd selfintersection-number. Blowing up the isolated fixed points of τ gives a surface $\tilde{X}(M,V)$ on which τ operates having no isolated fixed points. The exceptional curves on $\tilde{X}(M,V)$ obtained by this blowing up belong to the fixed point set of τ. The surface $\tilde{X}(M,V)/\tau$ is non-singular. On it we have a chain of rational curves mapping to ∞_τ. This is a resolution of ∞_τ. It need not be minimal. In fact, ∞_τ could be a regular point. In any case, the existence of this resolution by a chain of rational curves proves that ∞_τ is a quotient singularity [6], [1]. The above investigation of X(M,V) for M = M' is due to Karras [12] (Lemma 3.3). The fact that ∞_τ is a quotient singularity was proved earlier by H. Cohn and E. Freitag (see the literature quoted in [12]). Gundlach [5] has given necessary and sufficient conditions that ∞_τ is regular. Such symmetric cusps are called quasi regular.

THEOREM (Karras). A cusp given by (M,V) with M = M' is quasi regular if and only if its cycle $((b_0,b_1,\dots,b_{r-1}))$ is equal to one of the following cycles

i) $((3,\underbrace{2,\dots,2}_{m}))$ with $m \geq 0$

ii) $((4,\underbrace{2,\dots,2}_{m}))$ with $m \geq 0$

iii) $((\underbrace{2,\dots,2}_{n}, 3, \underbrace{2,\dots,2}_{m},3))$ with $m \geq n$

and if in iii) the two curves of selfintersection number -3 are interchanged under τ (which is automatic for m > n).

Consider the following curves in $\mathbb{C}^2$ (coordinates X,Y)

i) $(X+Y^2)(X^2+Y^{m+5}) = 0$ with $m \geq 0$

ii) $(X^2+Y^2)(X^2+Y^{m+3}) = 0$ with $m \geq 0$

iii) $(X^{n+3}+Y^2)(X^2+Y^{m+3}) = 0$ with $m \geq n \geq 0$

Let F(X,Y) = 0 be one of these curves. The double cover of $\mathbb{C}^2$ branced along F(X,Y) = 0 has the point above $(0,0) \in \mathbb{C}^2$ as isolated singular point whose minimal resolution is a cycle of rational curves with selfintersection numbers as given in the preceding theorem of Karras. This can be checked directly. By a theorem of Laufer [15] (see also [13]) a singularity whose resolution is a cycle of rational curves is determined up to biholomorphic equivalence by its cycle of selfintersection numbers. Therefore, the structure of the local rings $\mathcal{O}(M,V)$ of quasi regular cusps is now known ([12], Satz 3), namely

(2) $$\mathcal{O}(M,V) \cong \mathbb{C}[C,Y,Z]/(Z^2 = F(X,Y))$$

where F(X,Y) is the polynomial given in i), ii), iii) above and where τ corresponds to the natural involution of the double cover. See also H. Cohn as quoted in [12].

In the following examples a), b), c) of quasi regular cusps we indicate the fixed point set C of τ on X(M,V) by heavily drawn lines. Isolated fixed points of τ on X(M,V) do not occur in examples a), b), c).

a)

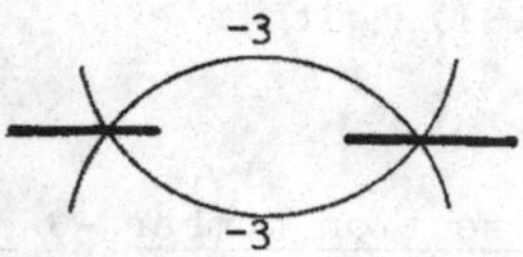

b)

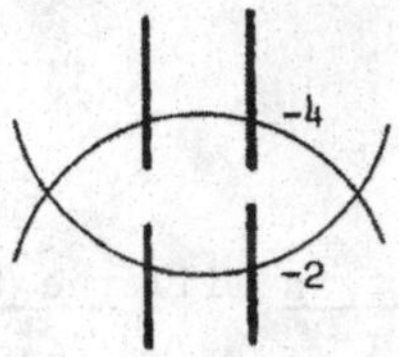

c)

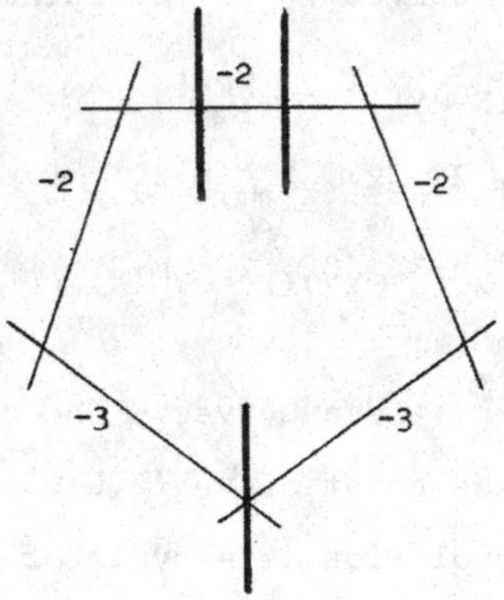

In example a) we have $K = \mathbb{Q}(\sqrt{5})$ with $M = \sqrt{5}.\mathcal{O}$ and $[U_M^+ : V] = 2$. (For a field K we denote its ring of integers by $\mathcal{O}$.) After dividing by τ (which interchanges the two (-3)-curves) we have in $X(M,V)/\tau$ the following situation

a)

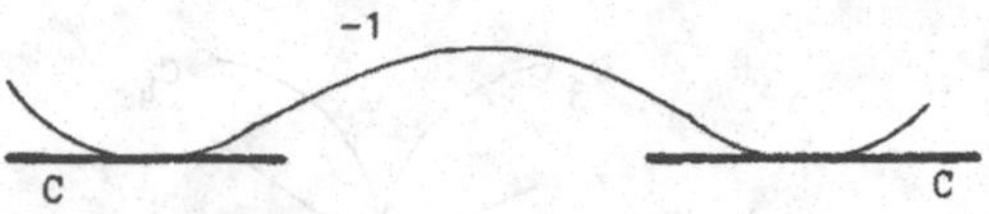

The non-singular rational (-1)-curve is the image of the two (-3)-curves. The image curve of C will also be denoted by C. It simply touches the (-1)-curve in two points. If we blow down the (-1)-curve we get $(H^2/G(M,V))/\tau \cup \{\infty_\tau\}$ which shows that ∞_τ is regular. After blowing down the (-1)-curve, the two components of C become singular. Each has a cusp (in the sense of curve singularities). The two cusps have separate tangents which checks with iii) $(m = n = 0)$. The structure of $\mathcal{O}(M,V)$ is given by (2). Therefore, there must exist three Fourier series f, g, h as in (1) generating $\mathcal{O}(M,V)$ and satisfying $h^2 = (f^3+g^2)(f^2+g^3)$.

In example b) we have $K = \mathbb{Q}(\sqrt{2})$ with $M = \mathcal{O}$ and $V = U_M^+$

b)

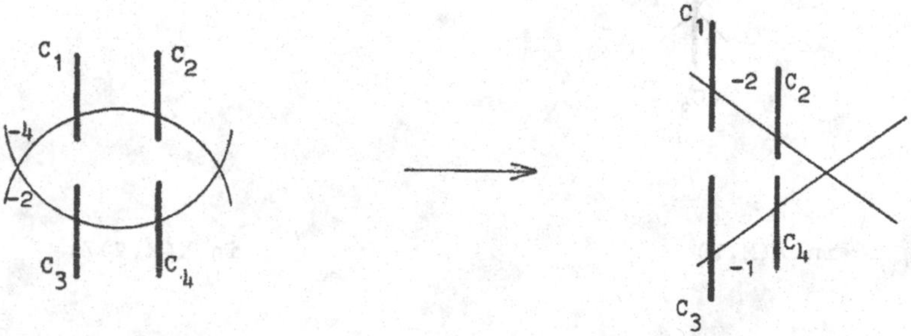

in X(M,V) in X(M,V)/τ

We have numbered the four branches of C.

In $X(M,V)/\tau$ we blow down the (-1)-curve, the (-2)-curve becomes a (-1)-curve and can be blown down also. The image of the two curves is ∞_τ, which is therefore a regular point. In $(H^2/G(M,V))/\tau \cup \{\infty\}$ the four branches of C in a neighborhood of ∞_τ behave as follows :

b)

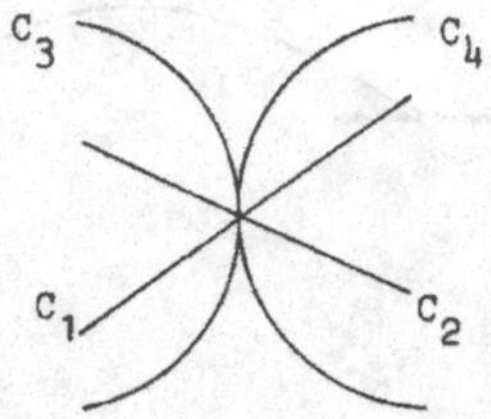

C_3, C_4 touch simply, all other intersections are transversal. This checks with ii) (m = 1).

In example c) we have $K = \mathbb{Q}(\sqrt{7})$ with $M = \sqrt{7}.\mathcal{O}$ and $V = U_M^+$.

c)

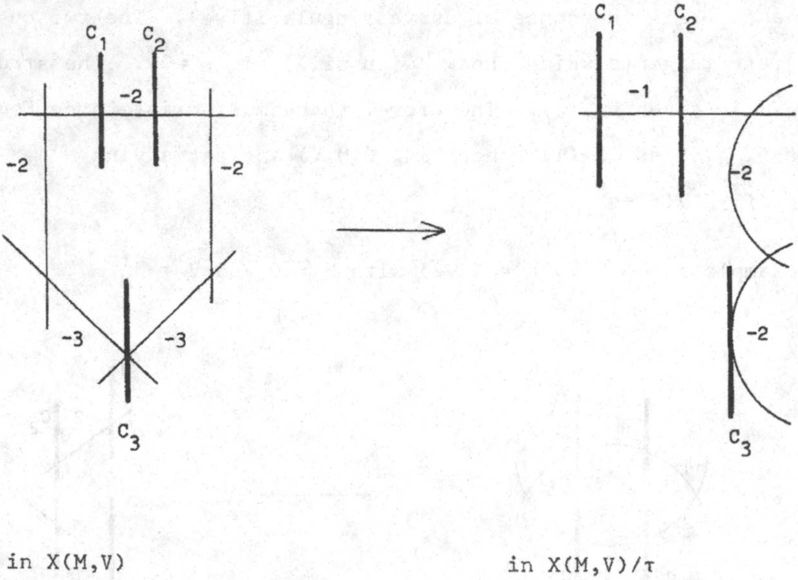

In $X(M,V)/\tau$ the (-2)-curve touches the component C_3 of C simply. Blowing down 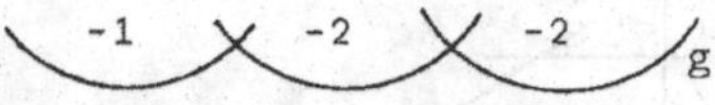

gives the regular point ∞_τ where C_1, C_2, C_3 behave locally like

$$(X^3 + Y^2)(X^2 + Y^6) = 0$$

with $X^3 + Y^2 = 0$ corresponding to C_3, and $X \pm i.Y^3 = 0$ to C_1 and C_2 respectively (compare iii), n = 0, m = 3).

The following symmetric cusp is not quasi regular.

d) 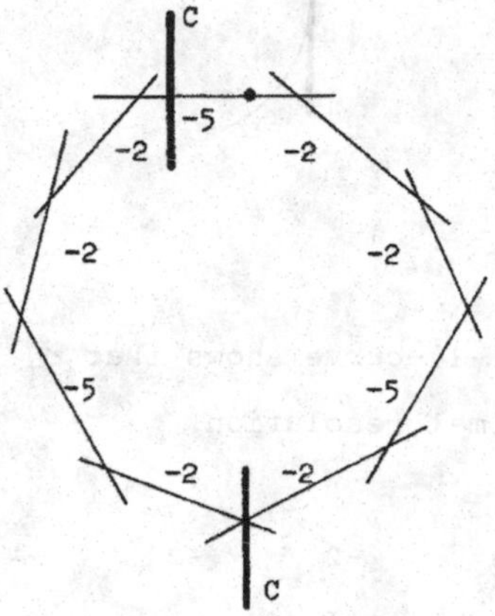

• = isolated fixed point of τ

We have $K = \mathbb{Q}(\sqrt{13})$ with $M = \mathcal{O}$ and $[U_M^+ : V] = 3$. Before dividing by τ we blow up the isolated fixed point. Then we divide by τ and obtain a configuration

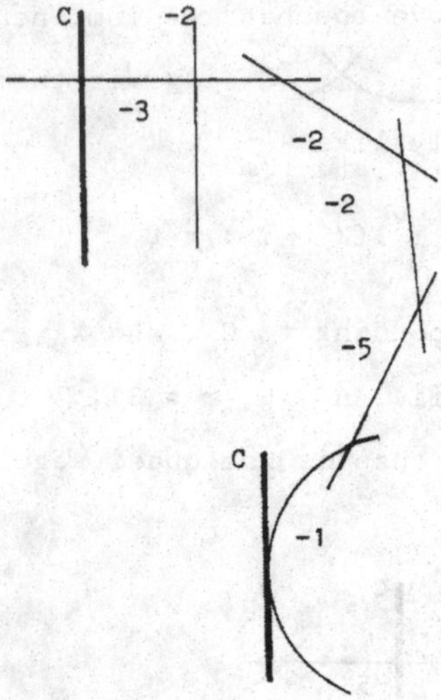

which after blowing down the (-1)-curve shows that ∞_τ is a quotien singularity admitting the minimal resolution

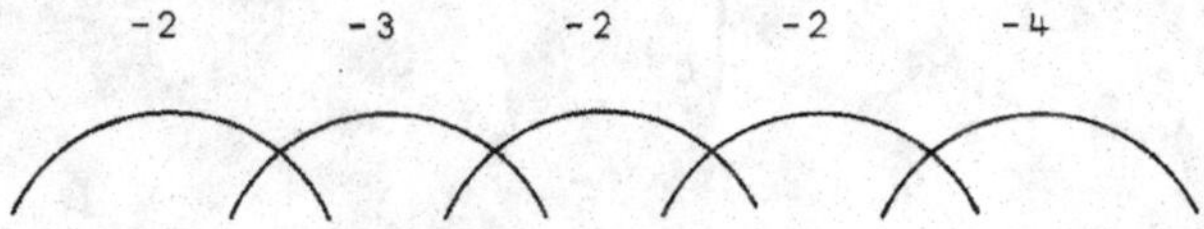

Thus it is the quotient singularity of type (36; 11,1), see [6].

§2. THE DIMENSION FORMULA FOR HILBERT CUSP FORMS.

Let K be a real quadratic field and $\mathcal{O}$ the ring of integers of K. The Hilbert modular group $SL_2(\mathcal{O})/\{\pm 1\}$ operates effectively on H^2. According to Siegel the volume of $H^2/SL_2(\mathcal{O})$ equals $2\zeta_K(-1)$. The volume is normalized such that if Γ is a subgroup of $SL_2(\mathcal{O})/\{\pm 1\}$ of finite index a which acts freely on H^2, then

$$\text{(3)} \qquad \operatorname{vol}(H^2/\Gamma) = 2\zeta_K(-1).a = e(H^2/G)$$

where $e(A)$ denotes the Euler number of the space A. (Though H^2/Γ is non-compact, the Euler number can be calculated by the volume, this is a special case of a result of Harder, see [7] and the literature quoted there.)

Let $S_k(\Gamma)$ be the complex vector space of cusp forms of weight k for Γ where Γ is a subgroup of $SL_2(\mathcal{O})/\{\pm 1\}$ of finite index.

The weight k of a form f is defined by the transformation law

$$f\left(\frac{az_1+b}{cz_1+d}, \frac{a'z_2+b'}{c'z_2+d'}\right) = (cz_1+d)^k(c'z_2+d')^k f(z_1,z_2)$$

This is well-defined also for k odd, because the expression $(cz_1+d)^k(c'z_2+d')^k$ does not change if $\begin{pmatrix} a & b \\ c & d \end{pmatrix}$ is replaced by $\begin{pmatrix} -a & -b \\ -c & -d \end{pmatrix}$.

THEOREM. *If Γ has index a in $SL_2(\mathcal{O})/\{\pm 1\}$ and acts freely on H^2, then for* $k \geq 3$

$$\text{(4)} \qquad \dim S_k(\Gamma) = \frac{k(k-2)}{2}\zeta_K(-1).a + \chi$$

$$= \frac{k(k-2)}{4} e(H^2/\Gamma) + \chi,$$

where $\chi = 1 + \dim S_2(\Gamma)$.

The formula (4) is found in the literature only for k even. But it seems to be known also for odd k (see Shimizu [17], p. 63, footnote 11). On my request, H. Saito has checked that (4) can be proved for odd k in the same way as Shimizu does it.

If Γ acts freely, then H^2/Γ is a non-singular complex surface which can be compactified by finitely many points, the cusps, to give a compact surface $\overline{H^2/\Gamma}$. The isotropy groups of the cusps are of the form $G(M,V)$. The cusps are singular points of $\overline{H^2/\Gamma}$ and can be resolved in the minimal canonical way as recalled in §1. The resulting surface is a non-singular algebraic surface $Y(\Gamma)$. It is a regular surface, i.e. its first Betti number vanishes, but it is not necessarily simply-connected. The cusp forms of weight 2 can be extended to holomorphic differential forms on $Y(\Gamma)$ (sections of the canonical bundle of $Y(\Gamma)$). Therefore, $\dim S_2(\Gamma)$ is the geometric genus p_g of $Y(\Gamma)$ and χ the arithmetic genus. The fact that the constant term in the Shimizu polynomial (4) is the arithmetic genus of $Y(\Gamma)$ was discovered by Freitag (compare [7], 3.6.).

§3. THE FIELD $K = \mathbb{Q}(\sqrt{5})$.

In the field $K = \mathbb{Q}(\sqrt{5})$ the ring $\mathcal{O}$ of integers consists of all linear combinations $a + b(1+\sqrt{5})/2$ with $a,b \in \mathbb{Z}$. To the prime ideal generated in $\mathcal{O}$ by $\sqrt{5}$ there belongs a principal congruence subgroup of $SL_2(\mathcal{O})$, which we denote by Γ.

$$\Gamma = \{\begin{pmatrix}\alpha & \beta\\ \gamma & \delta\end{pmatrix} \in SL_2(\mathcal{O}) \mid \alpha \equiv \delta \equiv 1 (\mathrm{mod}\ \sqrt{5}), \beta \equiv \gamma \equiv 0 (\mathrm{mod}\ \sqrt{5})\}.$$

Because $\begin{pmatrix}-1 & 0\\ 0 & -1\end{pmatrix} \notin \Gamma$, the group Γ can be regarded as a subgroup of the Hilbert modular group $G = SL_2(\mathcal{O})/\{\pm 1\} = PSL_2(\mathcal{O})$. The group Γ acts freely on H^2. The volume of H^2/G is equal to $2\zeta_K(-1) = 1/15$. The factor group G/Γ is isomorphic to $PSL_2(\mathbb{F}_5)$ because $\mathcal{O}/\sqrt{5}.\mathcal{O} \cong \mathbb{F}_5$. In its turn, $PSL_2(\mathbb{F}_5)$ is isomorphic to the alternating group A_5. Namely, A_5 is the group of automorphisms of the icosahedron and acts on the six axes of the icosahedron through its vertices in the same way as $PSL_2(\mathbb{F}_5)$ acts on the six points of the projective line $P_1(\mathbb{F}_5)$. We have

(5) $$e(H^2/\Gamma) = |A_5|.2\zeta_K(-1) = 60.\frac{1}{15} = 4 .$$

The space H^2/Γ is compactified by adding six cusps. Since the class number of $\mathbb{Q}(\sqrt{5})$ is 1, the action of G on $P_1(K) = K \cup \{\infty\}$ has only one orbit, while the action of Γ on $P_1(K)$ has six. This follows, because the isotropy group of G and Γ at ∞ satisfy $|G_\infty/\Gamma_\infty| = 10$. In fact, G_∞/Γ_∞ is the dihedral group of order 10, this will be used later. Two points α/δ and γ/δ in $P_1(K)$ with $\alpha,\beta,\gamma,\delta \in \mathcal{O}$ and $(\alpha,\beta) = (\gamma,\delta) = 1$ belong to the same orbit precisely when $\alpha \equiv \gamma \pmod{\sqrt{5}}$ and $\beta \equiv \delta \pmod{\sqrt{5}}$, that is when α/β and γ/δ represent the same point of $P_1(\mathbb{F}_5)$. The surface H^2/Γ, compactified by six points, is denoted by $\overline{H^2/\Gamma}$. This is an algebraic surface with six singular points corresponding to the six cusps. Since the action of G on H^2 induces an action of $A_5 \cong G/\Gamma$ on $\overline{H^2/\Gamma}$ which acts transitively on the cusps, these six singular points have the same structure, and it is sufficient to investigate the structure of the singularity at $\infty = 1/0$. The isotropy group of Γ at this point has the form

(6) $$\Gamma_\infty = \{\begin{pmatrix} \varepsilon & \mu \\ 0 & \varepsilon^{-1} \end{pmatrix} \mid \varepsilon \text{ unit in } \mathcal{O}, \varepsilon \equiv 1 \pmod{\sqrt{5}}, \mu \equiv 0 \pmod{\sqrt{5}}\}.$$

The fundamental unit of $\mathcal{O}$ is $\varepsilon_0 = (1+\sqrt{5})/2$. The condition $\varepsilon \equiv 1 \pmod{\sqrt{5}}$ means that ε must be a power of $-\varepsilon_0^2$. The group Γ_∞ can also be written as $G(M,V)$ where $M = \sqrt{5}.\mathcal{O}$ and V is generated by ε_0^4. Thus $[U_M^+ : V] = 2$ and $G(M,V)$ is as in example a) of §1.
On the surface $Y = Y(\Gamma)$ that arises from $\overline{H^2/\Gamma}$ by resolution of the six singular points there are six pairwise disjoint configurations

(7)

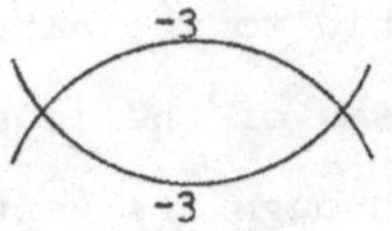

As a 4-dimensional manifold, Y can be obtained as follows : H^2/Γ has as deformation retract a compact manifold X whose boundary has six components. Each boundary component is a torus bundle over a circle. All boundary components are isomorphic. Every configuration (7) in Y has a tubular neighborhood having as boundary such a torus bundle. The manifold Y arises by glueing to X the tubular neighborhoods of these six configurations (7). Since the Euler number of each tubular neighborhood is 2, it follows from (5) and the additivity of e that

$$e(Y) = e(X) + 6.2 = e(H^2/\Gamma) + 12 = 16 \tag{8}$$

The action of A_5 on H^2/Γ described above induces an action on Y. The diagonal $z_1 = z_2$ of H^2 yields a curve in H^2/Γ, which can be compactified to a curve C in Y. The subgroup of Γ carrying the diagonal into itself is the ordinary principal congruence subgroup $\Gamma(5)$ of $SL_2(\mathbb{Z})$, which can also be regarded as subgroup of $SL_2(\mathbb{Z})/\{\pm 1\}$, the quotient group being A_5 again.

Therefore, each element of A_5 when acting on Y carries C to itself. The curve $H/\Gamma(5)$ has normalized Euler volume $-\frac{1}{6}.60 = -10$ and twelve cusps. The compactified curve $\overline{H/\Gamma(5)}$ has Euler number $-10 + 12 = 2$, thus is a rational curve which maps onto C. For reasons of symmetry, the curve C must pass through each of the six configurations (7) exactly twice. We now describe how the curve cuts a resolution (7) by reducing the question to the corresponding question for the diagonal in $H^2/G = (H^2/\Gamma)/A_5$. There is an exact sequence

(9) $$0 \to \mathcal{O}/\sqrt{5}.\mathcal{O} \to G_\infty/\Gamma_\infty \to U_M^+/V \to 1$$

The groups $\mathcal{O}/\sqrt{5}.\mathcal{O}$ and U_M^+/V are cyclic of order 5 and 2 respectively, and G_∞/Γ_∞ is a semi-direct product, namely the dihedral group of order 10.

To understand the formation of the quotient of the configuration (7) by this dihedral group, we check first that any non-trivial element g of $\mathcal{O}/\sqrt{5}.\mathcal{O}$ carries each of the two (-3)-curves to itself and has their intersection points as isolated fixed points. By blowing up these two points we come to the following configuration :

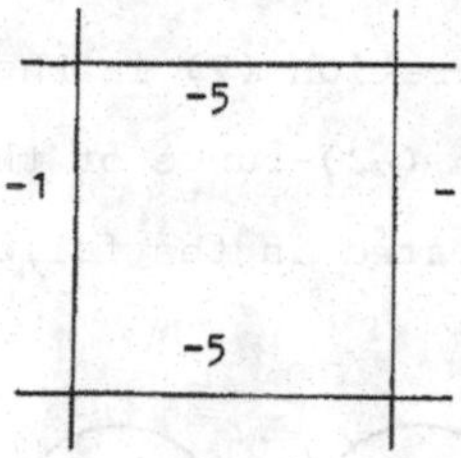

(the verticals are fixed lines for g)

After factorizing by $\mathcal{O}/\sqrt{5}.\mathcal{O}$ we obtain

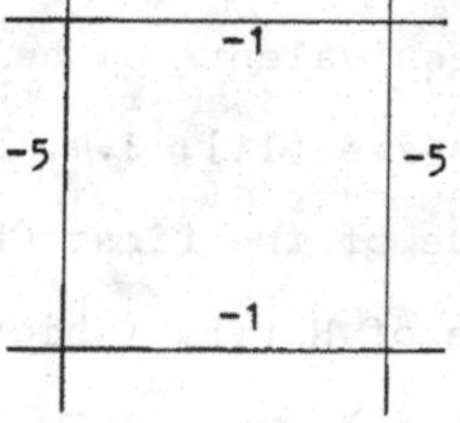

The group $U_M^+/V \cong \mathbb{Z}/2\mathbb{Z}$ acts on this quotient by "rotation", carrying each (-1)-curve to the other one, each (-5)-curve to the other one.
Factorization leads to

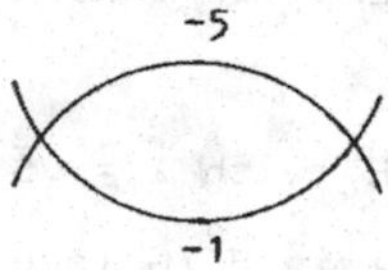

and blowing down the (-1)-curve gives a configuration consisting of a rational curve with a double point. This is the resolution of the cusp of $\overline{H^2/G}$. The curve in the desingularized compactification of H^2/G represented by $z_1 = z_2$ is usually called F_1 (see [10]). It passes transversally through the resolved cusp as follows

(10)  (see [7], §4.)

As explained the configuration (7) is a ten-fold covering of (10). We conclude that C passes through each configuration (7) in the two "corners" and meets in these two points each (-3)-curve of the configuration (7) transversally. This is illustrated in the following diagram

(11) 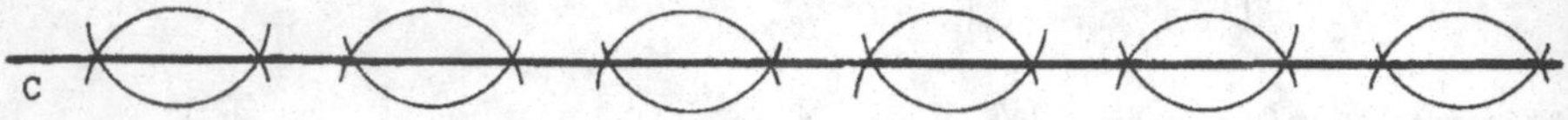

The curve C is non-singular, because of the described behaviour at the cusps of $\overline{H^2/\Gamma}$ and because two curves on H^2 equivalent to the diagonal $z_1 = z_2$ under $SL_2(\mathcal{O})$ cannot intersect in H^2 (see [11], 3.4. or [10]). Therefore $\overline{H/\Gamma(5)} \to C$ is bijective. The value of the first Chern class c_1 of Y on C equals twice the Euler volume of $H/\Gamma(5)$ (which is -10) plus 24 (see [7], 4.3. (19)). Thus we have in Y

(12) $c_1[C] = 4$ and $C.C = 2$ (by the adjunction formula).

Because Y is regular, this implies that Y <u>is a rational surface</u> (compare [9], [7]).

The curve $\lambda z_2 - \lambda' z_1 = 0$ in H^2 with $\lambda = \sqrt{5}.\varepsilon_0$ is a skew-hermitian curve which determines the curve F_5 in H^2/G (see [10]). The inverse image D of F_5 in H^2/Γ consists of 15 connectedness components. Namely, as can be checked, the subgroup of $A_5 = G/\Gamma$ which carries the curve in H^2/Γ given by $\lambda z_2 - \lambda' z_1 = 0$ to itself is of order 4. The curve F_5 passes through the resolved cusp of H^2/G as follows

Therefore D intersects each configuration (7) in the following way

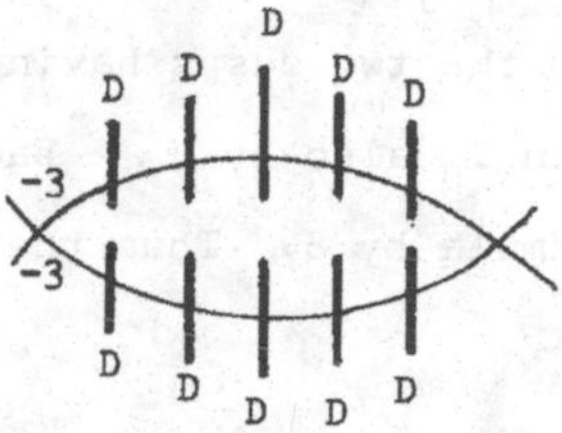

A component of D intersects exactly two of the configurations (7) and each in two points, one intersection point on each (-3)-curve. It is easy to see that each component of D is a non-singular rational curve.

The involution $(z_1,z_2) \to (z_2,z_1)$ on H^2 induces an involution τ on H^2/Γ, because $A' \in \Gamma$ if $A \in \Gamma$. The involution τ keeps every cusp of H^2/Γ fixed, because it operates on $P_1(K)$ by conjugation $(x \mapsto x')$ and the cusps can be represented by rational points. Each cusp is symmetric, Γ operates on each of the configurations (7) by interchanging the two (-3)-curves. The curve C is pointwise fixed under τ. In fact, C is the complete fixed point set. This can be seen as follows. The involution $(z_1,z_2) \mapsto (z_2,z_1)$ **induces an involution** on H^2/G which has $F_1 \cup F_5$ as fixed point set ([7], §4.).

Therefore, the fixed point set of τ on H^2/Γ is at most $C \cup D$. The inter section behaviour of such a component D_j with a configuration (7) shows that D_j is carried to itself under τ, but is not pointwise fixed.

The quotient Y/τ is a non-singular algebraic surface. We have

(13) $$e(Y/\tau) = \frac{1}{2}(e(Y) + e(C)) = \frac{1}{2}(16+2) = 9$$

By example a) in §1, the surface Y/τ has six exceptional curves. If we blow them down, we get an algebraic surface Y_0 with $e(Y_0) = 3$ and six distinguished points $P_1,\ldots,P_6$ resulting from the exceptional curves. Since Y is rational, Y/τ and Y_0 are rational. Thus Y_0 is the complex projective plane, and the image of C on Y_0 is a rational curve with a double cusp in each point P_j $(j=1,\ldots,6)$ and is otherwise non-singular. "Double cusp in P_j" means that the curve has two branches in P_j, each with a cusp, the two cusps having separate tangents. We denote the image of C in Y_0 also by C. Each double cusp reduces the genus in the Plücker formula by 6. Thus the degree n of C in $Y_0 = P_2(\mathbb{C})$ satisfies

$$\frac{(n-1)(n-2)}{2} - 6.6 = 0.$$

Therefore C is a curve of degree 10 in $P_2(\mathbb{C})$, as can also be infered from (12). The image of D in $Y_0 = P_2(\mathbb{C})$ (also denoted by D) is the union of the 15 lines joining $P_1,\ldots,P_6$, as can be proved in a similar way.

The involution τ operating on Y commutes with each element of $G/\Gamma \cong A_5$. This follows from the fact that matrices $A, A' \in SL_2(\mathcal{O})$ are equivalent mod $\sqrt{5}$. Therefore, A_5 acts effectively on Y/τ and on $Y_0 = P_2(\mathbb{C})$.

Every action of A_5 on $P_2(\mathbb{C})$ can be lifted to a 3-dimensional linear representation, because $H^2(A_5,\mathbb{Z}_3) = 0$.
[I. Naruki has shown me a proof that $H^2(G,\mathbb{Z}_3) = 0$ for a non abelian, finite simple group G whose order is not divisible by 9. Such results essentially can be found in Schur's papers.]

The lifting is unique, because A_5 is simple. The character table shows that there are exactly two equivalence classes of non-trivial 3-dimensional representations of A_5. They are related by an outer automorphism of A_5. Hence the action of A_5 on $P_2(\mathbb{C})$ which we have found is essentially the one whose invariant theory was studied by F. Klein [14]. We recall some of Klein's results.

The group A_5 is isomorphic to the finite group I of those elements of SO(3) which carry a given icosahedron centered at the origin of the standard Euclidean space $\mathbb{R}^3$ to itself. The group I operates linearly on $\mathbb{R}^3$ (standard coordinates x_0, x_1, x_2) and thus also on $P_2(\mathbb{R})$ and $P_2(\mathbb{C})$. We are concerned with the action on $P_2(\mathbb{C})$. A curve in $P_2(\mathbb{C})$ which is mapped to itself by all elements of I is given by a homogeneous polynomial in x_0, x_1, x_2 which is I-invariant up to constant factors and hence I-invariant, because I is a simple group. The graded ring of all I-invariant polynomials in x_0, x_1, x_2 is generated by homogeneous polynomials A,B,C,D of degrees 2,6,10,15 with $A = x_0^2 + x_1^2 + x_2^2$. The action of I On $P_2(\mathbb{C})$ has exactly one minimal orbit where "minimal" means that the number of points in the orbit is minimal. This orbit has six points, they are called poles. These are the points of $P_2(\mathbb{R}) \subset P_2(\mathbb{C})$ which are represented by the six axes through the vertices of the icosahedron. Klein uses coordinates

$$A_0 = x_0, \quad A_1 = x_1 + ix_2, \quad A_2 = x_1 - ix_2$$

and puts the icosahedron in such a position that the six poles are given by

$$(A_0, A_1, A_2) = (\sqrt{5}/2, 0, 0)$$

$$(A_0, A_1, A_2) = (\tfrac{1}{2}, \varepsilon^{\nu}, \varepsilon^{-\nu})$$

with $\varepsilon = \exp(2\pi i/5)$ and $0 \leq \nu \leq 4$.

The invariant curve $A = 0$ does not pass through the poles. There is exactly one invariant curve $B = 0$ of degree 6 which passes through the poles, exactly one invariant curve $C = 0$ of degree 10 which has higher multiplicity than the curve $B = 0$ in the poles and exactly one invariant curve $D = 0$ of degree 15. In fact, $B = 0$ has an ordinary double point (multiplicity 2) in each pole, $C = 0$ has a double cusp (multiplicity 4) in each pole and $D = 0$ is the union of the 15 lines connecting the six poles. Klein gives formulas for the homogeneous polynomials A,B,C,D (determined up to constant factors). They generate the ring of all I-invariant polynomials. We list Klein's formulas :

$$A = A_0^2 + A_1A_2$$

$$B = 8A_0^4A_1A_2 - 2A_0^2A_1^2A_2^2 + A_1^3A_2^3 - A_0(A_1^5+A_2^5)$$

$$\begin{aligned} C = {} & 320A_0^6A_1^2A_2^2 - 160A_0^4A_1^3A_2^3 + 20A_0^2A_1^4A_2^4 + 6A_1^5A_2^5 \\ & - 4A_0(A_1^5+A_2^5)(32A_0^4-20A_0^2A_1A_2+5A_1^2A_2^2) + A_1^{10} + A_2^{10} \end{aligned}$$

$$\begin{aligned} 12D = {} & (A_1^5-A_2^5)(-1024A_0^{10}+3840A_0^8A_1A_1 \\ & -3840A_0^6A_1^2A_2^2 + 1200A_0^4A_1^3A_2^3 \\ & -100A_0^2A_1^4A_2^4 + A_1^5A_2^5) \\ & + A_0(A_1^{10}-A_2^{10})(352A_0^4-160A_0^2A_1A_2+10A_1^2A_2^2) \\ & + (A_1^{15}-A_2^{15}) \end{aligned}$$

According to Klein the ring of I-invariant polynomials is given as follows

$$\mathbb{C}[A_0,A_1,A_1]^I = C[A,B,C,D]/(R(A,B,C,D) = 0) \tag{14}$$

The relation $R(A,B,C,D) = 0$ is of degree 30.

We have

$$\begin{aligned} & R(A,B,C,D) = \\ & -144D^2 - 1728B^5 + 720ACB^3 - 80A^2C^2B \\ & +64A^3(5B^2-AC)^2 + C^3 . \end{aligned} \tag{15}$$

The equations for B and C show that the two tangents of $B = 0$ in the pole $(\sqrt{5}/2,0,0)$ are given by $A_1 = 0$, $A_2 = 0$. They coincide with the tangents of $C = 0$ in that pole. Therefore the curves $B = 0$ and $C = 0$ have in each pole the intersection multiplicity 10. Thus they intersect only in the poles.

When we restrict the action of I to the conic $A = 0$, we get the well-known action of I on $P_1(\mathbb{C})$ (which can also be obtained via the isomorphism $SO(3) \cong PSU(2)$). The curves $B = 0$, $C = 0$, $D = 0$ intersect $A = 0$ tranversally in 12, 20, 30 points respectively. If one uses a suitable conformal map $S^2 \to P_1(\mathbb{C}) \cong \{A = 0\}$ these points correspond to the 12 vertices, 20 center points of the faces, 30 center points of the edges of the icosahedron (always projected from the origin of $\mathbb{R}^3$ to S^2). Putting $A = 0$, the relation $R(A,B,C,D) = 0$ gives a famous icosahedral identity.

We consider the uniquely determined double cover W of $P_2(\mathbb{C})$ branched along $C = 0$. The action of I can be lifted to the double cover.

The study of the Hilbert modular surface H^2/Γ led to an action of G/Γ ($\cong A_5$) on the complex projective plane. We also found the invariant curve $C = 0$. We use an isomorphism $G/\Gamma \cong I$ to identify G/Γ and the icosahedral group. Since the action of I on the projective plane is essentially unique and the invariant curve $C = 0$ well determined as curve of degree 10 with double cusps in the poles, we have proved the following result.

THEOREM. *Let Γ be the principal congruence subgroup of $SL_2(\mathcal{O})$ for the ideal $(\sqrt{5})$ in the ring $\mathcal{O}$ of integers of the field $\mathbb{Q}(\sqrt{5})$. Then the Hilbert modular surface H^2/Γ can be compactified by six points (cusps in the sense of modular surfaces) to give a surface $\overline{H^2/\Gamma}$ with these cusps as the only singular points. The surface $\overline{H^2/\Gamma}$ admits an action* of the icosahedral group I. It is I-equivariantly isomorphic to the double cover W of $P_2(\mathbb{C})$ branched along the Klein curve $C = 0$. This curve has singularities ("double cusps") in the six poles of the action I and otherwise no singularities. The double cover W has a singular point above each double cusp of C and no further singular points. Under the isomorphism these singular points correspond to the six singular points of $\overline{H^2/\Gamma}$. The involution of the double cover W corresponds to the involution of $\overline{H^2/\Gamma}$ induced by $(z_1,z_2) \to (z_2,z_1)$ on H^2. The surface W is rational.

We use this theorem to gain information on the modular forms relative to Γ. A modular form of weight k is a holomorphic function $f(z_1,z_2)$ on H^2 transforming under elements of Γ as recalled in §2. The form f is a cusp form if it vanishes in the cusps. The forms of weight 2r correspond bijectively to the holomorphic sections of K^r where K is the canonical bundle of H^2/Γ. A form is symmetric if $f(z_1,z_2) = f(z_2,z_1)$, skew-symmetric if $f(z_1,z_2) = -f(z_2,z_1)$. Let W' be the double cover W of $P_2(\mathbb{C})$ with the six singular points removed and $P_2'(\mathbb{C})$ the projective plane with the six poles removed. Let $\pi : W' \to P_2'(\mathbb{C})$ be the covering map, $\tilde{\gamma}$ the divisor in W' represented by the branching locus $C = 0$ and γ the divisor in $P_2'(\mathbb{C})$ given by $C = 0$. If L is a line in $P_2'(\mathbb{C})$, then $\tilde{\gamma} + \pi^*(-3L)$ is a canonical divisor of W'. Because $\pi^*\gamma = 2\tilde{\gamma}$, we conclude that $\pi^*(\gamma-6L)$ is twice a canonical divisor and also $\pi^*(4L)$ is twice a canonical divisor on W'. Therefore, under the isomorphism $H^2/\Gamma \to W'$, a homogeneous polynomial of degree 4r in A_0,A_1,A_2 defines a section of K^{2r} and thus a modular form relative to Γ of weight 4r. It can be proved, that the abelian group $\Gamma/[\Gamma,\Gamma]$ has a trivial 2-primary component. This implies that a homogeneous polynomial of degree k in A_0,A_1,A_2 defines a modular form relative to Γ of weight k. In fact, these modular forms are symmetric. There is a skew-symmetric form of weight 5, whose divisor is $\tilde{\gamma}$ (under the isomorphism $H^2/\Gamma \to W'$). We denote it by c. Thus we have obtained a graded subring

$$M'(\Gamma) = \sum_{k \geq 0} M'_k(\Gamma) = \mathbb{C}[A_0, A_1, A_2, c]/(c^2 = C)$$

of the full graded ring $M(\Gamma) = \sum M_k(\Gamma)$ of modular forms for Γ. (Here C is the Klein polynomial of degree 10.) We have

$$\dim M'_k(\Gamma) = \binom{k+2}{2} + \binom{k-3}{2} = k^2 - 2k + 7 \qquad \text{for } k \geq 3$$

$$\dim M'_2(\Gamma) = 6$$

$$\dim M'_1(\Gamma) = 3$$

The arithmetic genus χ of the non-singular model Y of $\overline{H^2/\Gamma}$ equals 1, because Y is rational. The dimension formula (§2 (4)) and §3 (5) imply that $M'_k(\Gamma) = M_k(\Gamma)$ for $k \geq 2$. We have to use that there exist six Eisenstein series of weight k (for $k \geq 2$) belonging to the six cusps which shows $\dim M_k(\Gamma) - \dim S_k(\Gamma) = 6$ for $k \geq 2$. Because the square of a modular form f of weight one belongs to $M_2(\Gamma) = M'_2(\Gamma)$, the zero divisor of f gives a line in $P_2(\mathbb{C})$. Thus $M_1(\Gamma) = M'_1(\Gamma)$. Of course, there are no modular forms of negative weight.

THEOREM. The ring of modular forms for the group Γ is isomorphic to

$$\mathbb{C}[A_0, A_1, A_2, c]/(c^2 = C).$$

The ring of symmetric modular forms for Γ is

$$\mathbb{C}[A_0, A_1, A_2]$$

The vector space of skew-symmetric forms is

$$c.\mathbb{C}[A_0, A_1, A_2]$$

The group $G/\Gamma = I$ = icosahedral group operates on these spaces by the

Klein representation of I *of degree* 3 *in terms of the coordinates* A_0, A_1, A_2 *of* $\mathbb{C}^3$.

We now consider the full Hilbert modular group $G = SL_2(\mathcal{O})/\{\pm 1\}$ for $\mathbb{Q}(\sqrt{5})$ and obtain in view of (14) and (15).

THEOREM. *The ring of modular forms for the group* G *is isomorphic to*

$$\mathbb{C}[A,B,c,D]/(144D^2 = -1728B^5 + 720Ac^2B^3 - 80A^2c^4B + 64A^3(5B^2 - Ac^2)^2 + c^6)$$

The ring of symmetric modular forms for G is isomorphic to

$$\mathbb{C}[A,B,C,D]/(R(A,B,C,D) = 0) \tag{16}$$

For the preceding theorems compare the papers of Gundlach [3] and Resnikoff [16] and also [8] where results on $\mathbb{Q}(\sqrt{5})$ where derived using the principal congruence subgroup of $SL_2(\mathcal{O})$ for the prime ideal (2). In [8] the relation R(A,B,C,D) = 0 was obtained in a different form connected to the discriminant of a polynomial of degree 5. The modular form D occurs in Grundlach's paper [3] as a product of 15 modular forms for Γ of weight 1 each cuspidal at 2 cusps and vanishing along the "line" between these 2 cusps. The zero divisors of the six Eisenstein series for Γ of weight 2 correspond to the six conics passing through 5 of the six poles. (Each Eisenstein series is cuspidal in five cusps.) In H^2/G the curve C = 0 becomses F_1 (given by $z_1 = z_2$). The restriction of B to F_1 gives a cusp form of weight 12 on $H/SL_2(\mathbb{Z})$, therefore must be Δ (up to a factor). The curves B = 0, C = 0 intersect only in the six poles of the action of I, in agreement with the fact that Δ does not vanish on H.

Remark. I. Naruki has given a geometric interpretation of the curve

$B = 0$. Let $S(5)$ be the elliptic modular surface in the sense of T. Shioda associated to the principal congruence subgroup $\Gamma(5)$ of $SL_2(\mathbb{Z})$. Choose a "zero section" σ of $S(5)$, then each regular fibre of $S(5)$ becomes a group (1-dim. complex torus). The binary icosahedral group $I' = SL_2(\mathbb{F}_5)$ is the group of automorphisms of $S(5)$ which carry σ to itself. The element $-1 \in I'$ acts as the involution which is $x \to -x$ on each regular fibre. Dividing $S(5)$ by this involution and blowing down 24 exceptional curves which come from the 12 singular fibres of $S(5)$ gives $P_1(\mathbb{C}) \times P_1(\mathbb{C})$ on which $I = I'/\{\pm 1\}$ operates. Dividing $P_1(\mathbb{C}) \times P_1(\mathbb{C})$ by the natural involution interchanging components yields $P_2(\mathbb{C})$ on which I acts by the Klein representation. Under this procedure $B = 0$ is the image of the curve in $S(5)$ containing all the points of the regular fibres of $S(5)$ which have precisely the order 4. A paper of Naruki (über die Kleinsche Ikosaeder-Kurve sechsten grades) will appear in Mathematische Annalen.

§4. THE FIELD $K = \mathbb{Q}(\sqrt{2})$.

In this field the ring $\mathcal{O}$ of integers consists of all linear combinations $a + b\sqrt{2}$ with $a, b \in \mathbb{Z}$. The fundamental unit is $\varepsilon_0 = 1 + \sqrt{2}$. We consider the principal subgroup $\tilde{\Gamma}(2)$ of $SL_2(\mathcal{O})$ for the ideal (2). The group $\tilde{\Gamma}(2)/\{\pm 1\}$ is a subgroup $\Gamma(2)$ of the Hilbert modular group $G = SL_2(\mathcal{O})/\{\pm 1\}$. The group $G/\Gamma(2)$ is an extension of the symmetric group S_4 by a group of order 2 (which is the center of $G/\Gamma(2)$). The non-trivial element in the center is represented by the matrix

$$\begin{pmatrix} \varepsilon_0 & 0 \\ 0 & \varepsilon_0^{-1} \end{pmatrix} = D_{\varepsilon_0}$$

of $SL_2(\mathcal{O})$. Let Γ be the subgroup of G obtained by extending $\Gamma(2)$ by

D_{ε_0}. Then $G/\Gamma \cong S_4$. The group Γ acts freely on H^2. We shall investigate Γ similarly as we treated the congruence subgroup with respect to $(\sqrt{5})$ in §3. Often details will be omitted an proofs only skecthed.

The Hilbert modular surface $\overline{H^2/\Gamma(2)}$ has six cusps, each resolved by a cycle of type ((4,2,4,2)). The non-singular surface thus obtained will be called Y_2. The curve F_1 in $\overline{H^2/G}$ is given by $z_1 = z_2$, the curve F_2 by $\lambda z_2 - \lambda' z_1 = 0$ with $\lambda = \sqrt{2}.\varepsilon_0$. The inverse images of F_1 and F_2 in Y_2 are also denoted by F_1 and F_2 respectively. F_1 has 8 and F_2 has 6 components in Y_2. The curves F_1 and F_2 in Y_2 pass through each of the six resolved cusps as follows

(17)

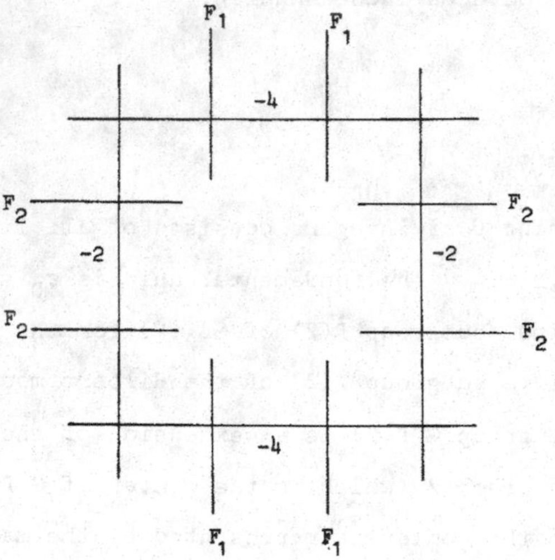

The 14 components of $F_1 \cup F_2$ are disjoint, non-singular rational curves. Each component of F_1 has selfintersection number -1, hence is an

exceptional curve. Each component of F_2 has selfintersection number -2. Because $2\zeta_K(-1) = \frac{1}{6}$, the Euler number of $H^2/\Gamma(2)$ is $48/6 = 8$, and we have (as in §3 (8))

$$e(Y_2) = 8 + 6.4 = 32$$

In fact, Y_2 <u>is a</u> K3-<u>surface with</u> 8 <u>points blown up</u>. This can be shown by the methods of [9], see [2]. The involution on Y_2 given by D_{ε_0} will be denoted by δ. It operates freely on Y_2. The non-singular model Y for $\overline{H^2/\Gamma}$ (obtained by resolving the six cusps) equals Y_2/δ. Therefore, Y has Euler number 16, <u>it is an Enriques surface with</u> 4 <u>points blown up</u>. (An Enrique surface can be defined as a surface with fundamental group of order 2 whose universal covering is a K3-surface.) Each cusp of $\overline{H^2/\Gamma}$ is resolved by a cycle of type ((4,2)) (type ((4,2,4,2)) divided by δ). The inverse image of F_1 and F_2 in Y are also called F_1, F_2. They have 4 or 3 components respectively, the four components of F_1 being exceptional curves. The curves F_1 and F_2 in Y pass through each of the six resolved cusps as follows

(18)

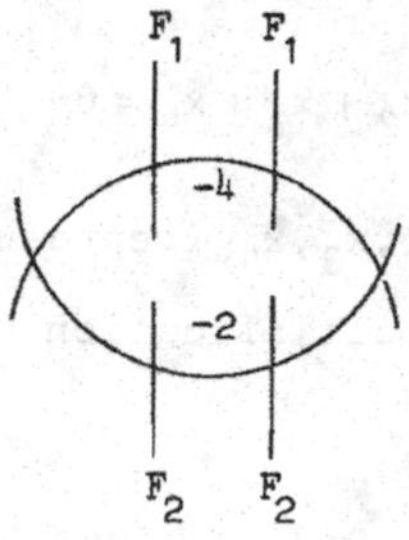

The involution $\tau : (z_1, z_2) \to (z_2, z_1)$ on H^2 induces an involution τ on Y, because $A \in \Gamma \leftrightarrow A' \in \Gamma$. It commutes with the action of every element of $G/\Gamma \cong S_4$ on Y, because A,A' are equivalent mod 2. The fixed point set of τ on Y is $F_1 \cup F_2$. We have $e(F_1) = 8$ and $e(F_2) = 6$.

Therefore

$$e(Y/\tau) = \frac{1}{2}(e(Y) + e(F_1) + e(F_2)) = \frac{1}{2}(16+8+6) = 15$$

We now look at example b) of §1 and see that, from each cusp, Y/τ has 2 curves to blow down successively. We blow down these 12 curves and obtain a surface Y_0 with $e(Y_0) = 3$. If A is a component of F_1 on Y_0 and B a component of F_2 on Y_0, then a simple calculation shows $c_1(A) = 3$ and $c_1(B) = 6$ where c_1 is the first Chern class of Y_0. Therefore Y_0 is rational and is in fact the projective plane $P_2(\mathbb{C})$, on which F_1 becomes a union of 4 lines intersecting in 6 points and F_2 a union of 3 conics with a contact point in each of the six points (compare example b) in §1). The group $G/\Gamma \cong S_4$ operates on $Y_0 = P_2(\mathbb{C})$ with $F_1 \cup F_2$ as an invariant curve of degree 10. The isomorphism $G/\Gamma \cong S_4$ is established by the permutation of the four components of F_1. There is up to projective equivalence only one projective representation of S_4 permuting four lines in general position. It can be lifted in 2 ways to a linear representation :

Embed $\mathbb{C}^3$ in $\mathbb{C}^4$ by

$$x_1 + x_2 + x_3 + x_4 = 0 \qquad (19)$$

Take the permutations of x_1, x_2, x_3, x_4 (representation ρ_1 of S_4) or the permutations followed by multiplication with their signs (representation ρ_1 of S_4).

Consider the projective plane with homogeneous coordinates x_1, x_2, x_3, x_4 subject to (19). Then

$$F_1 \text{ is given by } x_1 . x_2 . x_3 . x_4 = 0,$$

because this is the only invariant curve of degree 4 which has 4 lines as components. The six cusps of $\overline{H/\Gamma}$ correspond to the six intersection points $(0,0,1,-1)$ (and permutations) of the 4 lines. Furthermore,

F_2 is given by $(x_1x_2+x_3x_4)(x_1x_3+x_2x_4)(x_1x_4+x_2x_3) = 0$

because this is the only invariant curve of degree 6 passing through (0,0,1,-1) with 3 irreducible conics as components. Let σ_k be the kth elementary symmetric function of x_1,x_2,x_3,x_4 $(\sigma_1=0)$. The polynomial

$$(20) \qquad C = x_1x_2x_3x_4(x_1x_2+x_3x_4)(x_1x_3+x_2x_4)(x_1x_4+x_2x_3) = \sigma_4(\sigma_3^2-4\sigma_2\sigma_4)$$

of degree 10 describes the branch locus $F_1 \cup F_2$.

THEOREM. Let Γ be the extended principal congruence subgroup of $G = SL_2(\mathcal{O})/\{\pm 1\}$ for the ideal (2) in the ring $\mathcal{O}$ of integers of the field $\mathbb{Q}(\sqrt{2})$. Then $\overline{H^2/\Gamma}$ is isomorphic to the double cover W of $P_2(\mathbb{C})$ along the curve C = 0 of degree 10. This curve has exactly 6 singular points which give singular points of W corresponding to the six cusps of $\overline{H^2/\Gamma}$. Desingularizing W in the canonical way gives a surface Y which is an Enriques surface with 4 points blown up. (The exceptional points in Y come from the 4 linear components of C = 0.)

To gain information for the modular forms relative to Γ, one has to deal with difficulties arising from the fact that Γ has a non-trivial character $\Gamma \to \{1,-1\}$. If one compares with the result of Gundlach [4] where these "sign questions" were treated, one can prove as in §3 that the ring of modular forms for the group Γ is isormorphic to

$$(21) \qquad \mathbb{C}[x_1,x_2,x_3,x_4,c]/(\sigma_1=0,c^2=C)$$

This checks with the dimension formula (§2 (4)), because as in §3 we have $e(H^2/\Gamma) = 4$ and $\chi = 1$ (since Y is an Enriques surface). Comparing with Gundlach [4] shows in addition that $G/\Gamma \cong S_4$ operates on the ring (21) by the representation ρ_2. The ring of invariant polynomials for this representation is generated by $\sigma_2, \sigma_4, \sigma_3^2, \sigma_3\Delta$ where $\Delta = \prod_{i<j} (x_i - x_j)$ is the discriminant. We have a relation $R(\sigma_2, \sigma_4, \sigma_3^2, \sigma_3\Delta) = 0$ for these generators, namely

$$R(\sigma_2, \sigma_4, \sigma_3^2, \sigma_3\Delta) = 27(\sigma_3\Delta)^2 + [-4(\sigma_2^2+12\sigma_4)^3 + (27\sigma_3^2+2\sigma_2^3-72\sigma_2\sigma_4)^2]\sigma_3^2 \tag{22}$$

which can be taken from the formula for the discriminant of a polynomial of degree 4. It follows

<u>THEOREM</u>. <u>The ring of symmetric modular forms for the Hilbert modular group</u> $G = SL_2(\mathcal{O})/\{\pm 1\}$ <u>of the field</u> $\mathbb{Q}(\sqrt{2})$ <u>is isomorphic to</u>

$$\mathbb{C}[\sigma_2, \sigma_4, \sigma_3^2, \sigma_3\Delta]/(R(\sigma_2, \sigma_4, \sigma_3^2, \sigma_3\Delta) = 0).$$

This agrees with Gundlach [4], Satz 1. But there the relation was not determined. The ring of modular forms for G is obtained attaching the skew-symmetric form c of weight 5 satisfying

$$c^2 = C = \sigma_4(\sigma_3^2 - 4\sigma_2\sigma_4)$$

The modular forms $G, \tilde{H}, H, \theta$ (belonging to various characters of $SL_2(\mathcal{O})/\{\pm 1\}$) which Gundlach [4] mentions in his Theorem 1 find the following description in our set up (up to a factor). We also give the zero divisors.

$G = \Delta$, (six lines)

$\tilde{H} = \sigma_3$, (three lines)

$H = \sqrt{\sigma_3^2 - 4\sigma_2\sigma_4}$, (part of the branching locus; three conics)

$\theta = \sqrt{\sigma_4}$, (part of the branching locus; four lines)

The theory we have developed for $\mathbb{Q}(\sqrt{2})$ involves the symmetry group S_4 of a cube. Similar considerations for $\mathbb{Q}(\sqrt{3})$ are possible, but more complicated. Here the group A_4 (symmetry group of a tetrahedron) enters. Gundlach [4] has also investigated $\mathbb{Q}(\sqrt{3})$, but the translation into our geometric method must be done at some other occasion.

§5. ON THE FIELDS $\mathbb{Q}(\sqrt{7})$ AND $\mathbb{Q}(\sqrt{13})$.

In $\mathbb{Q}(\sqrt{7})$ there is no unit of negative norm. Therefore, we consider the extended group $GL_2^+(\mathcal{O})$ of all matrices $\begin{pmatrix}\alpha & \beta\\ \gamma & \delta\end{pmatrix}$ with $\alpha,\beta,\gamma,\delta \in \mathcal{O}$ and determinant a totally positive unit. For the prime ideal $(\sqrt{7})$ let $\Gamma^+(\sqrt{7})$ consist of all matrices of $GL_2^+(\mathcal{O})$ which are congruent to $\pm\begin{pmatrix}1 & 0\\ 0 & 1\end{pmatrix}$ mod $(\sqrt{7})$. Let D be the group of diagonal matrices $\begin{pmatrix}\varepsilon & 0\\ 0 & \varepsilon\end{pmatrix}$ with ε a unit. Since the fundamental unit ε_0 equals $8+3\sqrt{7}$, this diagonal group is contained in $\Gamma^+(\sqrt{7})$. The groups $GL_2^+(\mathcal{O})/D$ and $\Gamma^+(\sqrt{7})/D$ operate effectively on H^2. We denote them by G^+ and Γ respectively. G^+ is the extended Hilbert modular group with $[G^+ : G] = 2$ where $G = SL_2(\mathcal{O})/\{\pm 1\}$. We have

$$G^+/\Gamma \cong PSL_2(\mathbb{F}_7) = G_{168}$$

This is the famous simple group of order 168. The group Γ operates freely on H^2. The surface H^2/Γ is compactified by 24 points (cusps). Each cusp is resolved as in §1 (example c)). This gives a non-singular surface Y. Because $\zeta_{\mathbb{Q}(\sqrt{7})}(-1) = \frac{2}{3}$, we have

$$e(Y) = \frac{2}{3}\cdot 168 + 5.24 = 232$$

We consider the curves F_1,F_2,F_4 in $\overline{H^2/G^+}$. They are given by $z_1 = z_2$, $(3+\sqrt{7})z_2 - (3-\sqrt{7})z_1 = 0$ and $z_1 - z_2 = \sqrt{7}$ respectively. Their inverse images in Y will also be denoted by F_1,F_2,F_4. These are non-singular disjoint curves in Y. They pass through each of the 24 cusps as follows

(23)

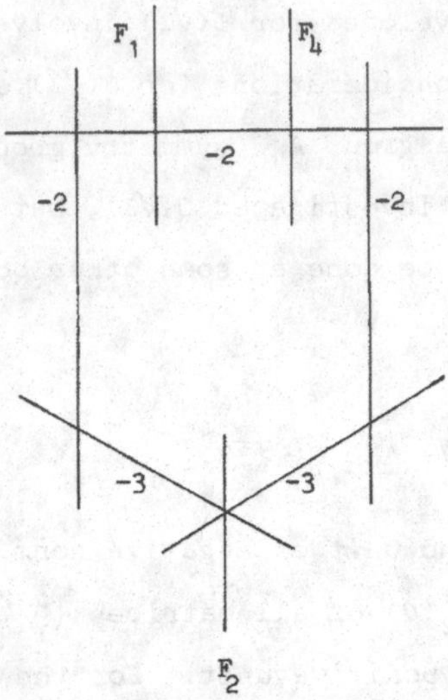

The Euler numbers of F_1, F_4, F_2 in Y are given by

$$e(F_1) = -\frac{1}{6}.168 + 24 = -4$$

$$e(F_4) = -\frac{1}{4}.168 + 24 = -18$$

$$e(F_2) = -\frac{1}{4}.168 + 24 = -18,$$

because $-\frac{1}{6}$, $-\frac{1}{4}$, $-\frac{1}{4}$ are the normalized Euler volumes of the curves F_1, F_4, F_2 in H^2/G^+.

The involution $(z_1, z_2) \to (z_2, z_1)$ of H^2 induces an involution τ of Y commuting with each element of $G^+/\Gamma \cong G_{168}$. The fixed point set of τ in Y is $F_1 \cup F_4 \cup F_2$. Therefore,

(24) $$e(Y/\tau) = \frac{1}{2}(232-4-18-18) = 96.$$

The example c) in §1 shows that each cusp gives rise to three curves which can be blown down successively. We obtain a surface Y_0 with

$$e(Y_0) = 96 - 3.24 = 24$$

The group G_{168} acts on Y_0. One can proof that Y_0 is rational. There is a famous action of G_{168} on $P_2(\mathbb{C})$, see [18], §88, §133-140. This action has an orbit consisting of 21 points. <u>Up to an equivariant isomorphism</u>

Y_0 <u>is obtained from</u> $P_2(\mathbb{C})$ <u>by blowing up these</u> 21 <u>points</u>. <u>The curves</u> F_1, F_4, F_2 <u>become invariant curves of degrees</u> 4,18,12.
This result has to be proved in some other paper. It should be used to investigate the structure of the ring of Hilbert modular forms relative to Γ and G^+.

Our last example concerns the field $\mathbb{Q}(\sqrt{13})$. It is due to van der Geer [2] who has proved many interesting results on the Hilbert modular surfaces of principal congruence subgroups. Let $\mathcal{O}$ be the ring of integers in $\mathbb{Q}(\sqrt{13})$. Let $\tilde{\Gamma}$ be the congruence subgroup of $SL_2(\mathcal{O})$ for the prime ideal 2 of $\mathcal{O}$. Then $\Gamma = \tilde{\Gamma}/\{\pm 1\}$ is a normal subgroup of $G = SL_2(\mathcal{O})/\{\pm 1\}$. The quotient group is $SL_2(\mathbb{F}_4) \cong A_5$. We consider the Hilbert modular surface $\overline{H^2/\Gamma}$. It has 5 cusps. Each is resolved as in §1, example d). Let Y be the non-singular surface obtained in this way. Since

$$2\zeta_{\mathbb{Q}(\sqrt{13})}(-1) = \frac{1}{3},$$

we have

$$e(Y) = \frac{1}{3}\cdot 60 + 5\cdot 9 = 65.$$

The inverse image in Y of the curve F_1 on $\overline{H^2/G}$ has 10 disjoint components which are non-singular rational curves of selfintersection number -1. (Proof as in [8]). The inverse image will also be denoted by F_1. It passes through each of the five cusps as follows

(25)

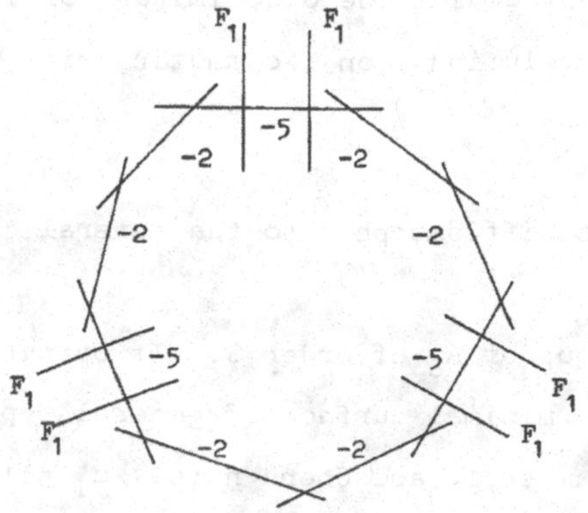

Each component of F_1 goes through 3 of the 5 cusps and is determined by these three cusps. We blow down the ten components of F_1 and obtain a surface Y_1 of Euler number 55. It has arithmetic genus $5 = \frac{1}{4}\, e(H^2/\Gamma)$, see [7]. Therefore $p_g = 4$. The surface Y_1 is a minimal surface of general type. The space of sections of the canonical bundle K of Y_1 is isomorphic to the space of cusp forms $S_2(\Gamma)$. The cusp forms define a "map"

$$\phi_K : Y_1 \to P_3(\mathbb{C}).$$

The action of $G/\Gamma \cong A_5$ on $S_2(\Gamma)$ is the standard action on $\mathbb{C}^4$ represented in $\mathbb{C}^5$ by

$$x_1 + x_2 + x_3 + x_4 + x_5 = 0$$

It turns out that ϕ_K is holomorphic of degree 1 and $\phi_K(Y_1)$ is given in $P_4(\mathbb{C})$ by

$$(26) \qquad \sigma_1 = 0, \quad \sigma_2\sigma_3 - 2\sigma_5 = 0,$$

where σ_k is the kth elementary symmetric function of $x_1,\ldots,x_5$. The surface (26) has 15 double points which are images under ϕ_K of the 15 configurations 

on Y_1 (see (25)). Otherwise ϕ_K is bijective. Because (26) gives a relation between the cusp forms of weight 2, it can be used to gain more information on the ring of modular forms for Γ (see [2]).

The ideal (2) does not divide the discriminant of $\mathbb{Q}(\sqrt{13})$. Therefore, we do not have an involution τ on Y commuting with G/Γ.

Remarks.

1) The surface Y_1 is diffeomorphic to the general quintic hypersurface in $P_3(\mathbb{C})$.
2) Consider a subgroup of A_5 of order 5. It operates freely on Y_1. The quotient is a minimal surface of general type with arithmetic genus 1, Euler number 11 and Chern number $c_1^2 = 1$. We recall that

Godeaux has studied free actions of groups of order 5 on quintic surfaces and considered the corresponding quotients (L. Godeaux, Les surfaces algébriques non rationelles de genres arithmétique et géometrique nuls, Paris 1934).

REFERENCES

[1] E. BRIESKORN, Rationale Singularitäten komplexer Flächen. Invent. Math. 4, 336-358 (1968).

[2] G.B.M. VAN DER GEER, On Hilbert modular surfaces of principal congruence subgroups. Dissertation, Rijksuniversiteit te Leiden, 1977.

[3] K.-B. GUNDLACH, Die Bestimmung der Funktionen zur Hilbertschen Modulgruppe des Zahlkörpers $Q(\sqrt{5})$. Math. Ann. 152, 226-256 (1963).

[4] K.-B. GUNDLACH, Die Bestimmung der Funktionen zu einigen Hilbertschen Modulgruppen. Journal f.d.r.u.a. Math. 220, 109-153 (1965).

[5] K.-B. GUNDLACH, Some new results in the theory of Hilbert's modular group. "Contributions to function theory", Tata Institute, pp. 165-180, Bombay (1960).

[6] F. HIRZEBRUCH, Über vierdimensionale Riemannsche Flächen mehrdeutiger analytischer Funktionen von zwei komplexen Veränderlichen. Math. Ann. 126, 1-22 (1953).

[7] F. HIRZEBRUCH, Hilbert modular surfaces. L'Enseignement Math. 19, 183-281 (1973).

[8] F. HIRZEBRUCH, Hilbert's modular group of the field $Q(\sqrt{5})$ and the cubic diagonal surface of Clebsch and Klein. Russian Math. Surveys 31:5, 96-110 (1976), from Uspekki Mat. Nauk 31:5, 153-166 (1976).

[9] F. HIRZEBRUCH and A. VAN DE VEN, Hilbert modular surfaces and the classification of algebraic surfaces. Invent. Math. 23, 1-29 (1974).

[10] F. HIRZEBRUCH and D. ZAGIER, Intersection numbers of curves on Hilbert modular surfaces and modular forms of Nebentypus. Invent. Math. 36, 57-113 (1976).

[11] F. HIRZEBRUCH and D. ZAGIER, Classification of Hilbert modular surfaces.
"Complex Analysis and Algebraic Geometry", Iwanami Shoten und Cambridge Univ. Press 1977, p. 43-77.

[12] U. KARRAS, Eigenschaften der lokalen Ringe in zweidimensionalen Spitzen,
Math. Ann. 215, 117-129 (1975).

[13] U. KARRAS, Klassifikation 2-dimensionaler Singularitäten mit auflösbaren lokalen Fundamentalgruppen.
Math. Ann. 213, 231-255 (1975).

[14] F. KLEIN, Weitere Untersuchungen über das Ikosaeder.
Gesammelte mathematische Abhandlungen Bd. II, Springer-Verlag, Berlin 1922 (Reprint 1973), pp. 321-380 (Math. Ann. 12, (1877)), see in particular pp. 339, 347, 354.

[15] H. LAUFER, Taut two-dimensional singularities,
Math. Ann. 205, 131-164 (1973).

[16] H.L. RESNIKOFF, On the graded ring of Hilbert modular forms associated with $Q(\sqrt{5})$,
Math. Ann. 208, 161-170 (1974).

[17] H. SHIMIZU, On discontinuous groups operating on the product of upper half planes,
Ann. of Math. 77, 33-71 (1963).

[18] H. WEBER, Lehrbuch der Algebra, 2. Aufl. Bd. II,
Friedrich Vieweg & Sohn, Braunschweig 1899.

ON ZETA-FUNCTIONS OF RANKIN TYPE ASSOCIATED WITH SIEGEL MODULAR FORMS

by A.N. ANDRIANOV

Nowadays, when the theory of Hecke operators on the spaces of modular forms of one complex variable has been so heavily exploited by so many prominent mathematicians, there is a quite natural and similar field where one can see at the moment more blank spots than cultivated areas: the theory of Hecke operators on the spaces of Siegel's modular forms. In this report we want to present some recent results in this field and especially some open questions. I'd like to thank Prof. Stefan, who helped me to improve the language of the manuscript.

1. Siegel's modular forms.

Here we collect some definitions and facts. For details, see [1].

Let

$$H_n = \{Z = X + iY \in M_n(\mathbb{C});\ {}^tZ = Z,\ Y > 0\}$$

be the Siegel upper halfplane of genus n and

$$\Gamma^n = Sp_n(\mathbb{Z}) = \{M \in M_{2n}(\mathbb{Z});\ {}^tMJ_nM = J_n\} ,$$

where $J_n = \begin{pmatrix} 0 & E_n \\ -E_n & 0 \end{pmatrix}$, be the Siegel modular group of genus n. A function $f(Z)$, holomorphic on H_n, is called a modular form of genus n and weight k (k is an integer, $k > 0$) if the following two conditions are fulfilled:

1) for every $M = \begin{pmatrix} A & B \\ C & D \end{pmatrix} \in \Gamma^n$ and $Z \in H_n$

$$f((AZ+B)(CZ+D)^{-1}) = \det(CZ+D)^k f(Z) ;$$

2) $f(Z)$ is bounded in regions of the form

$\{Z = X + i\,Y \in H_n,\ Y \geq c\,E_n (c > 0)\}$.

We denote by $\underline{M}_k^n$ the $\mathbb{C}$-linear space of all modular forms of genus n and weight k . The space $\underline{M}_k^n$ has finite dimension over $\mathbb{C}$. Each form $f \in \underline{M}_k^n$ can be expanded into a Fourier series

$$f(Z) = \sum_{N \in \underline{N}_n} a(N) \exp(2\pi i \operatorname{Tr}(NZ)) \tag{1.1}$$

where

$$\underline{N}_n = \{N = (a_{ij}) \in M_n(\mathbb{Q});\ {}^tN = N \geq 0\ ,\ a_{ii},\ 2a_{ij} \in \mathbb{Z}\}$$

is the set of all symmetric semi-definite semi-integral matrices of order n . It is easy to see that

$$a(U\,N\,{}^tU) = a(N) \qquad (N \in \underline{N}_n,\ U \in SL_n(\mathbb{Z})) \tag{1.2}$$

The example of the theta-series of genus n (see [2]) shows that the Fourier coefficients $a(N)$ of a modular form f can have important arithmetical interpretation in the terms of integral solutions of some systems of quadratic equations. That was the original reason to introduce these modular forms and that is why the theory of Hecke operators we are going to discuss is so concerned with the properties of the Fourier coefficients of modular forms.

2. The Hecke operators.

For details, see [3] and [4] .

Let $H \subset G$ be two multiplicative groups such that for each $g \in G$ the sets $H \backslash HgH$ and HgH/H are finite. Denote by $L(H, G)$ the free $\mathbb{Q}$-module ($\mathbb{Q}$ is the field of rationals)generated by the left cosets (Hg) $(g \in G)$. The group H acts on $L(H, G)$ by multiplication from the right:

$$h : \Sigma_i\, a_i (Hg_i) \longrightarrow \Sigma_i\, a_i\, (Hg_i h) \qquad (h \in H) .$$

Denote by $D(H, G)$ the submodule of all H-invariant elements in $L(H, G)$. $D(H, G)$ is an associative ring with respect to the multiplication:

$$(\Sigma_i a_i(Hg_i))\cdot(\Sigma_j b_j(Hg_j)) = \Sigma_{i,j} a_i b_j (Hg_i g_j) \quad .$$

The ring $D(H,G)$ is called the Hecke ring of the pair (H, G) (over $\mathbb{Q}$).

Let

$$S = S^n = \{M \in M_{2n}(\mathbb{Q});\ {}^tM J_n M = r(M)J_n ,\quad r(M) \in \mathbb{Q} ,\ r(M) > 0\}$$

and for a prime number p let

$$S_p = S_p^n = \{M \in S^n \cap M_{2n}(\mathbb{Z}[p^{-1}]);\quad r(M)^{\pm 1} \in \mathbb{Z}[p^{-1}]\}$$

where $\mathbb{Z}[p^{-1}]$ is the ring of all rational numbers of the form a/p^r with $a \in \mathbb{Z}$, $r \in \mathbb{Z}$. S and S_p are groups, $\Gamma = Sp_n(\mathbb{Z}) \subset S_p \subset S$ and for each $M \in S$ the sets $\Gamma\backslash\Gamma M\Gamma$ and $\Gamma M\Gamma/\Gamma$ are finite. So we can define the Hecke rings

$$L = L^n = D(\Gamma, S) ,\ L_p = L_p^n = D(\Gamma, S_p) .$$

The rings L, L_p are commutative integral domains and L is the tensor product of L_p where p runs over the set of all prime numbers.

If

$$f \in \underline{M}_k^n \quad \text{and} \quad X = \Sigma_i a_i(\Gamma M_i) \in L ,$$

then the function

$$f|X = f|_k X = \Sigma_i a_i f|_k M_i ,$$

where for $M = \begin{pmatrix} A & B \\ C & D \end{pmatrix} \in S$ we set

$$f|_k M = r(M)^{nk-n(n+1)/2} \det(CZ + D)^{-k} f((AZ + B)(CZ + D)^{-1}) ,$$

does not depend on the choice of the respresentatives M_i in the left cosets ΓM_i and again belongs to the space $\underline{M}_k^n$. In this way we get a linear representation of the ring L on the space $\underline{M}_k^n$. The corresponding operators on $\underline{M}_k^n$ are called the Hecke operators.

As it was proved by Maass and Jarkovskaya (see [4], § 1.3) each space $\underline{M}_k^n$ has a basis $\{f_i\}$ such that each f_i is an eigenfunction of all Hecke operators from L :

$$f_i|X = \lambda_i(X)f_i \qquad (X \in L) \quad .$$

We shall therefore consider below only the eigenfunctions of all Hecke operators from L (or L_p).

If $f \in \underline{M}_k^n$ is an eigenfunction of all Hecke operators from L_p (p is a prime number):

$$f|X = \lambda(X)f \qquad (X \in L_p) \ ,$$

then the map $X \to \lambda(X)$ is a non-zero homomorphism of L_p into $\mathbb{C}$. The set of all non-zero homomorphisms of L_p into $\mathbb{C}$ can be described as follows:

Let $A = (\alpha_o, \alpha_1, \ldots, \alpha_n) \in (\mathbb{C}^*)^{n+1}$ $(\mathbb{C}^* = \mathbb{C} - \{0\})$ and let $X = \Sigma_i\, a_i(\Gamma M_i) \in L_p$. Each representative M_i in the left coset ΓM_i can be chosen in the "triangular" form

$$M_i = \begin{pmatrix} p^{d_{io}}{}^tD_i^{-1} & B_i \\ 0 & D_i \end{pmatrix}, \text{ where } D_i = \begin{pmatrix} p^{d_{i1}} & * & \ldots & * \\ 0 & p^{d_{i2}} & \ldots & * \\ \ldots & \ldots & \ldots & \ldots \\ 0 & 0 & \ldots & p^{d_{in}} \end{pmatrix} .$$

We set

$$\psi_A(X) = \Sigma_i\, a_i \prod_{j=0}^{n} (\alpha_j\, p^{-j})^{d_{ij}}$$

The map $X \to \psi_A(X)$ is a non-zero homomorphism of L_p into $\mathbb{C}$ and each such homomorphism has the form ψ_A for some $A \in (\mathbb{C}^*)^{n+1}$, and $\psi_A = \psi_{A'}$ if and only if $A' = wA$, $w \in W$, where W is the finite group of transformations of $(\mathbb{C}^*)^{n+1}$

generated by all permutations in $\alpha_1,\dots,\alpha_n$ and by transformations

$$\alpha_o \to \alpha_o\,\alpha_i\ ,\ \alpha_i \to \alpha_i^{-1}\ ,\ \alpha_j \to \alpha_j\ (j \neq 0,\ i) \qquad (i = 1,\dots,n)\ .$$

Each W-invariant polynomial in $\alpha_i^{\pm 1}$ can be expressed as a polynomial in $\psi_A(X_j)$ for some finite set $\{X_j\} \subset L_p$ (see [3]).

If $f \in \underline{M}_k^n$ is an eigenfunction of all operators $X \in L_p$ with the eigenvalues $\lambda_f(X)$ and $\lambda_f(X) = \psi_A(X)$, where $A = A_f(p) = (\alpha_o(p), \alpha_1(p),\dots,\alpha_n(p)) \in (\mathbb{C}^*)^{n+1}$, we shall call the numbers $(\alpha_o(p),\dots,\alpha_n(p))$ the p-parameters of f. Considering the action on $\underline{M}_k^n$ of the element $(\Gamma\ p\ E_{2n}) \in L_p$ it is easy to see that the p-parameters $(\alpha_i(p))$ of an eigenfunction $f \in \underline{M}_k^n$ satisfy the equation

$$\alpha_o^2(p)\,\alpha_1(p)\dots\alpha_n(p) = p^{nk - n(n+1)/2} \qquad (2.1)$$

3. Eigenvalues of Hecke operators and Fourier coefficients of eigenfunctions.

The problem of finding and studying the relations between the Fourier coefficients of an eigenfunction of the Hecke operators and the corresponding eigenvalues is of significant importance for the arithmetical theory of modular forms. On one hand, the relations would help us to understand the multiplicative properties of the Fourier coefficients, which would have significance in arithmetics; on the other hand, without such relations we can not find analytical relations between the eigenfunction and associated Euler products constructed by the eigenvalues and therefore investigate analytical properties of the Euler products.

If $n=1$ the solution of the problem (given by Hecke) is very simple: let $f \in \underline{M}_k^1$ be an eigenfunction of all Hecke operators from L^1, in particular

$$f|T(m) = \lambda_f(m)\ f,\ (m = 1,2,\dots)$$

where

$$T(m) = \sum_{M \in \Gamma^1 \backslash M_2(\mathbb{Z}),\ \det M = m} (\Gamma^1 M) \in L^1 \qquad (m = 1, 2, \ldots)$$

then

$$a(m) = a(1)\,\lambda_f(m)\ , \quad (m = 1, 2, \ldots) \tag{3.1}$$

where $a(0), a(1), \ldots$ are the Fourier coefficients of f.

The relations (3.1) allow to investigate the analytical properties of two types of the Euler products (zeta-functions) associated with the eigenfunction f:

Let for each prime p $(\alpha_o(p), \alpha_1(p))$ be the p-parameters of f. The coefficients of the polynomials

$$Q_{p,f}(t) = (1-\alpha_o(p)t)(1-\alpha_o(p)\alpha_1(p)t)\ ,$$

$$Q^{(2)}_{p,f}(t) = (1-\alpha_1(p)t)(1-\alpha_1^{-1}(p)t)$$

are invariant with respect to the group W (see § 2), and therefore they can be expressed in the terms of the eigenvalues. Define the Euler products

$$Z_f(s) = \prod_p [Q_{p,f}(p^{-s})]^{-1}\ ,$$

$$Z_f^{(2)}(s) = \prod_p [Q^{(2)}_{p,f}(p^{-s})]^{-1}\ .$$

The Euler products converge absolutely and uniformly if $\mathrm{Re}\ s$ is sufficiently large. From (3.1) and properties of the Hecke operators follow the relations

$$\sum_{m=1}^{\infty} \frac{a(m)}{m^s} = a(1)\ Z_f(s)\ , \tag{3.2}$$

$$\sum_{m=1}^{\infty} \frac{a(m^2)}{m^s} = a(1) \cdot \prod_p \left(1 + \frac{1}{p^{s-k+1}}\right) \cdot Z_f^{(2)}(s-k+1) \tag{3.3}$$

The series on the left hand of (3.2) can be written by the Mellin integral transform of f ; this fact permitted Hecke to prove that $Z_f(s)$ has an analytical continuation over all s-plane and satisfies to a functional equation. The zeta-function $Z_f^{(2)}(s)$ was investigated originally by Rankin [5] by means of another relation. Rankin's results were improved by Shimura [6], who has used the relation (3.3) and the fact that the series on the left hand of (3.3) can be expressed by means of an integral convolution of the product of f by a theta-series with an Eisenstein series for a congruence subgroup of Γ^1.

For $n > 1$ one can hardly expect that such simple relations as (3.1) ever exist. However, relations similar to (3.2) and (3.3) can be obtained for all n :

Let $f \in \underline{M}_k^n$ be an eigenfunction of all Hecke operators from L^n and let, for each prime p, $(\alpha_o(p), \alpha_1(p),\dots,\alpha_n(p))$ be the p-parameters of f. Define the polynomials

$$Q_{p,f}(t) = (1-\alpha_o(p)t) \prod_{r=1}^{n} \prod_{1\leq i_1<\dots<i_r\leq n} (1 - \alpha_o(p)\alpha_{i_1}(p)\dots\alpha_{i_r}(p)t) ,$$

$$Q^{(2)}_{p,f}(t) = \prod_{i=1}^{n} (1 - \alpha_i(p)t)(1 - \alpha_i^{-1}(p)t) .$$

The coefficients of these polynomials are invariant with respect to the group W, and therefore they can be expressed in the terms of the f-eigenvalues of the Hecke operators from L_p^n. The Euler products

$$Z_f(s) = \prod_p [Q_{p,f}(p^{-s})]^{-1} ,$$

$$Z_f^{(2)}(s) = \prod_p [Q^{(2)}_{p,f}(p^{-s})]^{-1}$$

converge absolutely and uniformly if $\operatorname{Re} s$ is sufficiently large. We shall call these Euler products the zeta-function of the Hecke type and of the Rankin type, respectively.

The following result of Jarkovskaya (Math. Sbornik, 1975) gives a generalization of (3.2) for arbitrary $n \geq 1$:

Suppose that $f \in \underline{M}_k^n$ is an eigenfunction of all Hecke operators from L^n with the Fourier expansion (1.1), and let $Z_f(s)$ be the associated zeta-function of the Hecke type. For each given $N \in \underline{N}_n$ we have the identity

$$\sum_{m=1}^{\infty} \frac{a(m\,N)}{m^s} = R_{f,N}(s) \cdot Z_f(s) , \qquad (3.4)$$

where $R_{f,N}(s)$ is a Dirichlet series which is 2^n-restricted (we call the Dirichlet series

$$\sum_{m=1}^{\infty} \frac{c(m)}{m^s}$$

d-restricted if $c(m) = 0$ as soon as $m \equiv 0 \pmod{p^d}$ for some prime p).

As to the zeta-function of the Rankin type, we have proved the following result

<u>Theorem</u> 1. If $f \in \underline{M}_k^n$ is an eigenfunction of all Hecke operators from L^n , with the Fourier expansion (1.1), then for each given $N \in \underline{N}_n$ we have

$$\sum_{M \in SL_n(\mathbb{Z}) \backslash M_n^+(\mathbb{Z})} \frac{a(MN\,{}^tM)}{(\det M)^s} = R_{f,N}^{(2)}(s) \; Z_f^{(2)}(s-k+1) , \qquad (3.5)$$

where M runs over a representative system for the left cosets, by $SL_n(\mathbb{Z})$, of the set of integral matrices of order n with positive determinant, and $R_{f,N}^{(2)}(s)$ is a Dirichlet series which is (n+1)-restricted.

To prove the relations like (3.4) and (3.5) it is sufficient to prove the corresponding "local" relations for each prime p . For example (3.5) is a consequence of the following result.

<u>Theorem</u> 1'. Let $f \in \underline{M}_k^n$ be an eigenfunction of all Hecke operators from L_p^n, where p is a prime number, and let (1.1) be the Fourier expansion of f. Then, for each given $N \in \underline{N}_n$, the formal power series

$$Q_{p,f}^{(2)}(p^{k-1}t) \cdot \left\{ \sum_{d=0}^{\infty} \left(\sum_{\substack{M \in SL_n(\mathbb{Z}) \backslash M_n^+(\mathbb{Z}) \\ \det M = p^d}} a(M\, N\, {}^tM) \right) t^d \right\}$$

is a polynomial of degree not more than n.

This result as well as all other known results of this kind are specialisations of the following general theorem.

Denote for a prime p by $H_p = H_p^n$ the Hecke ring over $\mathbb{Q}$ of the pair $(SL_n(\mathbb{Z}), T_p^n)$, where

$$T_p^n = \{M \in M_n(\mathbb{Z})\ ;\ \det M = p^d\ ,\ d = 0, 1, \ldots\}$$

(the same definition as above, although T_p^n is not a group but a semigroup) and let $H_p^* = H_p[v]$ be the polynomial ring in one variable over H_p. Define a representation of the ring H_p^* on the space of all function $a : \underline{N}_n \to \mathbb{C}$ which satisfy to (1.2): if $y = \Sigma_i\, c_i\, v^{d_i}(SL_n(\mathbb{Z})\cdot M_i) \in H_p^*$ and $a(N)$ is a function we set

$$(a|y)(N) = \Sigma_i\, c_i\, a(p^{d_i} M_i\, N\, {}^tM_i)\ , \quad (N \in \underline{N}_n)\ , \ b_i = p^{-d_i n(n+1)/2} (\det M_i)^{k(n-1)-n(n+1)}$$

In these notations, we have

<u>Theorem</u> 2. Let $f \in \underline{M}_k^n$ with the Fourier expansion (1.1) be an eigenfunction of all Hecke operators from L_p^n for a prime p. Suppose that $y_0, y_1, \ldots$ is a sequence of elements from H_p^* such that the formal power series

$$Y(t) = \sum_{d=0}^{\infty} y_d\, t^d$$

is rational in the sense that there is a polynomial $q(t)$ over H_p^*, $q(0)=1$, such that

$q(t)\,Y(t)$ is a polynomial. Then there is a polynomial $Q(t) \neq 0$ over $\mathbb{C}$, which depends only on $q(t)$ and the f-eigenvalues of the Hecke operators from L_p^n, and whose coefficients can be effectively expressed in the terms of the eigenvalues, such that for each given $N \in \underline{N}_n$, the formal power series over $\mathbb{C}$

$$Q(t)\{\sum_{d=0}^{\infty} (a|y_d)(N)\, t^d\} \qquad (3.6)$$

is a polynomial.

This theorem (together with its proof) contains in particular all known relations between Fourier coefficients of eigenfunctions and eigenvalues of Hecke operators. The proof is based on a study of the relations between the Hecke rings L_p and H_p^*, realised as subrings of a Hecke ring of the group

$$\Gamma_o^n = \{\begin{pmatrix} A & B \\ C & D \end{pmatrix} \in \Gamma^n,\ C = 0\}. \qquad (3.7)$$

The proofs will be published in Matem. Sbornik.

As to the computations of the series $R_{f,N}(s)$ and $R_{f,N}^{(2)}(s)$ in (3.4) - (3.5), or more generally, of the polynomials (3.6) for arbitrary n, some new ideas will have to be found. For $n = 2$ the series $R_{f,N}(s)$ can be computed without difficulty from the results of [4], Ch. 2; the series $R_{f,N}^{(2)}(s)$ are computed in [7].

4. On integral representations of Euler's products.

The theorem 2 allows us to get a lot of relations like (3.4), (3.5) with different kinds of Euler products. We restrict ourselves to the zeta-functions of the Hecke type and of the Rankin type, because even for n=1 they are the only types of the Euler products whose analytical properties can be investigated at the moment.

Because of the relations (3.4) and (3.5) to get some "good" integral representations of the zeta-functions in the terms of the modular form f it is enough, provided of course that we can compute the series $R_{f,N}(s)$, $R^{(2)}_{f,N}(s)$, to do it for the series on the left hand of the relations. As to (3.4), for n=1 it is the Mellin transform of f , for n=2 the situation is far more complicated: it was shown in [4] that the series can be obtained as the result of an integral convolution of the restriction of f on the symmetric space of the group $SL_2(\mathbb{C})$ imbeded in H_2 , with an Eisenstein series for a discrete subgroup of $SL_2(\mathbb{C})$ of the Picard type. This allows to prove that in this case the zeta-function $Z_f(s)$ has an analytical continuation and satisfies a functional equation; for $n > 2$ nothing is known in the most interesting case when f is a cusp form (if f is not a cusp form, $Z_f(s)$ can be expressed through the similar function for a modular form of genus n-1). The situation is much better for the relations (3.5): the series on the left hand side has a "good" integral representation for all n. For example, if n is even, $f \in \underline{M}^n_k$ is a cusp form and $N \in \underline{N}_n$, $N > 0$, the following integral representation can be proved: if Re s is sufficiently large, we have

$$2\pi^{n(n-1)/4}(\det 4\pi N)^{-s/2}\{\prod_{i=1}^{n}\Gamma(\frac{s-i+1}{2})\}\cdot\sum_{M\in SL_n(\mathbb{Z})\backslash M_n^+(\mathbb{Z})}\frac{a(MN{}^tM)}{(\det M)^{s-h}} =$$

$$= \int_{D(\Gamma_o(q))} f(Z)\,\overline{\Theta^{(h)}_{2N}(Z)}\,(\det Y)^{s+n+1/2} \times \tag{4.1}$$

$$\times\left\{\sum_{\binom{A\ B}{C\ D}\in\Gamma_o\backslash\Gamma_o(q)}\frac{\chi_N(\det D)\det(CZ+D)^{k-n/2-h}}{|\det(CZ+D)|^{s-2h+1}}\right\}d^*Z ,$$

where $h=0$ if k is even and $h=1$ otherwise, q is the smallest natural number such that the matrix $q(2N)^{-1}$ has rational-integral coefficients and even coefficients on the main diagonal,

$$\Gamma_o(q) = \Gamma_o^n(q) = \{\begin{pmatrix} A & B \\ C & D \end{pmatrix} \in \Gamma^n , \ C \equiv 0 \pmod q\} ,$$

Γ_o is the group (3.7), $D(\Gamma_o(q))$ is a fundamental domain of $\Gamma_o(q)$ on H_n,

$$d^*Z = (\det Y)^{-(n+1)} \prod_{1\leq i\leq j\leq n} dx_{ij}\, dy_{ij} \qquad (Z = X + iY \in H_n)$$

is the invariant measure on H_n, $\Gamma(s)$ is the gamma-function,

$$\Theta_{2N}^{(h)}(Z) = \sum_{M\in M_n(\mathbb{Z})} (\det M)^h \exp(2\pi i \operatorname{Tr}(MN{}^tMZ)) \qquad (4.2)$$

is a theta-series of the matrix $2N$, and χ_N is a Dirichlet character mod q which can be defined by

$$\chi_N(d) = (\operatorname{sign} d)^{\frac{n}{2}} \left(\frac{(-1)^{\frac{n}{2}} \det 2N}{|d|}\right) ,$$

where $(-)$ is the generalized Legendre symbol.

For the proof for $n=2$, see [7]. The proof in the general case is the same.

Accordingly to [7], § 5, the theta-series (4.2) is a modular form of the weight $(\frac{n}{2} + h)$ with the character χ_N with respect to the group $\Gamma_o^n(q)$. Therefore in order to study the analytical properties of the integral in (4.1) it is actually sufficient to study the Eisenstein series under the integral. For $n = 2$ this can be done following the ideas of the Maass' work [8], which allow us to prove that the integral has an analytical continuation over the entire s-plane. Unfortunately, we could not obtain a functional equation in this way. As to the case of general n , now everybody refers for everything connected with Eisenstein series to the famous Langland's preprint. I wonder, however, whether it is easy (or possible) to extract the properties of the Eisenstein series in (4.1) from this preprint. Here I mean just this question: what are the explicit gamma and zeta-factors to get a holomorphic function with a functional equation?

Finally we should like to mention that some examples suggest that if $f \in \underline{M}_k^n$ is a cusp form, then to get a holomorphic function we have to replace $Z_f^{(2)}(s)$ by the product

$$\zeta(s)\, Z_f^{(2)}(s)$$

where $\zeta(s)$ is the Riemann zeta-function.

The Leningrad Branch of the
Steklov Mathematical Institute.

Bibliography

[1] Séminaire H. Cartan 1957/58, Fonctions automorphes, Secrétariat mathématique, Paris, 1958.

[2] A.N. Andrianov, G.N. Maloletkin, Behavior of theta series of degree n under modular substitutions, Math. USSR Izvestija Vol.9 (1975), N° 2, 227-241.

[3] I. Satake, Theory of spherical functions on reductive algebraic groups over p-adic fields, Publ. Math., IHES N° 18 (1963).

[4] A.N. Andrianov, Euler products associated with the Siegel modular forms of genus 2, Uspehi Mat. Nauk., vol. 29 N° 3 (1974), 43-110.

[5] R.A. Rankin, Contributions to the theory of Ramanujan's function $\tau(n)$ and similar arithmetical functions, II, Proc. Cambridge Philos. Soc. 35 (1939), 357-372.

[6] G. Shimura, On the holomorphy of certain Dirichlet series, Proc. London Math. Soc., 31, N° 1 (1975), 79-98.

[7] A.N. Andrianov, Symmetric squares of zeta-functions of Siegel's modular forms of genus 2, Trudy Mat. Inst. Steklov, 142 (1976), 22-45.

[8] H. Maass, Dirichletsche Reihen und Modulformen zweiten Grades, Acta Arithmetica, 24 (1973), 225-238.

ADRESSES OF AUTHORS

A.N. ANDRIANOV,	Math. Institute, Fontanka 25,Leningrad 191011/USSR
H. COHEN,	U.E.R. de Mathématique et Informatique,Université de Bordeaux, 351 Cours de la Libération, 33405 Talence / France
M. EICHLER,	Mathematisches Institut der Universität Basel, Basel / Switzerland
S. GELBART,	Dept. of Mathematics, Cornell University, Ithaca N.Y. 14853 / USA
F. HIRZEBRUCH,	Mathematisches Institut der Universität Bonn, Wegelerstrasse 10, D-5300 Bonn / BRD
S. KUDLA,	Dept. of Mathematics, University of Maryland, College Park, Maryland 20742 / USA
C. MORENO,	Dept. of Mathematics, University of Illinois at Urbana Champaign, Urbana, Illinois 61801 / USA
J. OESTERLÉ,	Université de Paris-Sud, bat.425, F-91405 Orsay/ France
M. RAZAR,	Dept. of Mathematics, University of Maryland, College Park, Maryland 20742 / USA
J-P. SERRE,	Collège de France, 11 place Marcelin Berthelot, F-75231 Paris Cedex 05 / France
H.M. STARK,	Dept, of Mathematics, M.I.T., Cambridge Mass. 02139 / USA
M-F. VIGNÉRAS,	École Normale Supérieure de Jeunes Filles, 1 rue Maurice Arnoux,F-92120 Montrouge / France
D. ZAGIER,	Mathematisches Institut der Universität Bonn, Wegelerstrasse 10 , D-5300 Bonn / BRD